An Introduction to Power Electronics

An Introduction to Power Electronics

B. M. BIRD

University of Bristol

K. G. KING

Westinghouse Brake & Signal Co Ltd
Chippenham, Wiltshire

JOHN WILEY & SONS

Chichester · New York · Brisbane · Toronto · Singapore

Reprinted July 1984
Reprinted May 1985

Library of Congress Cataloging in Publication Data:

Bird, B. M.
An introduction to power electronics.

Bibliography: p.
Includes index.
1. Power electronics. I. King, K. G. II. Title.
TK7881.15.B57 1983 621.31'7 82-13664
ISBN 0 471 10430 2 (cloth)
ISBN 0 471 90051 6 (paper)

British Library Cataloguing in Publication Data:

Bird, B.M.
An introduction to power electronics.
1. Electric current convertors 2. Diodes
3. Thyristors
I. Title II. King, K. G.
621.31'37 TK2796

ISBN 0 471 10430 2 (cloth)
ISBN 0 471 90051 6 (paper)

Phototypeset by Macmillan India Ltd., Bangalore.
Printed at The Bath Press, Avon

Preface

The intention in writing this book has been to steer a course between the extremes of academic theory on the one hand and unenlightened empiricism on the other, neither pursuing theoretical analysis for its own sake nor attempting to produce a technician's handbook. Accordingly, we have included as much theory and simple mathematics as seems necessary for a rational and consistent understanding of the principles which underlie at least the greater part of practical power electronic technology, and have tried to avoid constructing the text around an arbitrary selection of particular techniques. We hope the result is a book suitable both as a textbook for students engaged in university and polytechnic courses and as an aid to sound design for practising engineers.

Practical aims notwithstanding, we have thought fit, in the interests of clarity and brevity, to set limits to what might be termed secondary aspects of the subject. Thus, while we are aware that the characteristics of a controlled rectifier may be significantly modified if the d.c. output current is not continuous and smooth, we have for most purposes assumed that it is, on the grounds that such an assumption provides a reasonable starting point, in a work of limited scope, for a basic understanding which the reader can extend, if he wishes, by referring to more specialized texts. Equally, where it does not materially affect the argument, we have ignored imperfections in components—voltage drops and leakage currents in semiconductor devices, magnetizing current, leakage reactance and resistance in transformers, and so on. We think it will be clear from the context whether such simplifying assumptions are being made, or whether a feature which is of generally secondary importance is of primary significance to the particular matter under discussion.

The fact that the book is written in terms of thyristors should not be taken to indicate that we do not recognize the established and growing importance of power transistors in power electronics, or even that we are biassed in favour of thyristors. The discussions of forced-commutation systems, to the extent that they are applicable in detail (excluding, for example, the special considerations of thyristor commutation circuits), are equally valid whether the switching elements are thyristors, transistors or devices as yet unknown, and the relatively extended treatment of thyristor ratings and characteristics may, we hope, be construed as an object lesson in the consideration of power semiconductor devices generally.

We should like to record our appreciation of the assistance and advice we have received in preparing the subject matter from colleagues and friends in the Westinghouse Brake and Signal Company Ltd. (Westcode Semiconductors).

January, 1983

B. M. Bird
K. G. King

Contents

Symbols

a	coefficient, tolerance factor
A	area
b	coefficient
c	specific heat
C	capacitance
C_{th}	thermal capacity
d	length (thickness), factor $R/2\omega_0 L$
e	e.m.f.
E	energy
f	frequency
f_0	resonant frequency
i	instantaneous current
I	current, r.m.s. current
$\hat{I}$	peak current
$\check{I}$	minimum instantaneous current
$\bar{I}$	mean (half-cycle) current
k	factor, constant, thermal conductivity
L	inductance
m	number of phases
M	mass
n	number, order of harmonic
p	number of pulses per cycle
P	power (mean)
Q	reactive power, quality factor, electrical charge
r	positive integer, incremental resistance
R	resistance
R_{th}	thermal resistance
S	apparent power
t	time
T	time interval
v	instantaneous voltage
v_0	threshold voltage
V	voltage, r.m.s. voltage
$\hat{V}$	peak (crest) voltage
$\bar{V}$	mean (half-cycle) voltage

w	instantaneous dissipation (power loss)
$\overline{W}$	mean dissipation (power loss)
$\hat{W}$	peak dissipation (power loss)
X	reactance
y	ordinate
Z	impedance
Z_{th}	thermal impedance
Z_t	transient thermal impedance
α	angle of delay, common-base current gain, parameter of commutation circuit
β	angle of advance, angular pulse duration
γ	switching ratio
δ	damping factor, per-unit increase
Δ	per-unit deviation
ε	emmisivity
ε_X	per-unit source (transformer) reactance
θ	temperature
θ_A	absolute temperature
λ	angular pulse duration
μ	overlap angle
τ	time, time-constant
ϕ	electrical angle, phase angle
ψ	electrical angle
ω	angular frequency
ω_0	resonant angular frequency

Subscripts

a	alternating, a.c. circuit, interphase, air
a, b, c	of defined periods
A	absolute, anode, of (point) A
b	base (small-signal), base (mechanical)
B	base (large-signal), blocking, of (point) B
BO	break-over
c	collector (small-signal), commutation, critical, cooler, carrier
C	collector (large-signal), of capacitor
d	direct, d.c. circuit, delay
D	diode
e	emitter (small-signal)
E	emitter (large-signal)
f	fall, forced-commutation, form
F	forward
G	gate
h	harmonic (distortion)
i	input
ipr	of interphase reactor
j	junction
k	clamping
K	cathode
L	of inductance (inductor)
L	latching
m	magnetizing, modulation
mtg	mounting
M	maximum
n	n th-harmonic
n	natural convection
N	of neutral point
o	output
p	primary, periodic, power
ph	phase
q	turn-off
r	reverse, rise, ripple
rr	reverse recovery

R	of resistance (resistor)
R	reverse
s	secondary, sharing, initial, string
t	transient
T	of transformer, triggering (-erred)
Th	of thyristor
u	utilization, ultimate
V	voltage
w	water (liquid)
X, Y, Z	of points X, Y, Z
0	threshold
1	fundamental (-frequency)
2, 3, . . .	2nd, 3rd, . . . harmonic
I, II, III	of phase I, II, III

CHAPTER 1

Power Semiconductor Devices

Power electronics is for the most part based on semiconductor switching devices: that is, devices which by virtue of their inherent characteristics or their mode of use exhibit, under different conditions, electrical properties approximating to complete blocking or unrestricted conduction, but not intermediate values of resistance except, briefly, during transient switching periods. These devices—power diodes, transistors and thyristors—all embody elements of a monocrystalline semiconductor material (disregarding polycrystalline selenium, whose use in this field is nowadays confined to cheap, low-power rectifier equipment and surge absorbers), and in virtually all cases the material is silicon.

In the authors' opinion, a physicist's understanding of the internal workings of semiconductor devices is not necessary to the engineer primarily concerned with their application in equipment, at least in the orthodox kinds of equipment considered in this book. The devices are accordingly discussed here, as far as possible, in terms of their external characteristics, with such references to the underlying physics only as may be of real assistance in understanding certain important aspects of their characteristics and behaviour. In this context, however, it is helpful to appreciate the basic concept, or at least the vocabulary, of electrical conduction in a semiconductor such as silicon, and the way in which it is modulated in the action of the three devices considered.

Pure silicon (the same exposition would apply to germanium) conducts only slightly at normal temperatures, owing to the very limited availability of electrons that are free to move within the crystal structure; high conductivity, as required in, say, a forward-biassed diode, is brought about by the addition, in very small concentrations, of certain impurities, generally known as doping.

Two modes of conduction are possible as a result of adding impurities. Essentially, an electric current is conceived as a movement of electrons, but if it is postulated that during a certain interval of time an electron moves from a point A to a space at B, leaving A vacant, the space may be considered to have moved from B to A. Thus a current may also be conceived as a movement of spaces, more commonly termed holes, and as the electron is a negative charge-carrier, so the hole, which in this context represents the absence of an electron, may be treated as a positive charge-carrier, with a charge equal and opposite to that of the electron. In a pure (intrinsic) semiconductor, the movement of electrons is in fact mirrored by the movement of holes, but in the doped material conduction occurs principally as a flow of electrons or of holes, according to the nature of the

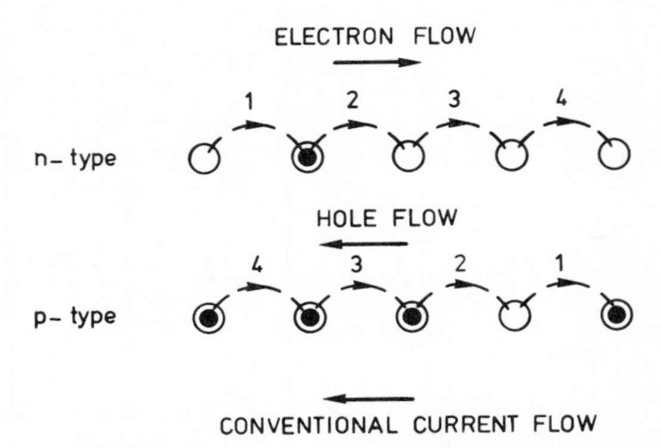

Figure 1.1 Current flow by movement of electrons in n-type material and of holes in p-type material.

impurity; in the first case the negative electrons are termed the majority carriers, and the material is said to be of n-type, while in the second the majority carriers are positive, and the material is of p-type. Figure 1.1 illustrates diagrammatically the actual movement of electrons in the two cases.

THE p–n JUNCTION

Power diodes, transistors and thyristors, at least as far as this book is concerned, are so-called bipolar devices, signifying that their action depends basically on the properties of an intimate junction between dissimilar forms of the semiconductor material, p-type and n-type. The junction may be formed in various ways, but principally by introducing the required impurities in turn into a single piece of material either by an alloying process or by diffusion.

The characteristics of the p–n diode, insofar as they approximate to the ideal rectifier characteristics, can be simply, if incompletely, explained by the reactions of the majority carriers to applied voltages. If the diode, comprising adjacent regions populated by majority carriers in the form of holes and electrons respectively, as depicted in Figure 1.2(a), is subjected to a voltage such that the 'p' terminal is made positive with respect to the 'n' terminal, the effect is to propel holes and electrons towards the junction, as in Figure 1.2(b), where they combine, while the consequent loss of charge carriers is made up by replenishment from the voltage source. Inasmuch as this can be envisaged as a continuous process, there is little impediment to the flow of current, and the result is a low-impedance conducting characteristic in what is referred to as the forward direction of current flow. If the applied voltage is reversed, as in Figure 1.2(c), holes in the p region, and electrons in the n region, are attracted away from the junction, so that no combination is possible, and the device is depleted of majority carriers; it thus presents a high-impedance blocking characteristic in this direction of applied voltage, which is termed the reverse direction.

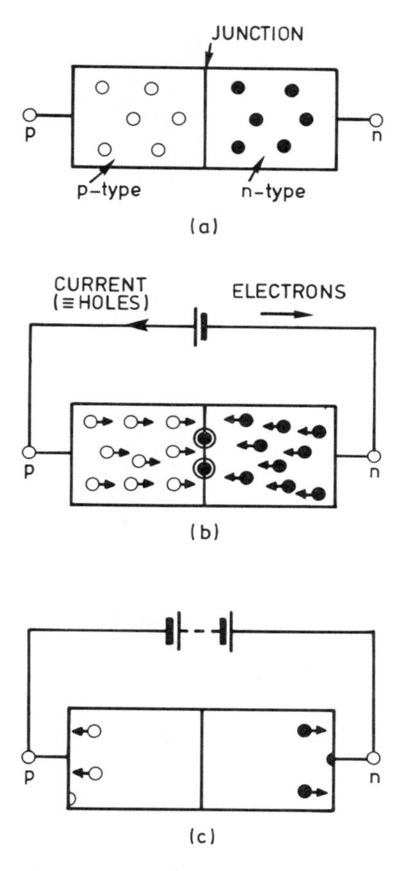

Figure 1.2. Diagrammatic representation of a p-n junction diode: (a) with no applied voltage; (b) with forward applied voltage; (c) with reverse applied voltage.

The behaviour of an actual power diode is somewhat more complex than the above description would suggest, in particular because of the modifying effects of the external contacts.

The characteristics of a practical diode show several departures from the ideals of zero forward and infinite reverse impedance, as illustrated by the typical representation of Figure 1.3. In the forward direction, a potential barrier associated with the distribution of charges in the vicinity of the junction, together with other effects, leads to a voltage drop which in the case of silicon is usually of the order of one volt for currents in the normal range. In reverse, within the normal operating range of voltage, a very small current flows which is largely independent of the voltage.

According to basic theory, the forward and reverse characteristics should form a single exponential function, of the form $i = a(e^{bv} - 1)$, but in practice this does

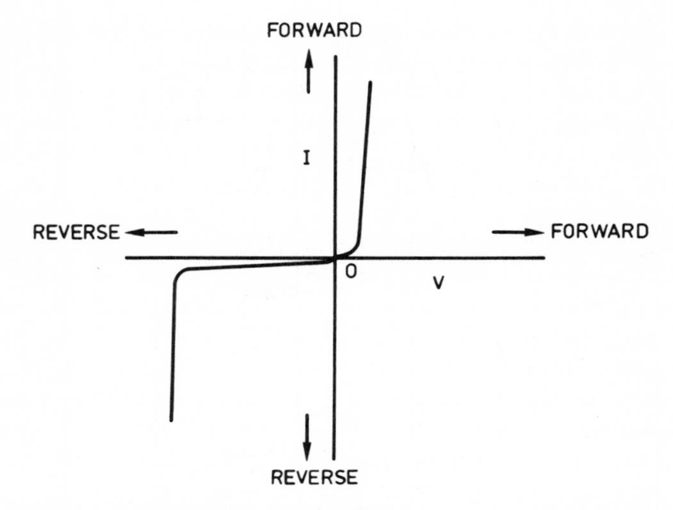

Figure 1.3 Typical static characteristic of a silicon diode (forward and reverse to different scales).

not fully account for all the effects involved, which include, in the forward direction, ohmic resistance, and in the reverse direction, the possibilities of an uneven distribution of current through the area of the junction, due to non-uniformity in the material, and of leakage current across the surface of the semiconductor element. For practical purposes the static characteristics may more conveniently be represented in the manner shown in Figure 1.4, with the forward characteristic expressed as a threshold voltage v_0 and a linear incremental or slope

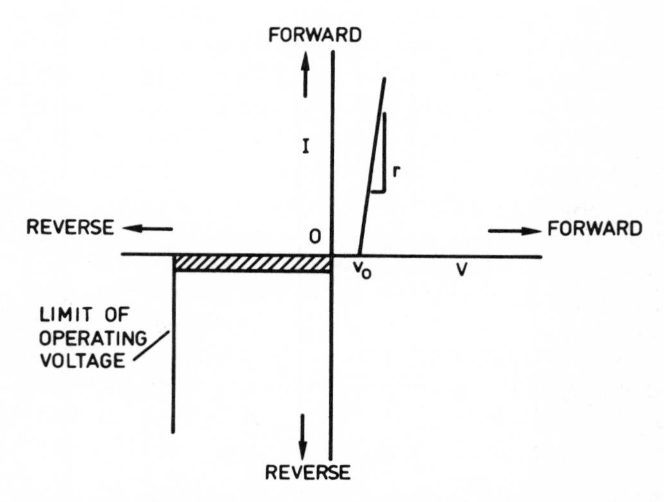

Figure 1.4 Practical representation of the static characteristics of a silicon diode.

resistance r, and the reverse characteristic as a range of possible leakage currents irrespective of voltage within the normal working range.

At a certain level of reverse voltage (characteristic of a particular diode) the characteristic changes more or less abruptly to a very low incremental resistance, i.e. a line of almost constant voltage, as a result of the so-called avalanche effect, whereby a proliferation of mobile charge carriers is produced by collisions within the crystal. A substantial flow of current in the avalanche mode is not inherently destructive, but damage is nevertheless likely to occur at quite low reverse current, since it is not always practicable, at least in a large diode, to ensure a sufficient degree of uniformity in the junction to prevent a concentration of current, and therefore of dissipation, in a very small part of the total area; neither can the possibility always be ruled out that a destructive current may flow at the surface of the semiconductor element at a voltage below the avalanche voltage of the material. So-called 'avalanche diodes' are characterized and tested to withstand a useful amount of reverse current as a means of protection.

THE TRANSISTOR

The elementary explanation given above of the rectifying properties of the p–n junction diode, albeit lacking in rigour, illustrates the general dependence of the conductance of the device upon a sufficiency of mobile charge carriers. In the transistor, the element of control which the diode lacks is introduced by providing a means of regulating the availability of charge carriers.

The three-layer structure of the junction transistor element may be thought of as a special conjunction of two diodes sharing a common layer; the alternating forms of the semiconductor material are thus arranged either p–n–p or n–p–n, as illustrated in Figure 1.5, where the terminals are labelled in accordance with common usage. In use, the collector junction is biassed in reverse, i.e. a positive voltage is applied to the collector in the case of the n–p–n transistor, so that, if no connection were made to the emitter, only a small leakage current would flow between the collector and the base. The base region is, however, made very thin; in consequence, if a forward current is made to flow in the emitter junction, most of

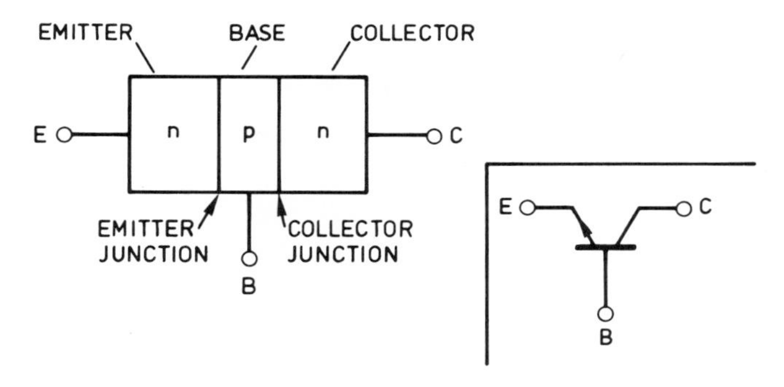

Figure 1.5 n–p–n transistor. (Inset: normal symbol.)

the majority carriers that flow from the emitter to the base overshoot into the collector region, and thereby enable a current to flow between collector and base to the extent permitted by the available carriers. Since the total flow of majority carriers from the emitter represents, approximately, the emitter current, the resulting collector current is somewhat less than the emitter current (the effects of leakage currents are ignored here), as expressed by the ratio $\alpha = i_C/i_E$, where normally $0.9 < \alpha < 1$.

The transistor thus affords a means of controlling the current in the collector circuit by varying the emitter current, as in Figure 1.6, with a possible power gain, since the collector voltage may be much greater than the forward emitter-to-base voltage, but no current gain. This mode of operation, with the emitter current as the input signal, is known as the common-base mode. Usually, the transistor is used in the common-emitter mode illustrated in Figure 1.7, where the input signal is the base current; this may be regarded as a variation of the common-base mode in which 100% positive feedback is provided by allowing the collector current to flow back to the emitter, and the result is to reduce the input current required to that excess of emitter current over collector current dictated by the basic transistor characteristic, expressed by α. That is,

$$i_B = i_E - i_C = i_C\left(\frac{1-\alpha}{\alpha}\right) \tag{1.1}$$

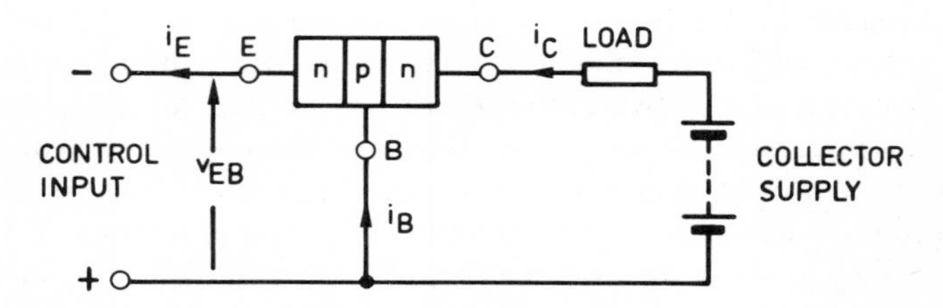

Figure 1.6 n–p–n transistor operating in the common-base mode.

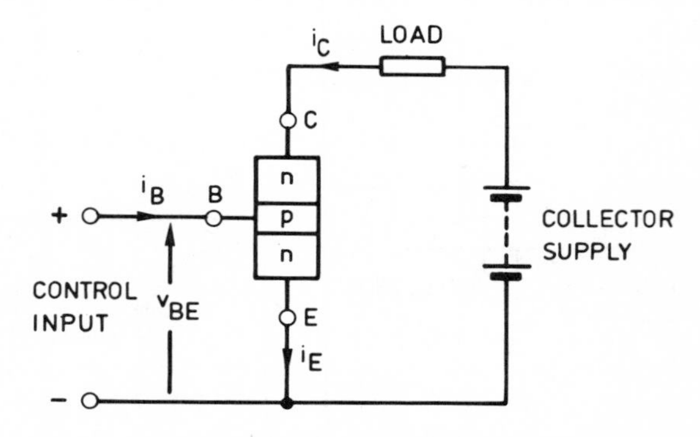

Figure 1.7 n–p–n transistor operating in the common-emitter mode.

Since α is close to unity (in the useful range of collector current) this represents a considerable current gain, and consequently a much greater power gain may be realized than in the common-base mode.

Visualizing the operation of the transistor as dependent upon the injection of control current into the base region assists an understanding of the practical limitations upon the current and voltage ratings of the device. In general terms, a high current rating requires a large element area in order to keep the resistive components of loss acceptably low and to avoid excessive concentration of the loss, while a high voltage rating necessitates a relatively thick base region to avoid a direct breakdown between collector and emitter. In a large area, it is difficult to achieve uniform penetration of base current from the connection at the edge, owing to the lateral resistance of the material; this effect becomes more marked as the base current is increased, because of the non-linearity of the base–emitter junction, and contributes, with other factors, to a sharp reduction in current gain at high collector currents. To offset this, it would be appropriate to increase the thickness of the base region, but this is in conflict with the necessity for a thin base to enable the charge carriers from the emitter to cross over to the collector without an excessive reduction in numbers, and results in a reduction in α, and hence an increase in base current such that the reduction in base resistance is of little benefit. Similarly, to obtain a high voltage rating by increasing the base thickness tends to lead to a low α and hence is in conflict with gain and, indirectly, current rating.

To overcome as far as possible the problem of base-current penetration, the base and emitter connections in large transistors are arranged in relatively complex interdigitated patterns in order to keep the current paths within the base region as short as possible. Nevertheless, both the voltage and the current ratings of transistors have generally been restricted, so far, to something like an order less than those attainable in silicon diodes.

THE THYRISTOR

The thyristor is essentially a development of the transistor in which a fourth layer is added on the collector side to form a second emitter. The direct effect of this is to enable it to sustain alternative stable states, conducting and non-conducting, and to operate in a triggering mode; indirect effects are to overcome the difficulties associated with the transistor in regard to high current and voltage ratings, and also to enable it to withstand a high voltage in the reverse direction and hence function as a rectifier.

Figure 1.8 shows the thyristor with a forward voltage applied to it; the diode pairs formed by the p_1 and n_1 layers and by the p_2 and n_2 layers offer no impedance to voltage in this direction, and the applied voltage is therefore supported, if the thyristor is in a non-conducting, or blocking, condition, by the reverse-biassed middle junction j_2. On the other hand, if the thyristor is to conduct, an adequate supply of charge carriers must be produced to cross the middle junction; this is achieved by the combined effects of the two emitters, p_1 and n_2, the first producing holes and the second electrons, and since the effects are

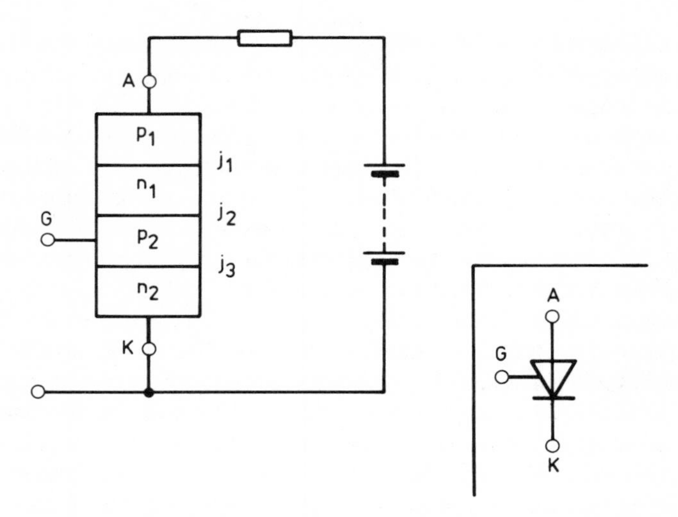

Figure 1.8 Thyristor with forward voltage applied. (Inset: general triode thyristor symbol.)

directly additive, the total quantity of carriers made available may easily be at least equivalent to the current flowing across the junction. The result is that forward current, once established, is self-sustaining over the whole of the element area, and not dependent on any externally supplied base current. There is thus no essential limit to the area, or to the current rating. At the same time, the fact that the two emitter–base pairs (p_1–n_1 and n_2–p_2) need be designed only for an average value of α of 0.5, to produce the effective combined α of unity required to make conduction self-sustaining, makes it possible to use relatively thick base layers, and hence to obtain a much greater voltage-blocking capability than is practicable in a transistor.

At a forward current of the order of the leakage current, the value of α associated with each emitter falls, as in a normal transistor, and under this condition the sum may be made less than unity; in the absence of externally supplied base current, therefore, an alternative stable state exists in which the thyristor will block forward voltage, within a certain range, passing only a small leakage current comparable with the reverse current of a diode. To trigger the thyristor from the blocking to the conducting state, sufficient base current is injected—normally at the p_2 terminal, which is referred to as the gate—to increase the current in the main circuit by the modest amount required to raise the combined α to something greater than unity; this effect, however, is necessary only in a small area of the element, from which the state of self-sustaining conduction will spread at a finite rate across the whole area, and neither the problems of designing the gate structure nor the magnitude of the gate current required are directly related to the size of the element. This means that the effective current gain can be very high, while the fact that in principle a pulse of gate current lasting only a few microseconds is all that is necessary to initiate a stable conduction period of any required duration enables average power gains of the order of 10^6 to be achieved without difficulty.

The thyristor can also be triggered from the forward blocking to the conducting state by an enhanced leakage current resulting from an increase in applied voltage beyond a certain level termed the breakover voltage; albeit not necessarily destructive, this mode of triggering is not normally permitted. A reverse voltage (negative at p_1) applied to the thyristor is blocked, if not excessive, by the anode junction, j_1 (the gate junction, j_3, has very limited voltage-blocking capability), and the reverse characteristic is similar to that of a diode.*

For comparison with the diode characteristic shown in Figure 1.3, Figure 1.9 illustrates typical forward and reverse characteristics of a thyristor. In the blocking condition ('off state'), i.e. when it has not been triggered, the thyristor exhibits a forward characteristic resembling the reverse characteristic of a diode, and a similar order of leakage current so long as the applied voltage is below the breakover voltage. If a steadily increasing current is applied to the gate, the effect is gradually to reduce the breakover voltage until a conducting ('on state') characteristic is obtained which is generally similar to the forward characteristic of a diode, albeit with a somewhat higher voltage drop at a given anode current density; the transitional phase between the fully blocking and the fully conducting states is, however, avoided in normal usage, and the thyristor is triggered from the former state to the latter by an adequate gate current at any required instant of time.

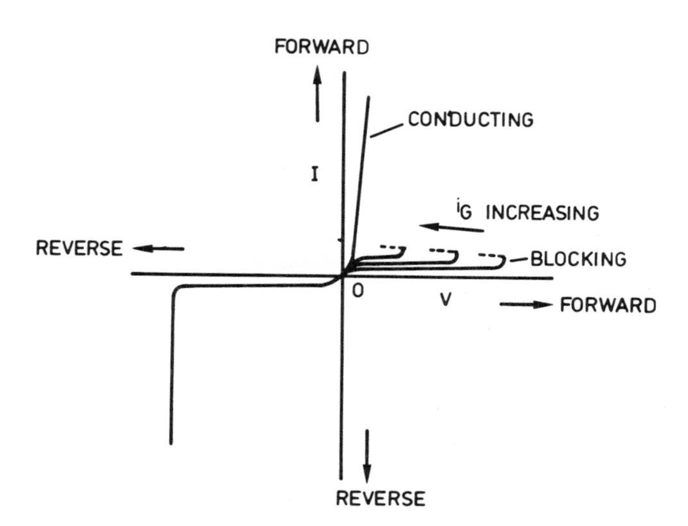

Figure 1.9 Characteristics of a reverse-blocking triode thyristor (not to scale).

* While other forms of thyristor exist—some are referred to at the end of this chapter—the word 'thyristor' used in this book without qualification denotes what is properly termed a reverse-blocking triode thyristor.

Once the thyristor is carrying forward current in the fully conducting state, it will continue to do so, without a continuous gate current, so long as the anode current does not fall below a critical low level known as the holding, or sustaining, current. For the conducting state to be established, such that it will persist after the removal of gate drive, it is necessary that the anode current should exceed a higher critical level, the latching current.

Mechanical construction

The two or more layers in the element of a power diode or thyristor are typically contained within a thickness of the order of 0.2–0.3 mm of silicon, while the diameter of the element may be anything up to 75 mm or so. The gating area in a simple thyristor is preferably located at the centre of the element, to minimize the spreading time (it can be at the edge), and the connection is made to the gate through an aperture in the cathode layer, as illustrated in Figure 1.10. The element is mounted in one of two ways: in a so-called single-ended device, it is mounted in contact with a substantial metal (normally copper) base, which together with a ceramic or metal-and-glass cover constitutes a hermetically sealed enclosure and also serves to conduct heat away from the element to a cooling device, to which it is attached by a threaded stud, or, in the case of a flat-based device, by screws or a clamp. The element may be soldered to the base, or it may be held in intimate contact with it by spring pressure; the pressure-contact assembly is widely favoured for large-area devices for a number of reasons, among the most important being that it avoids problems of high temperatures and disparate coefficients of expansion which arise when the assembly is subjected to a hard-soldering process; soft soldering is unsuitable for large-area devices on account of its liability to thermal fatigue, and a consequent deterioration in thermal conductance, in cyclic operation. The base also serves as one electrical connection—normally the anode in a thyristor or the collector in a transistor, but either the anode or the cathode of a diode—and the other main connection is made by soldering or by pressure contact to the opposite face of the element and

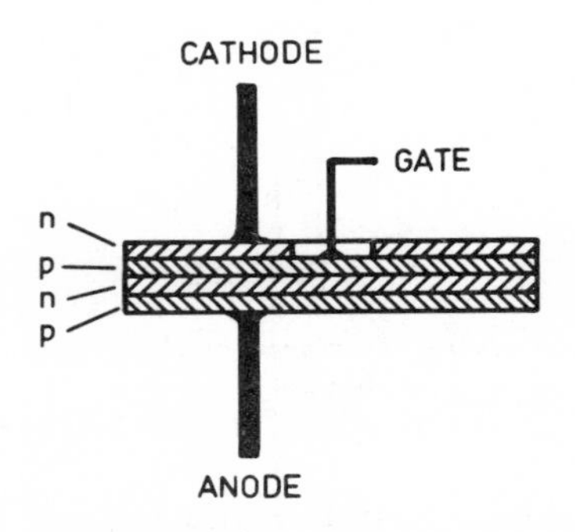

Figure 1.10 Connections to a thyristor with a centre gate.

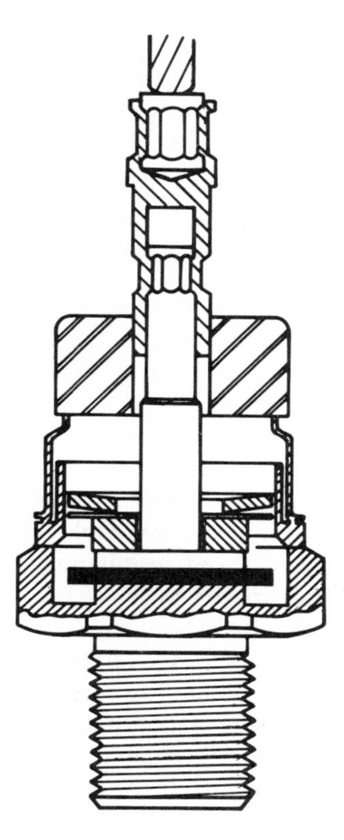

Figure 1.11 Typical construction of a stud-based (pressure-contact) diode.

brought out through a top cap. The typical construction of a pressure-contact diode is illustrated in Figure 1.11.

In a capsule device, the element is sandwiched between two substantial copper discs, which are attached by slightly flexible diaphragms to an insulating ring, usually ceramic, as illustrated in Figure 1.12. The element and the copper discs are held in intimate contact by heavy external pressure, which also clamps the device against the cooling surface or surfaces. The capsule form of construction offers the

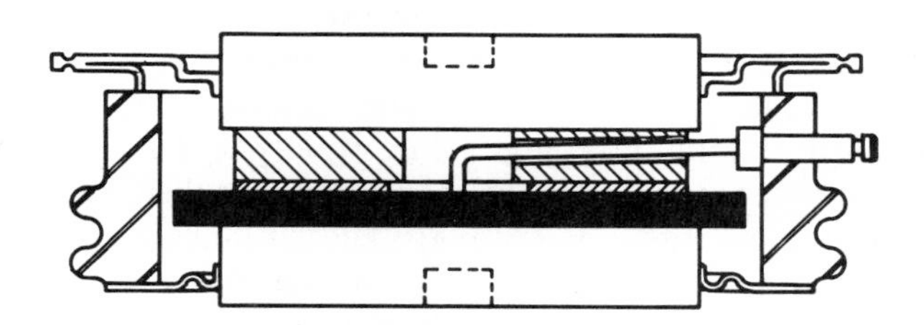

Figure 1.12 Typical construction of a capsule thyristor.

advantage that coolers can be applied to both sides, enabling the overall thermal resistance to be reduced to about half that obtainable with single-ended construction; a further advantage is that the possibility of mounting the capsule either way round leads to greater flexibility in the arrangement of coolers and electrical connections in multiple assemblies.

THE SWITCHING BEHAVIOUR OF DIODES AND THYRISTORS

A diode may be considered to be a switching device since it will change rapidly from a conducting to a virtually non-conducting state according to the polarity of the voltage applied to it. The onset of conduction, when forward voltage is applied, albeit not strictly instantaneous, is virtually so, and calls for no consideration here. The reversion to the blocking state upon the application of a reverse voltage is more complicated and will be discussed in some detail, in relation equally to the thyristor, which in this respect behaves in the same way.

The basic merit of the thyristor as a controllable switching device, that the state of forward conduction can be made self-sustaining independently of, and over an area remote from, the gate, carries with it two disadvantages: firstly, an appreciable time is required for the state of conduction to spread over the whole area of the element, signifying in effect a relatively slow turn-on process, and secondly (some special devices excepted) conduction, once established, cannot be terminated by any action of the gate; the normal thyristor can be turned off only through an interruption of the anode current by some external means, which can in some circumstances lead to an unwelcome circuit complexity, significant circuit losses and a consumption of time which limit the frequency at which the thyristor can be used.

The turn-on process in thyristors

The turning-on of a thyristor in response to a gate firing signal may be considered to take place in three phases. First, a delay occurs before any kind of significant response is apparent. Secondly, conduction is established in a small area adjacent to the gate and under the more-or-less direct influence of the gate current; the proportion of the total area that this represents depends upon the design of the cell, but in a large thyristor of simple construction it could be as little as 1 or 2%. Finally the conducting area spreads with a fairly constant velocity, no longer influenced by the gate, typically of the order of 0.1 mm/μs, until the whole area is conducting or the period of conduction ends.

Corresponding to the increase in the conducting area, the resistance of the cell falls from the near-infinite 'blocking' value, rapidly to a relatively high value while conduction is concentrated in the gate region, and then relatively slowly to the steady-state value (time permitting). The instantaneous voltage drop, assuming an anode current rising fairly rapidly to a constant level, thus follows some such waveform as that illustrated in Figure 1.13, resulting in an instantaneous power dissipation $w = v_{AK} i_A$ as shown.

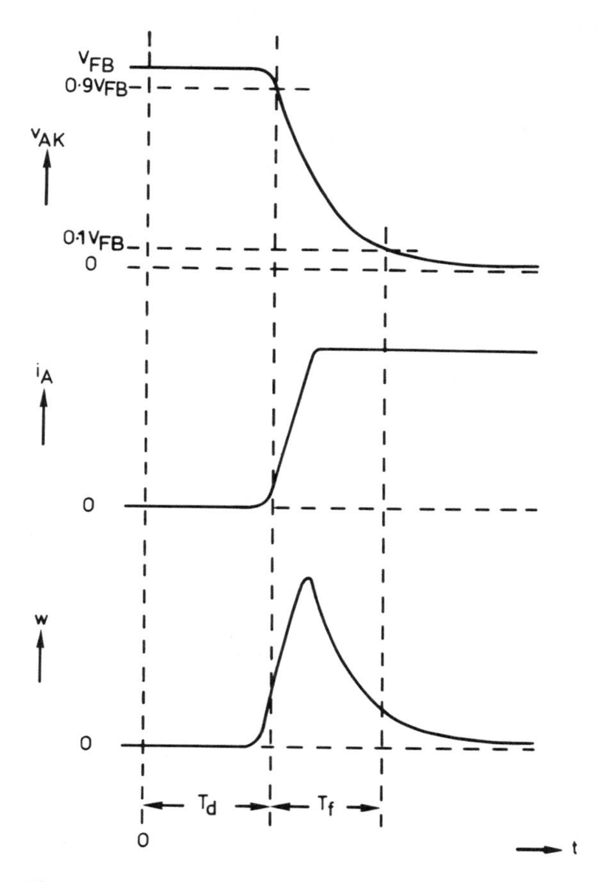

Figure 1.13 Typical transient waveforms of voltage, current and dissipation in a triggered thyristor. (Thyristor fired at $t = 0$.)

Conventionally, as shown in Figure 1.13, the total switching time is divided into the delay time and the fall time, these periods being arbitrarily delimited, for the purposes of measurement, at the instants when the voltage across the thyristor falls to 90% and 10% of the initial blocking voltage. An increasing level of gate drive reduces the delay time, but has relatively little effect on the fall time (Figure 1.14). In practice, since the switching process may be related to an infinite variety of anode current waveforms, as well as a range of firing currents, the switching waveforms are in a general sense indeterminate, and the conventional definition of turn-on time is somewhat meaningless unless related to specific conditions. Figure 1.15 illustrates the effect on the dissipation waveform of varying rise time in an anode current waveform of constant amplitude.

The energy loss associated with turning on, corresponding to the area under the dissipation waveform, can make a considerable difference to the average power

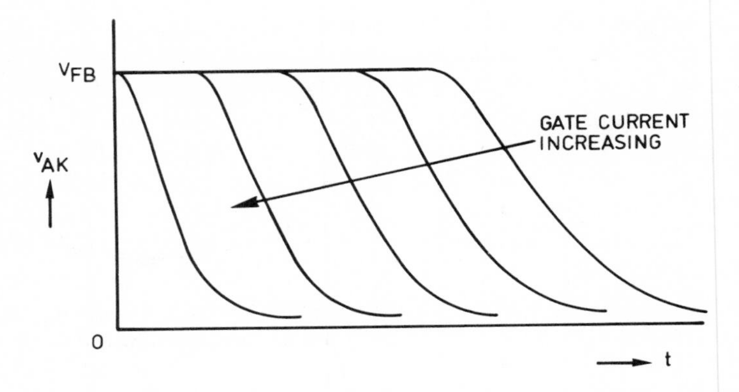

Figure 1.14 Effect of gate current amplitude on the turn-on time of a thyristor.

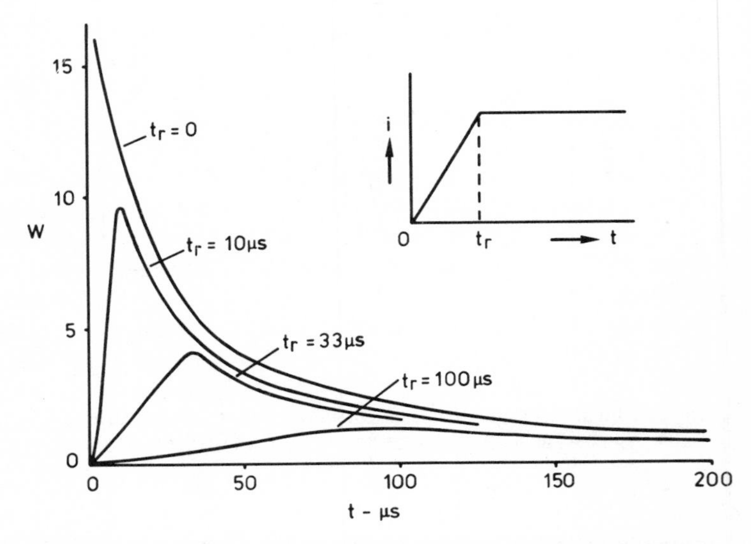

Figure 1.15 Effect of anode current rise time on the turn-on loss of a thyristor. (Thyristor fired at $t = 0$: instantaneous dissipation w in arbitrary units.)

loss in a thyristor operating at high frequency, when the spreading time—possibly 100 μs or more in a large thyristor—may easily constitute a large proportion of the total conduction period, or even exceed it, so that the average forward voltage drop is much higher than would be expected from the steady-state characteristics.

Also of considerable significance, if the anode current is fast-rising, is the high dissipation associated with the initial conduction in the gate region, both because of its contribution to the total energy loss and because of the concentration of heat dissipation around the gate, which implies a transiently increased thermal, as well as electrical, resistance. The turn-on loss is reduced, in these circumstances, if the area which is rendered conducting directly by the gate action is increased. This may be achieved, up to a point, by increasing the magnitude of the gate current pulse (given an appropriately short rise time). Thyristors intended for high-frequency operation are commonly designed with special gate configurations to enhance the initial effectiveness of the gate, by enlarging its perimeter, and to reduce the distance over which conduction has to spread; by these means the turn-on loss may well be reduced by an order of magnitude under onerous conditions. The methods employed include the use of multiple gates or a peripheral ring gate instead of, or in addition to, the central disc; in the extreme, the gate and cathode may be interdigitated in somewhat the same way as the base and emitter of a power transistor. A larger gate area generally implies the need for a correspondingly higher gate current (see 'Amplifying Gate Thyristors', below).

Small thyristors inherently tend to have more favourable ratios of gate area to total element area than large ones, and the need for special gate structures is therefore more pronounced in large cells. By the same token, using a group of parallel-connected small thyristors in preference to one large one can be advantageous in high-frequency circuits.

Turn-off processes

Since the thyristor, generally speaking, embodies no means of interrupting a current flowing in it, the process whereby, having once been triggered into conduction, it reverts to the non-conducting state is of fundamental importance in the application of the device. In some circuits, conduction ceases naturally as a result of the phenomenon known as natural commutation, in which the current is diverted to an alternative path by the influence of the voltages operating in the circuit—generally alternating supply voltages—and the thyristor is biassed onto its reverse blocking characteristic for a relatively long period. If, as is usually the case in circuits supplied from a d.c. source, natural commutation does not occur, an additional circuit is required whereby the current may be diverted from the thyristor, and a reverse voltage applied to it, for a brief but sufficient period to enable it to recover its forward blocking capability before forward voltage is reapplied; this process is referred to as forced, or artificial, commutation.

Reverse recovery

An important aspect of commutation is the transient behaviour of the conducting p–n junction when a reverse voltage is suddenly applied to it; this is illustrated in Figure 1.16. Some impedance has to be assumed in the source of reverse voltage, and the inductance L in Figure 1.16(a) represents stray inductance, or a small inductor included to limit the peak current and di/dt. The voltage V_R is here assumed to be constant, representing the process of natural commutation; the following discussion, albeit in terms of the thyristor, applies equally to diodes.

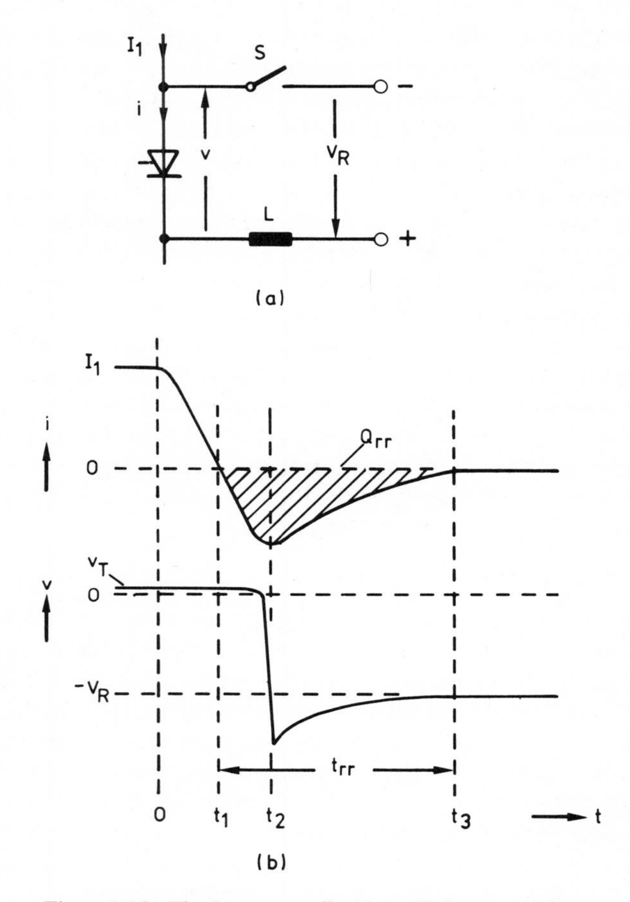

Figure 1.16 Thyristor turn-off with applied reverse voltage.

Upon the closure of switch S (Figure 1.16) to turn off the conducting thyristor (or diode), initially carrying current I_1 from the external circuit, the thyristor does not assume its reverse blocking state immediately but passes a considerable reverse current, which falls subsequently to the steady-state leakage value. The switch being closed at $t = 0$ (Figure 1.16 (b)), the thyristor current i falls from its initial value I_1 at a rate determined by V_R and L. At t_1 the current passes through zero and reverses, and in the period t_1–t_2 it continues to change at an unaltered rate while the thyristor offers no impedance to reverse current by virtue of the excess of available charge carriers at the anode junction (j_1, Figure 1.9) surviving from the period of forward conduction. This excess diminishes as the charge carriers are removed by the current or recombine, and at t_2 they become inadequate to support the current; at t_2, therefore, the current becomes dependent on the behaviour of the thyristor rather than the external circuit, and reverse voltage appears across the cell. The reverse voltage may exceed the applied voltage

V_R owing to the reversed di/dt in the inductance, although excessive overshoot is normally prevented by a suppression network.

The area under the reverse current waveform, hatched in Figure 1.16(b) (leakage current is ignored), represents a stored (recovered) charge Q_{rr} of which a substantial fraction is removed before the thyristor blocks reverse voltage at t_2: the remainder is dissipated with virtually the full reverse voltage across the cell, and is thereby associated with an energy loss $\int_{t_2}^{\infty} vi\,dt$, which is roughly proportional to $V_R Q_{rr}$. Given a rapid transition from forward conduction of the thyristor to reverse blocking, which Figure 1.16 illustrates, Q_{rr} increases with increasing forward current, and, since it is diminished by natural recombination during the recovery process, decreases with decreasing di/dt; it can vary widely according to the construction of the thyristor, and is subject to a considerable spread among nominally similar cells. Typically Q_{rr} may be of the order of 0.5 μC per ampere of forward current. The interval which elapses between the instant when the thyristor current passes through zero (t_1) and the instant at which the reverse current falls again to zero at t_3 is the reverse recovery time t_{rr}.

FORCED COMMUTATION

In a forced-commutation system the reverse voltage required to turn off the thyristor is supplied by a capacitor, which is charged to a suitable potential and discharges at a rate which is related to the current flowing initially in the thyristor. For the sake of economy, circuit efficiency and other considerations, the capacitor is made no larger than necessary, and the duration of the reverse voltage is therefore, generally speaking, the minimum that will ensure that the thyristor is reliably turned off.

Forced-commutation circuits fall into two broad categories—those in which the capacitor is switched (actually or effectively) directly in parallel with the thyristor, which is thereby subjected in reverse to the voltage to which the capacitor is initially charged, and those in which the applied reverse voltage is the forward voltage drop across a parallel-connected diode, which carries the discharge current from the capacitor (more accurately, part of it), controlled by an inductor. Reverse voltage is necessary, from a practical point of view, to extract stored charge from the thyristor in order to avoid an excessive and unpredictable prolongation of the turn-off process: the voltage drop of the diode in the second category of circuit mentioned is adequate in many cases (particularly with thyristors of recent design) although with some cells possibly as much as five volts may be necessary to reduce the reverse recovery time to a minimum.

Forced commutation with substantial reverse voltage

The operation of a forced-commutation circuit in which the capacitor voltage appears across the thyristor is illustrated in Figure 1.17. When the switch S is closed, the capacitor first supplies the reverse recovery charge demanded by the thyristor, and thereafter discharges at a rate depending on the capacitance and the

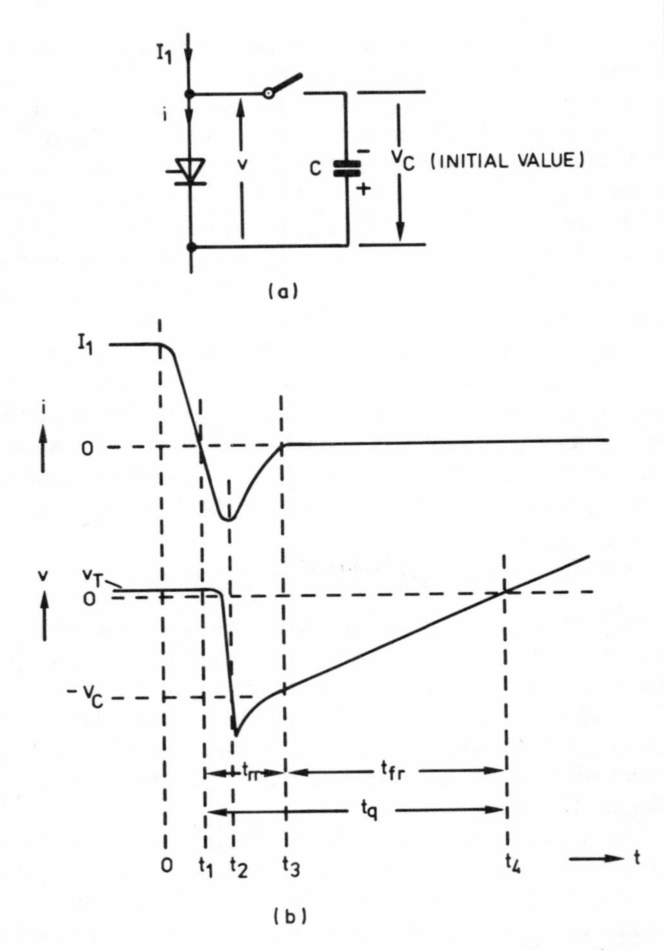

Figure 1.17 Thyristor turn-off with capacitor as reverse voltage source.

external circuit current, so that the waveform of the voltage across the thyristor is generally of the form shown in Figure 1.17(b) (where I_1 is assumed to be constant during the period of interest).

The voltage waveform in Figure 1.17 is drawn on the assumption that the thyristor blocks forward voltage after t_4, at the expiry of the reverse voltage period. For this it is necessary that the centre junction (j_2, Figure 1.8) shall by that time have recovered, i.e. become substantially free of available charge carriers; the only process that can bring this about is natural recombination, since any assistance from the external circuit would have to be in the form of a forward current, which it is the object of the turn-off process to prevent, and in consequence this so-called forward recovery takes considerably longer than the reverse recovery, and its commencement is moreover delayed until reverse recovery is nearly complete. The turn-off time t_q of a particular thyristor is the shortest period measured from t_1, the instant of zero current, to t_4 (Figure 1.17) that will permit it to block the subsequently applied forward voltage, and is substantially the sum of the reverse recovery time t_{rr} and the forward recovery time t_{fr}.

For the commutation circuit to operate effectively, it must provide a turn-off interval T_q in excess of the thyristor turn-off time t_q. Thyristors intended for use in forced-commutation circuits are accorded a value of t_q as a characteristic, which is quoted in relation to specific operating conditions, making due allowances for manufacturing tolerances. The best figures obtainable range from about 4 to several hundred microseconds according to voltage rating, t_q being roughly proportional to the square of the blocking voltage. In a given thyristor, the turn-off time varies appreciably with junction temperature, typically decreasing by about 0.2 μs for every 1°C decrease in θ_j in the region of the maximum rated temperature, and slightly with forward current, rate of change of current and rate of change of applied forward voltage.

The dependence of t_q upon junction temperature has a special significance in circuits in which the thyristor carries very short current pulses of high amplitude, in that the dissipation of energy, concentrated in the gate region, associated with turn-on at the beginning of the pulse may produce a considerable transient temperature rise, and thus result in an appreciably longer turn-off time than would be expected normally.

Determination of commutating capacitance

If the effects of reverse recovery and finite di/dt are ignored, the capacitance required for a given turn-off interval T_q is easily estimated in terms of the external circuit current i_1 and the initial voltage V_C to which it is charged (Figure 1.18):

$$C = \frac{1}{-V_C} \int_0^{T_q} i_1 \, dt \tag{1.2}$$

More simply, if the external circuit current can be considered constant at I_1 for the duration of the commutation period,

$$C = \frac{I_1 T_q}{-V_C} \tag{1.3}$$

Strictly speaking, this estimation of capacitance is slightly optimistic, in that it does not allow for the additional charge Q_{rr} which has to be supplied by the capacitor, and a safer result would be given by

$$C = \frac{I_1 T_q + Q_{rr}}{-V_C} \tag{1.4}$$

In practice Q_{rr}/I_1, being generally only a fraction of 1 μC/A, as noted above, is small in comparison with t_q, and ignoring Q_{rr} introduces no serious error (a safety

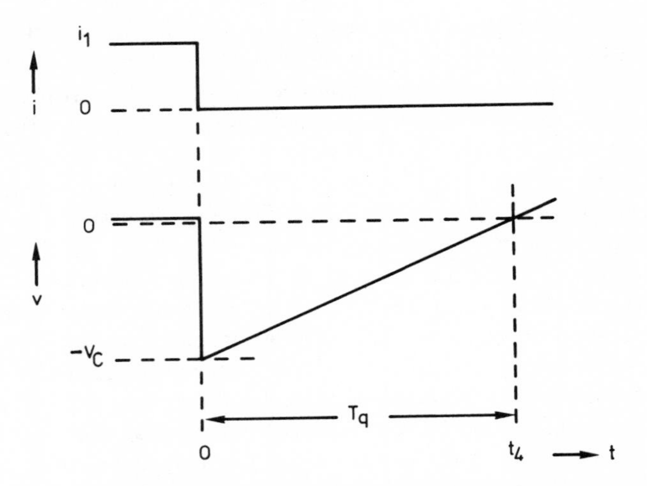

Figure 1.18 Simplified thyristor turn-off waveforms.

factor of at least 1.5 would normally be incorporated in T_q to cover various imponderables).

Commutation with inverse-parallel diode

Where a diode is connected across the thyristor, restricting the reverse voltage developed across it to something of the order of one volt, the discharge current from the capacitor is controlled during the turn-off period by an inductor which is chosen, in conjunction with the capacitor, to give an oscillation half-period somewhat greater than the required turn-off interval.

In Figure 1.19, the switch S is closed at $t = 0$ with the capacitor charged to a voltage V_C. The capacitor thereupon starts to discharge through the inductor L with a substantially sinusoidal current, so that

$$i_C \approx \frac{V_C}{\sqrt{(LC)}} \sin \omega_0 t$$
$$v_C \approx V_C \cos \omega_0 t \tag{1.5}$$

where $\omega_0 = 1/\sqrt{(LC)}$. By design, the peak discharge current $V_C/\sqrt{(L/C)}$ is considerably in excess of the external circuit current I_1 flowing initially in the thyristor (here assumed constant). The capacitor current initially represents a diminution of the thyristor current i_{Th}, which falls to zero at t_1, when $i_C = I_1$. Between t_1 and t_2, the instant when i_C falls again to equal I_1, the difference current $(i_C - I_1)$ flows in the diode; at t_2 the diode current falls to zero and thereafter i_C remains constant and equal to I_1, assuming that the thyristor has by that time recovered its forward blocking capability, and since di_C/dt thereby becomes zero, the voltage across L also becomes zero, and the capacitor voltage, now of reversed

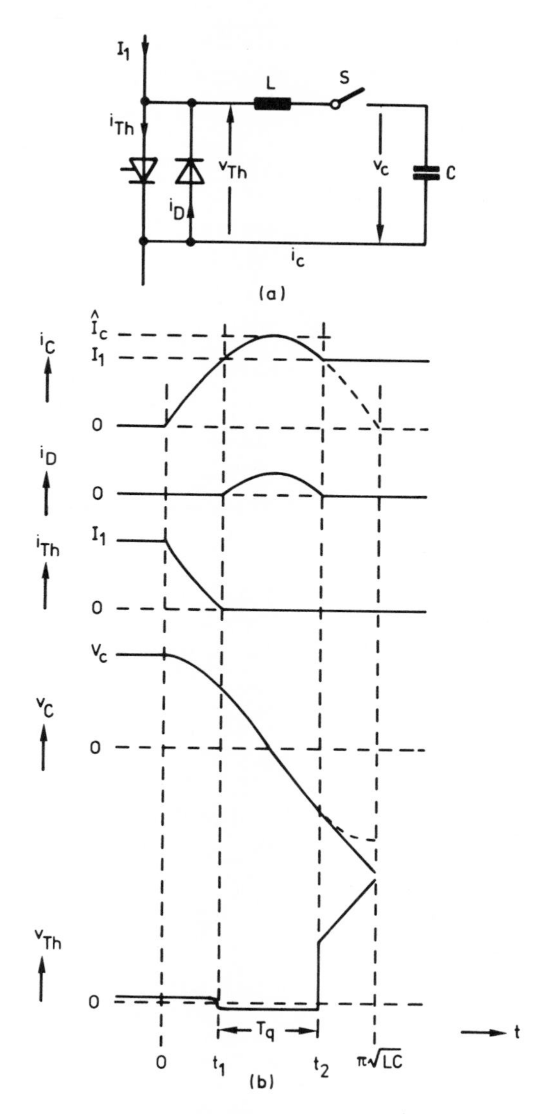

Figure 1.19 Forced commutation with inverse parallel diode.

polarity, therefore appears directly across the thyristor in the forward direction (Figure 1.19(b)).

The turn-off interval T_q is thus the period t_1–t_2. The very low value of reverse voltage in this method of commutation can result in an appreciable increase in the reverse recovery time of the thyristor, and hence in t_q, but this depends on the construction of the thyristor, and in many cases the effect is not noticeable.

Given I_1 and V_C, values of L and C required to produce a stipulated turn-off interval can be calculated from the basic relationship, illustrated by the waveforms of Figure 1.19(b),

$$\frac{I_1}{\hat{I}_C} = \cos\frac{\omega_0 T_q}{2} \tag{1.6}$$

where $\omega_0 = 1/\sqrt{(LC)}$

$$\hat{I}_C = \frac{V_C}{\sqrt{(L/C)}}$$

Hence
$$\frac{I_1\sqrt{(L/C)}}{V_C} = \cos\frac{T_q}{2\sqrt{(LC)}} \tag{1.7}$$

So long as $\pi\sqrt{(LC)}$ exceeds the required value of T_q, the above relationship may be satisfied by a range of values of L and C. The combination that gives the lowest value of C, and hence the minimum stored energy, is found as follows:

From (1.7),
$$\cos\frac{\omega_0 T_q}{2} = \frac{I_1}{V_C\omega_0 C}$$

or
$$\frac{1}{C} = \frac{\omega_0 V_C}{I_1}\cos\frac{\omega_0 T_q}{2} \tag{1.8}$$

Then for the minimum value of C

$$\frac{d}{d\omega_0}\omega_0\cos\frac{\omega_0 T_q}{2} = 0$$

from which
$$\tan\frac{\omega_0 T_q}{2} = \frac{2}{\omega_0 T_q} \tag{1.9}$$

This gives the optimum relationships for the half-period of oscillation and the peak discharge current:

$$\left.\begin{aligned} \pi\sqrt{(LC)} &= 1.83\,T_q \\ \frac{V_C}{\sqrt{(L/C)}} &= 1.53\,I_1 \end{aligned}\right\} \tag{1.10}$$

CHARACTERISTICS AND RATINGS OF DIODES AND THYRISTORS

In the following sections, as elsewhere, the exposition will be mainly in terms of thyristors; it is to be assumed that it is equally valid, where applicable, to diodes.

It is proper to make a distinction between ratings and characteristics. According to the usage employed here, the ratings of a device are the declared limits of its permissible operating conditions, with the implication that operation beyond these limits involves at least a risk that the device will be damaged, or that its life will be impaired. The characteristics, on the other hand, are descriptions, normally in the form of limits, of the operation of the device under stated conditions. However, since ratings are invariably dependent in some way upon characteristics, and since in many cases what appears for practical purposes as a rating may in fact be a limiting condition of operation for which a declared characteristic is guaranteed, or represent simply the limit of available test information, it is more convenient here to consider them together than, as in a data sheet, separately. It should be observed that quoted characteristics and ratings are limits applicable to types of cell, rather than to individual cells of the type in question, which in general display considerable variations, due to lack of precise control in manufacture, without transgressing the type limits. The characteristics and ratings of semiconductor devices are mainly electrical and thermal.

Effects of temperature

Temperature affects semiconductor devices in two ways: firstly it affects their electrical characteristics, and secondly it may have a direct effect on the materials used or inadvertently included in their construction.

Inasmuch as the temperature of a semiconductor device may not be uniform, electrical characteristics are defined in relation to the temperature of the operative part of the semiconductor element, namely the p–n junction. This is hardly a rigorous concept, since the junction itself may not be at a uniform temperature, and there may be more than one junction, but it is nevertheless a usable one, more accurately to be thought of in terms of a 'virtual junction', which is a point within the element at a temperature equal to the uniform temperature at which the element would exhibit the observed characteristics. In ordinary parlance, the virtual junction is referred to simply as 'the junction'. The specific effects of junction temperature on the various characteristics are considered below.

The direct effects of temperature, apart from drastic ones associated with severe fault conditions, are typically mechanical stresses caused by differential expansion in dissimilar materials, or possibly undesirable effects such as the migration or condensation of contaminants, relaxation in crimps, or thermal fatigue in solders. Such effects lead to ultimate limits of permissible operating temperature irrespective of electrical characteristics, but usually such forms of construction are employed that these limits are not restrictive in relation to the range of temperature over which the electrical characteristics are satisfactory, or over which operation is actually required. A partial exception to this may be seen in the use, for reasons of economy, of soft solder in some cells, which may place a restriction on the maximum permissible operating temperature under conditions of frequent cyclic junction temperature variations, if thermal fatigue of the solder, and the increase in the internal thermal resistance of the cell which it causes, are to be avoided.

The dissipation of losses in a semiconductor device leads to a rise in its internal temperature which is a function of the amount of the dissipation and the thermal resistance of the cell. The latter is normally defined (except for very small wire-ended devices) as between the virtual junction and the mounting base or stud, so that the temperature rise considered from the base to the junction is

$$\theta_j - \theta_b = \delta\theta_{j-b} = WR_{j-b} \tag{1.11}$$

where R_{j-b} is the thermal resistance and W is the dissipation. Similarly, the base assumes a temperature above that of its surroundings by an amount depending on the total thermal resistance external to the cell. If R_{b-a} is the thermal resistance from the base to the ambient fluid or heat sink, the total temperature rise is

$$\theta_j - \theta_a = \delta\theta_{j-a} = W(R_{j-b} + R_{b-a}) \tag{1.12}$$

The above is a slight oversimplification in that a small proportion of the heat produced is lost directly from the housing of the cell, or along the flexible top connection in the case of a single-ended cell. More significant in some circumstances is that a calculation based on thermal resistance yields essentially a steady-state solution, whereas for short periods the junction temperature rise is time-dependent, owing to the thermal capacities of various parts of the cell. This aspect, and also that of cooling semiconductor devices, are dealt with in Chapter 6.

Voltage ratings

Absolute limits are set to the permissible operating voltage by the breakover voltage in the forward direction, and by the sharp increase in reverse current in the avalanche region. In practice a maximum instantaneous reverse voltage rating is usually determined by reference to a particular value of leakage current, based on the known spread of characteristics for a particular type of cell; this constitutes a peak transient or non-repetitive rating, related to a specified duration: e.g. the amplitude of a 10 ms half-sine-wave. Lower values of voltage, arrived at by applying an empirical ratio or testing to a lower current limit, may be specified as repetitive instantaneous or d.c. ratings, limiting the average or continuous reverse loss to a reasonable, low level.

A repetitive voltage rating does not signify an appropriate nominal operating voltage for a semiconductor device, which can only be arrived at (unless by trial and error) through a knowledge of possible transient voltages, repetitive or non-repetitive, superimposed on the normal supply.

The effects of an increasing junction temperature on the reverse characteristic of a silicon diode or thyristor are to increase the leakage current (ideally according to an exponential law but in practice not necessarily, owing to leakage at the edge of the element) and to increase the avalanche voltage. The general effect is thus as illustrated in Figure 1.20. Figure 1.20 also shows the effects on the forward blocking characteristic of a thyristor, which are, again, to increase the leakage current, but, in consequence, to reduce the breakover voltage.

Relative to measurements at normal room temperature (25°C for the purposes of standardization), the reverse voltage rating as determined by measurement at a particular current may be higher at the maximum rated junction temperature,

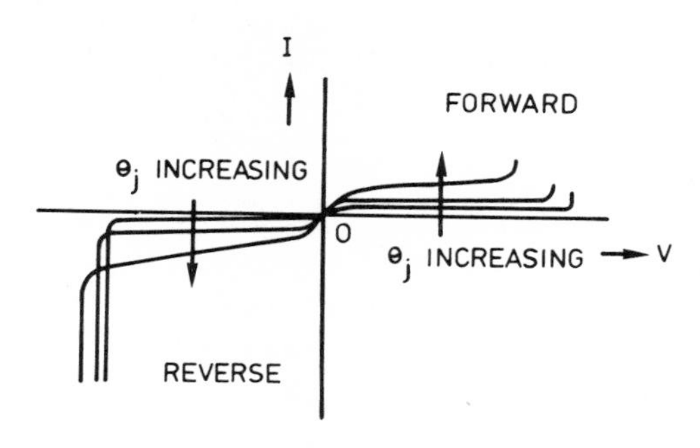

Figure 1.20 Effects of junction temperature on the blocking characteristics of a thyristor.

because of the increased avalanche voltage, or lower, because of the increased leakage current, depending on the shape of the characteristic; these alternative possibilities are illustrated in Figure 1.21. At low temperatures, the avalanche effect is normally predominant, and the voltage rating therefore reduced. The overall voltage ratings of a particular thyristor, which have to conform to a standardized relationship, are based on the least favourable of all the forward and reverse tests, at room temperature and at the maximum junction temperature, allowing small safety margins for possible errors in measurement.

An overriding factor in assigning a voltage rating to a semiconductor device is the need to avoid thermal instability, whereby the increase in reverse leakage current with increasing temperature and the increase in temperature resulting from increased current are in such a relationship that current and temperature both rise at an increasing rate until the cell is destroyed. If the reverse voltage is V_R and the thermal reistance of the cell is R_{th}, a small increase in reverse current δi_R results in an increase in junction temperature

$$\delta\theta_j = \delta i_R \, V_R \, R_{th} \tag{1.13}$$

This increase in junction temperature in turn causes an increase in reverse current

$$\delta i_R = \delta\theta_j \frac{di_R}{d\theta_j} = \delta i_R \, V_R \, R_{th} \frac{di_R}{d\theta_j} \tag{1.14}$$

i_R will thus increase unstably if $V_R \, R_{th} \, di_R/d\theta_j$ is greater than unity.

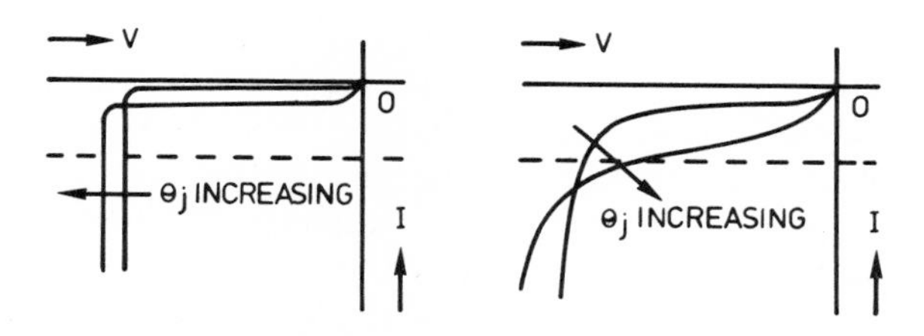

Figure 1.21 Possible effects of junction temperature on measured reverse voltage rating.

The likelihood of thermal instability increases with increasing reverse voltage and with increasing junction temperature, as may be illustrated as follows for the ideal relationship of reverse current to temperature. The latter may be expressed as

$$i_{\mathrm{R}} = a\mathrm{e}^{b\theta_{\mathrm{j}}} \tag{1.15}$$

where b is a constant and a is a quantity which is constant or increases with increasing voltage. Then

$$\frac{\mathrm{d}I_{\mathrm{R}}}{\mathrm{d}\theta_{\mathrm{j}}} = ab\mathrm{e}^{b\theta_{\mathrm{j}}} = bI_{\mathrm{R}} \tag{1.16}$$

The condition for instability in this case is

$$V_{\mathrm{R}} R_{\mathrm{th}} ab\mathrm{e}^{b\theta_{\mathrm{j}}} = 1$$

$$\text{or} \qquad V_{\mathrm{R}} R_{\mathrm{th}} b i_{R} = 1 \tag{1.17}$$

In general. thermal instability is not a serious problem with silicon devices, although it can be significant in certain cases where R_{th} and i_{R} tend to be relatively high, for example in the case of a small cell used without a cooler. It should be noted that the value of R_{th} to be considered in relation to equation (1.14) is somewhat higher than the normal one if the reverse current is not uniformly distributed throughout the area of the element.

Forward current ratings

There are in the main three considerations which may set limits to the permissible forward current, namely

(i) the requirement that the rated junction temperature should not be exceeded as a result of losses,
(ii) the safe current rating of the external connections, particularly flexible conductors and crimps,
(iii) the temperature gradients within the cell, which should not be such as to cause excessive stress through differential expansion.

Since the junction temperature depends on the internal temperature rise and the base temperature, the limiting current determined by (i) is a function of base temperature. On the other hand, (iii) is largely independent of base temperature, and (ii) is commonly defined without reference to temperature. A graph of current rating against base temperature is therefore normally in the form of a region of constant current adjoining a region in which the current falls with increasing base temperature, as illustrated in Figure 1.22.

If the reverse loss is so small as to be considered negligible, as is usually the case in a sizeable silicon device, the current rating falls to zero at a base temperature equal to the rated continuous junction temperature, as in Figure 1.22(a):

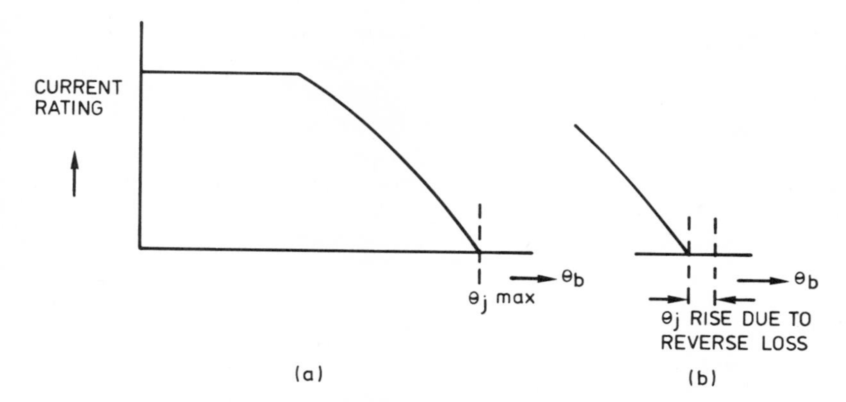

Figure 1.22 Variation of forward current rating with base temperature: (a) with negligible reverse loss; (b) with significant reverse loss.

otherwise a margin of temperature is left below that level corresponding to the effect of the reverse loss alone, as in Figure 1.22(b).

In most applications of diodes and thyristors, periodic variations in the instantaneous losses give rise to periodic fluctuations in junction temperature. Strictly speaking, therefore, equations (1.11) and (1.12) are valid only in terms of average loss and temperature rise: thus,

$$\left.\begin{aligned} \overline{\theta}_{\mathrm{j}} - \overline{\theta}_{\mathrm{b}} &= \overline{W} R_{\mathrm{j-b}} \\ \overline{\theta}_{\mathrm{j}} - \theta_{\mathrm{a}} &= \overline{W}(R_{\mathrm{j-b}} + R_{\mathrm{b-a}}) \end{aligned}\right\} \quad (1.18)$$

The forward conducting loss due to a given current I is calculated on the basis of the simplified characteristic illustrated in Figure 1.4:

$$\overline{W}_{\mathrm{F}} = \overline{I} v_0 + I^2 r = \overline{I} v_0 + \overline{I}^2 k_{\mathrm{f}}^2 r \quad (1.19)$$

where k_{f} is the current form factor.

Ignoring reverse loss, equation (1.19) is used, assuming a limit forward characteristic, to derive loss data for a range of idealized current waveforms, such as part-sinusoids and square waves of various conduction angles, and assuming a worst combination of forward characteristic and thermal resistance, to compute the thermal limits of current rating, for the same waveforms, as functions of base temperature. A small margin of temperature is normally allowed to take account of the periodic fluctuations referred to above; the effect may be calculated with fair accuracy using the techniques described in Chapter 6.

It is not normally necessary to base the ratings on a combination of the limit forward characteristic and the limit thermal resistance, since the improbability of approaching both together in one cell makes it possible to take a more optimistic view without causing a significant number of rejects in a temperature-rise test.

As the conduction angle in a current waveform of a given type decreases, its form factor increases. This may have two effects: firstly, the mean current rating at a particular base temperature decreases, as k_{f} in equation (1.19) increases, and secondly the r.m.s. current may rise to the limiting value determined by the connections, or some other factor, and thus determine the mean current rating directly. The mean current rating, for a given base temperature, therefore tends to vary with conduction angle in the way illustrated by Figure 1.23. More conveniently, for practical purposes, the losses and the permissible base temperature are presented as functions of current for various conduction angles in the manner of Figure 1.24.

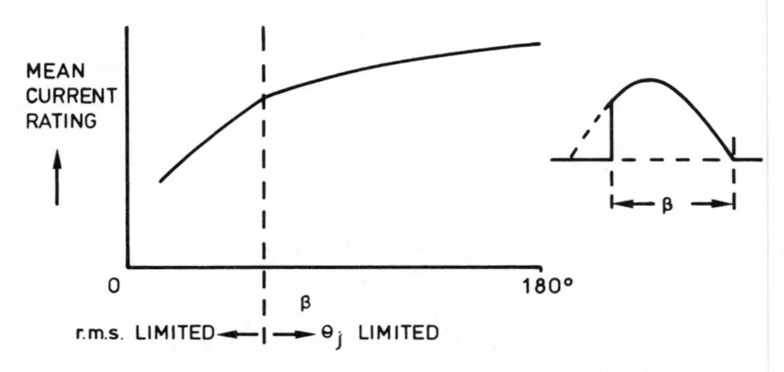

Figure 1.23 Variation of the forward current rating of a thyristor with conduction angle.

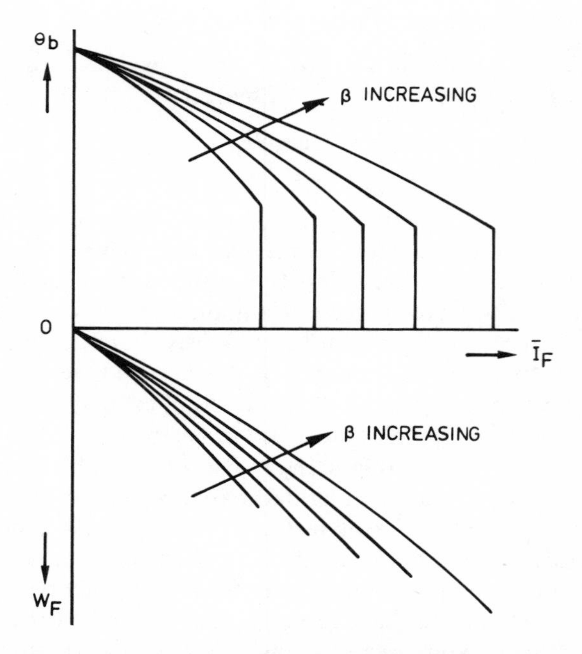

Figure 1.24 Practical presentation of the conduction loss and current rating of a thyristor.

Gate firing characteristics and gate ratings

The direct effect of applying a positive gate current to a thyristor can be considered to be a reduction in the breakover voltage, typically as illustrated by the graph of Figure 1.25. In a practical application, the minimum effective firing current is that which will reduce the breakover voltage below the voltage applied to the cell, and to cover the majority of applications it is defined with respect to a suitably low anode-to-cathode voltage—typically five volts. The figure quoted is the maximum limit of what is normally a fairly wide production spread, at a particular temperature, together with the corresponding maximum gate firing voltage. In addition, it is normal to quote the maximum gate voltage that is guaranteed not to fire any thyristor at its quoted minimum breakover voltage, typically 0.25 V.

The gate firing current and voltage are both markedly temperature dependent. The way in which they vary with temperature is itself variable, and for practical purposes is best illustrated by the boundaries of possible values over the range as in Figure 1.26, rather than by typical characteristics of individual cells.

At any particular temperature, the limit firing current and voltage are defined without regard to time. As stated above, the delay which occurs before appreciable anode current starts to flow depends considerably on the level of gate current, becoming very long—possibly some tens of microseconds—at the minimum firing current of a particular cell and approaching zero if the current is increased sufficiently (Figure 1.14). A current in the region of five times the quoted minimum firing current is commonly employed when fast switching is desirable. In addition to the magnitude, the rise time of gate current also has a substantial effect on the delay time, unless it is considerably shorter than the delay time.

The current–voltage characteristic of a thyristor gate generally resembles that of a silicon diode, albeit a poor one; it is not normally subject to any expressed limits. The gate ratings normally quoted define the maximum peak forward and reverse voltages, the maximum peak forward current and the maximum mean and

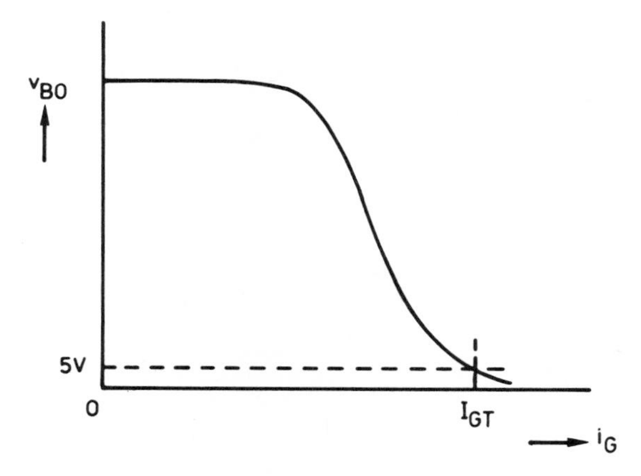

Figure 1.25 Variation of the break-over voltage of a thyristor with gate current.

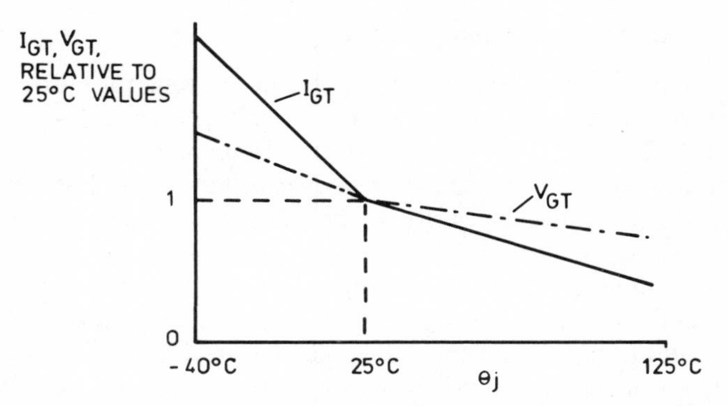

Figure 1.26 Typical limits of variation of I_{GT} and V_{GT} of a thyristor with junction temperature.

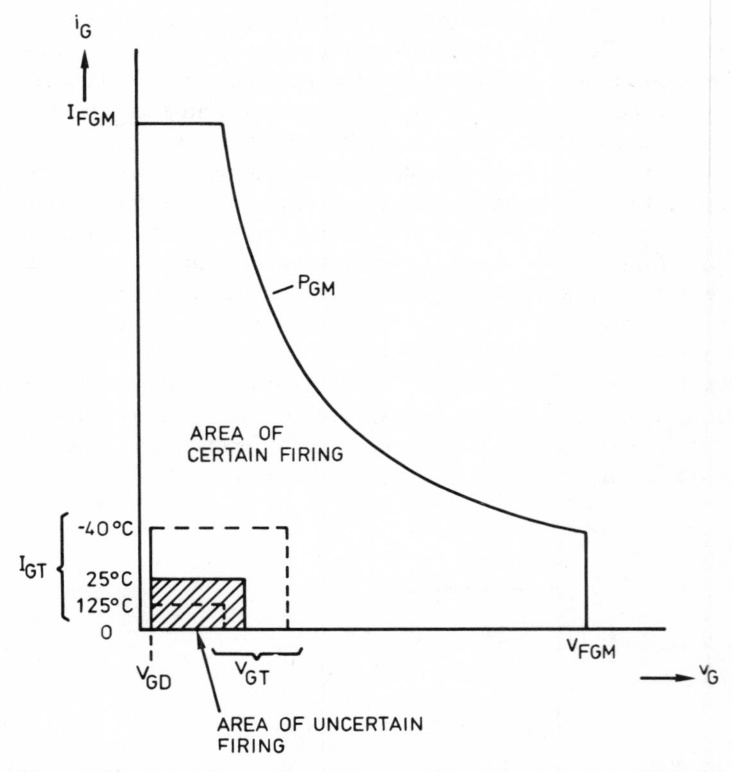

Figure 1.27 Form of quoted gate characteristics and ratings for a thyristor.

peak dissipation. These limits, with the characteristics described above, are illustrated in Figure 1.27.

Holding and latching currents

In the great majority of applications, latching current (I_L) is much more important than holding current, and it is therefore common to quote only one figure to cover both. Somewhat unexpectedly, a high level of gate drive does not reduce I_L, but increases it; by the same token I_L is to some extent dependent on the rate of fall of gate current when the gate drive is terminated.

d*i*/d*t*

Apart from the switching loss that occurs during the spreading phase of the thyristor turn-on process, as discussed above, a high rate of rise of forward current in the initial stage, before any appreciable spreading of conduction has taken place, is especially significant. The almost instantaneous release of energy in the limited area of conduction adjacent to the gate, if excessive, causes damage which may result in immediate destruction of the cell; in some cases of repetitive operation the effect may not be immediately catastrophic, but cumulative, leading to failure after a relatively short period of service, so that the apparent ability of a thyristor to withstand a high rate of rise of current for a brief period of observation is in itself no guarantee that its operating conditions are safe.

The rate of rise of current, commonly referred to as 'di/dt', for which a thyristor can safely be rated varies greatly according to its operating conditions, being reduced by factors which increase the energy dissipated and increased by those which improve the thyristor's switching characteristics. Thus the rating is reduced for

(a) an increasing forward blocking voltage immediately before switching;
(b) an increasing peak forward current;
(c) an increasing designed voltage rating, which generally implies a thicker silicon element and a higher forward voltage drop;

and is increased by

(d) an increased gate overdrive—i.e., the ratio between the gate current and the firing current limit for the cell (since the latter varies with temperature, the minimum operating junction temperature is also a factor)
(e) a reduced gate current rise time.

For practical reasons, however, this degree of complication is not normally reflected in published data, particularly since the widespread adoption of amplifying gates (q.v.) for high-current thyristors, which to a large extent masks the variations in operating conditions.

A common case to which the di/dt ratings described above cannot be applied is that of a small capacitor in series with a resistor connected across a thyristor for the purpose of limiting the rate of rise of forward voltage (see below) or suppressing transient overvoltages; the effectiveness of the suppression circuit is impaired if the resistor is not of a non-inductive type, in which case the di/dt when the thyristor is fired, discharging the capacitor, is unlimited except by stray inductance and, ultimately, by the switching speed of the thyristor itself. A mitigating factor is present, however, in the limited amount of energy stored in the capacitor, and the case is covered simply by specifying the maximum capacitance and minimum resistance that may be used without deliberate limitation of di/dt in a particular range of voltage.

Since di/dt effects are associated purely with the gate region of the thyristor, they bear no direct relation to the overall size of the element, and small cells often have much better di/dt ratings, relative to their continuous current ratings, than large ones. There may thus be an advantage, in circuits in which high di/dt is unavoidable, in using a number of small cells in parallel (provided the current is adequately shared during the initial switching period) rather than one large one. Special thyristors for high di/dt duty are made with a large gate periphery, and fired with very large gate currents or provided with amplifying gates.

The ability of a thyristor to withstand high di/dt increases greatly once the area of conduction has increased appreciably, and the di/dt rating of an ordinary cell can be increased by an order of magnitude if the circuit is modified in such a way that the steep rise of current is delayed until a short time after the thyristor is fired. This entails connecting in series with the thyristor a saturable reactor which in its unsaturated state passes sufficient 'priming' current to enable the thyristor to turn on normally, and after perhaps ten microseconds saturates to permit the current to rise rapidly to its full value. It is necessary to provide in some way for resetting of the flux in the reactor core between consecutive operations.

It will be observed that the conditions conducive to a high di/dt rating are generally similar to those which minimize the turn-on loss.

dv/dt

A thyristor may be triggered as a result of a high rate of rise of forward voltage, even though no gate current is applied. The effect may be thought of as due to capacitive current within the element performing the same function as a gate current, albeit not necessarily in the region of the gate. The rate of rise, commonly 'dv/dt', at which a thyristor may break over is basically a characteristic, but, as in the case of breakover due to excessive forward voltage, it may effectively constitute a rating, in that the cell may be damaged as a result of the irregular mode of switching.

As with gate current, the effect of dv/dt may be regarded as a reduction of the breakover voltage, the effect becoming greater with increasing junction temperature. Thus, dv/dt ratings are normally given for the most restrictive case of maximum junction temperature, and with some qualification as to the applied voltage, for example, a certain proportion of the quoted breakover voltage.

No ambiguity is possible when dv/dt is defined in terms of a linear ramp rising to a stated voltage, as illustrated in Figure 1.28. If the final voltage is equal to the nominal breakover voltage V_{BO}, i.e. the voltage at which the thyristor may break over regardless of dv/dt, the dv/dt rating tends in principle to zero; on the other hand, if the voltage is below a certain level the thyristor will withstand an infinite dv/dt without breaking over, because the capacitive current that flows represents an insufficient charge to accomplish the triggering process. There is thus a

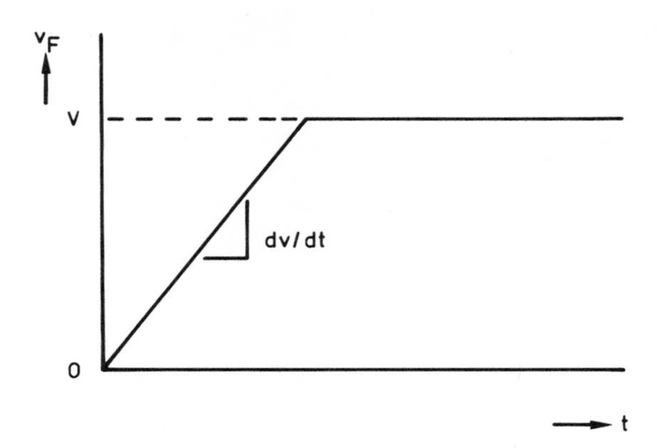

Figure 1.28 Linear rise of forward voltage applied to a thyristor to a stated level V.

variation of dv/dt rating with voltage generally of the form illustrated in Figure 1.29.

dv/dt ratings are sometimes expressed in relation to an exponential rise of voltage. The dv/dt figure is then defined conventionally as the slope of a linear voltage ramp which would reach a proportion $(1 - 1/e)$ of the asymptotic voltage in a time equal to the time constant of the exponential, which is $(1 - 1/e)$ times the initial rate of rise; this is illustrated in Figure 1.30. In comparison with a linear ramp with nominally the same dv/dt, the exponential ramp exhibits a higher slope up to $1/e$ of the final voltage and a lower slope thereafter, and there is no direct relationship between the dv/dt ratings assigned to a particular thyristor under the two conventions; generally speaking, the exponential figure is higher than the linear one. The exponential rating is preferred in some quarters on the grounds that it goes some way towards representing voltage waveforms frequently occurring in circuits which incorporate RC suppression networks, and that it is easily checked with simple apparatus; the linear approach, on the other hand, is

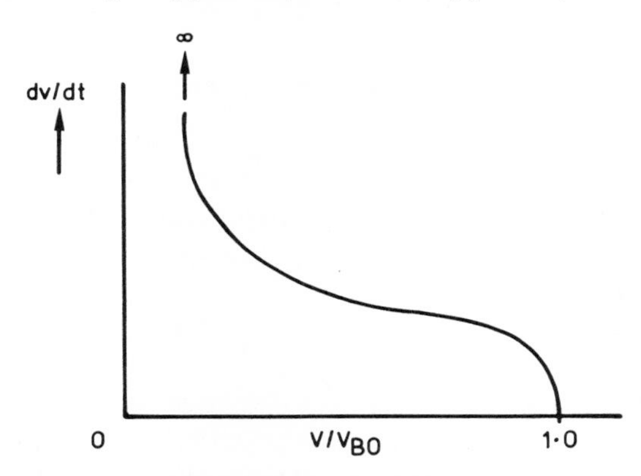

Figure 1.29 Variation of the dv/dt rating of a thyristor with final blocking voltage.

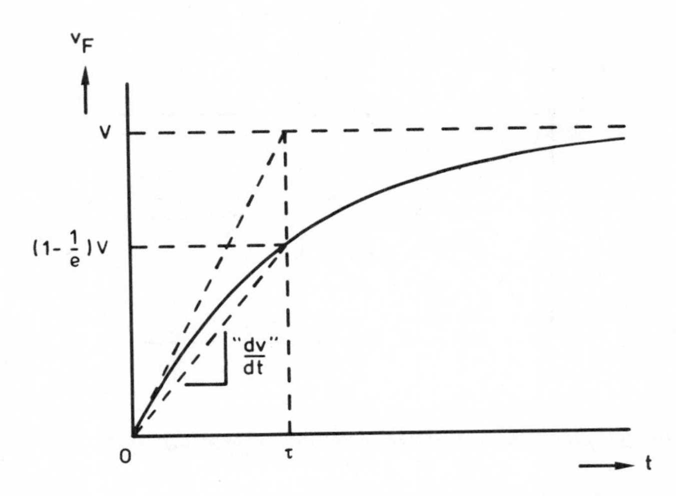

Figure 1.30 Conventional definition of exponential dv/dt.

more appropriate to comprehensive test methods, and is reasonably representative at least of some kinds of circuit waveforms.

Two different dv/dt ratings may be given for a thyristor, depending on whether or not it has been in a conducting state shortly before the application of forward voltage. That is to say, if the thyristor is turned off by a reverse voltage pulse of a duration not significantly greater than its turn-off time, in the manner illustrated in Figure 1.17 or Figure 1.19, its ability to accept a high dv/dt following the turn-off interval tends to be reduced in comparison with its dv/dt rating in a naturally commutating circuit, due to incomplete removal of charge carriers. This adverse effect on dv/dt capability is not usually significant in the situation represented by Figure 1.17, because the rate of change of voltage is determined at a relatively low level by the peak reverse voltage and the turn-off interval, but it can be in circuits that behave as depicted in Figure 1.19.

SPECIAL THYRISTOR STRUCTURES

The basic thyristor structure described above, while it still forms the basis for the design of a large proportion of the thyristors manufactured, has been elaborated or modified in various ways to overcome limitations in ratings and to improve performance. Some of the more important developments are as follows.

Shorted emitter

If the effect of excessive dv/dt is regarded as a capacitive current having an effect similar to that of an externally applied gate current, some improvement in dv/dt capability might be expected if a low resistance is connected between the gate and cathode terminals to provide an alternative path and so reduce the

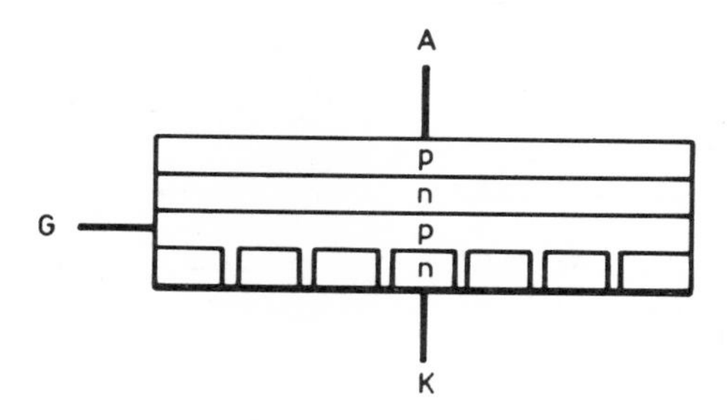

Figure 1.31 Shorted-emitter thyristor structure.

current flowing to the cathode (emitter). In fact the improvement is not very great, because shunting the gate has little effect over the remoter parts of the element area.

The shorted-emitter construction (Figure 1.31) provides a much more effective shunt, and hence a much better dv/dt capability, by incorporating multiple short-circuits distributed throughout the area. An inevitable side effect is a tendency for the firing current to be increased; this may, however, be alleviated to some extent by the careful proportioning of resistances within the gate-cathode structure.

Amplifying gate

Reference has been made above to the desirability of high gate firing current, with appropriately short rise-time, in order to minimize turn-on loss and obtain a satisfactory di/dt rating, and also to the need for higher firing current that arises with an enlarged gate perimeter in thyristors intended for high-frequency operation, and with the shorted-emitter structure. To avoid the necessity for firing circuits with very high outputs, many large thyristors are provided with a so-called amplifying or regenerative gate, whereby, in effect, an auxiliary triggering device, integral with the main structure, is used to apply a very large gate current to the main part of the element. The main firing current is derived from the anode circuit, and the auxiliary device requires only a firing pulse of normal magnitude—sufficient, that is, to ensure an acceptably short delay time. (The delay in the main part of the thyristor is virtually zero, because of the high firing current.)

A typical arrangement is illustrated diagramatically in Figure 1.32. The firing current pulse being applied to the gate terminal, a flow of current is initiated from the anode (assuming it to be at a positive potential) through the area adjacent to the gate. In flowing to the cathode connection, this produces a lateral potential gradient across the resistance of the cathode 'n' layer, and, as a result, a substantial proportion of the anode current is deflected to the main gate area, rapidly promoting full conduction in the main current path. This kind of principle may be associated with a wide variety of gate configurations.

Amplifying-gate thyristors have the advantage, among others, of removing one variable—the nature of the gate drive—from the conditions which affect the

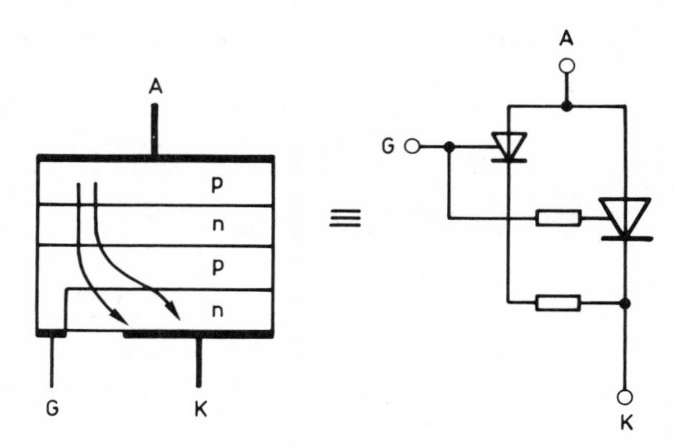

Figure 1.32 Amplifying-gate thyristor structure.

ratings, so that, for example, the di/dt rating can be not only higher, in general, but less complicated.

Asymmetrical thyristor

The term 'asymmetrical thyristor' is used to denote a triode thyristor with little or no reverse blocking capability (but also, as a rule, no useful reverse conducting capability). Given that thyristors are produced to the highest standards permitted by the available technology, it is a truism that particular characteristics can be improved only at the expense of an impairment of others. In the asymmetrical thyristor, sacrificing the reverse characteristic, which in many forced-commutation circuits is not required (the thyristor being in fact shunted by an inverse-parallel diode), facilitates a useful reduction in the turn-off time.

In some cases an inverse-parallel diode is incorporated into the thyristor element, eliminating the need (in appropriate circuits) for a separate diode and offering a technical advantage, in that the thyristor and the diode are in the closest possible proximity and therefore separated, electrically, by the minimum possible stray inductance.

Gate turn-off thyristor

In a small thyristor, the technique of gate–cathode interdigitation can be used to extend the influence of the gate to such effect that a reasonably large anode current can be turned off by a pulse of negative gate current without the need for an external forced-commutation circuit. In principle, this is an extremely useful facility, permitting a considerable simplification, and an improvement in efficiency, in many types of circuit.

However, applying this principle to thyristors of any size presents formidable problems, largely as a result of inevitable slight non-uniformity in the element, which tends to lead to a 'current-crowding effect', whereby the turn-off gate pulse diverts current progressively from those areas where the current gain is highest to the less sensitive areas, until ultimately the whole current is concentrated in such a small area, at such a high current density, that the turn-off capability disappears completely. As a result of this and other problems, gate turn-off thyristors have hitherto made little impact on power converter technology.

Bi-directional thyristors

The bi-directional thyristor, or 'triac',* with a triggering characteristic, similar to the forward characteristic of a normal reverse-blocking thyristor, with applied voltage of either polarity, replaces two parallel-connected thyristors in alternating-current circuits. In domestic and low-power industrial equipment, for such purposes as lighting, heating or motor control, the resulting simplification of both the power and control circuits, and of the mechanical structure, can bring about a useful saving in cost.

The triac consists essentially of two p–n–p–n structures connected in inverse parallel within a single element, as illustrated diagramatically in Figure 1.33. The ability of the device to function in this arrangement despite the fact that from the diagram the outer n_2 and n_4 layers appear to be short-circuited by the adjacent p layers is accounted for by the relatively high lateral resistance of these thin layers in the actual structure. Thus far, the structure is symmetrical in form, and since the terminals appear as either anodes or cathodes according to the direction of current flow, they are arbitrarily labelled as 'main terminal 1' (MT_1) and 'main terminal 2' (MT_2).

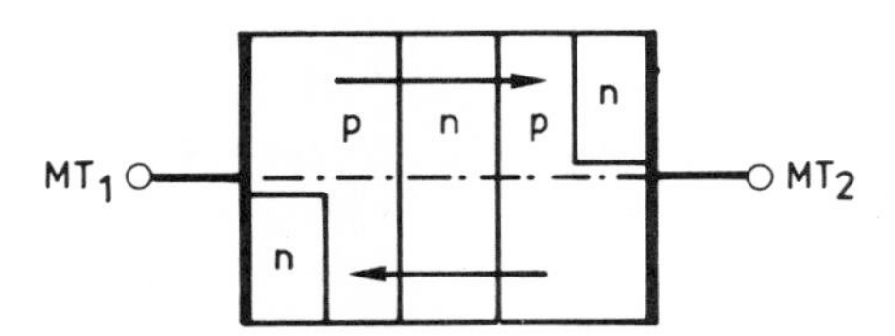

Figure 1.33 Essential structure of a bi-directional thyristor. (Arrows indicate directions of 'forward' conduction.)

The remaining aspects of the basic design are concerned with the means whereby it can be triggered from a single gate terminal irrespective of the polarity of the voltage applied to the main terminals. The arrangement commonly adopted permits firing with either direction of gate current, whichever the polarity of the main-terminal voltage, thus catering for a variety of economical firing systems. It is not necessary here to describe the complex structure of the triac in detail.

* A General Electric Co. (U.S.A.) acronym: TRIode thyristor for A.C. Control.

CHAPTER 2

Naturally Commutating Converters

Definitions

The rectifier and inverter circuits discussed in this chapter function by virtue of their connection to an alternating-current supply, which is generally assumed to have a capacity greatly in excess of the loading of the circuit in question. The significance of 'natural commutation' in such a circuit will become apparent in the course of the discussion; meanwhile, the following definitions should be noted.

A *rectifier* is an apparatus for deriving d.c. power from an a.c. source.

An *inverter* is an apparatus for deriving a.c. power from a d.c. source.

Converter is a general term embracing rectifiers and inverters (as well as other kinds of apparatus for converting power in one form to another form), but is often used specifically to denote an apparatus capable of functioning both as a rectifier and as an inverter.

Apart from the simplifying assumptions referred to in the introduction in regard to components and supply systems, it is generally assumed in this chapter that d.c. circuits include appreciable inductance, firstly because in most cases this is necessary for their efficient utilization, and secondly because, while there are circumstances in which the behaviour of a circuit is significantly altered if its inductance is low, the assumption nevertheless leads to a rationale which can if necessary be used as a starting point from which to proceed to other cases.

THE SINGLE-PHASE HALF-WAVE RECTIFIER

A single diode connecting an a.c. supply to a d.c. load, as in Figure 2.1, constitutes the simplest possible rectifier arrangement. For reasons which will become apparent, it is of little practical use, but serves as an introduction to the subject, and is included for the sake of completeness.

If the load is purely resistive ($L = 0$) the diode conducts when the supply voltage biasses it in the forward direction—that is, in what may be termed positive half-cycles of the supply—and blocks reverse voltage in the alternate, negative,

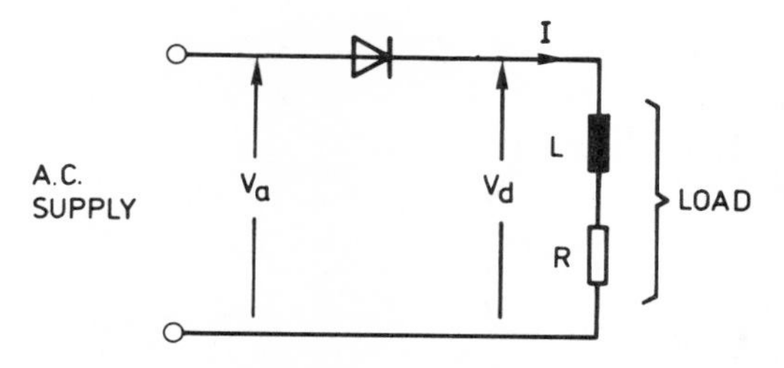

Figure 2.1 Single-phase half-wave rectifier with LR load.

half-cycles: hence the term 'half-wave'. The voltage applied to the load is thus as shown in Figure 2.2, and has a mean value

$$\bar{V}_d = \frac{\hat{V}_a}{\pi} = V_a \frac{\sqrt{2}}{\pi} \tag{2.1}$$

Power rectifiers are seldom, if ever, required in practice to supply purely resistive loads. If at the other extreme the load inductance is assumed to be infinite, any current that flows in the diode must be completely smooth, and therefore continuous; continuous current in the diode, however, would mean that the a.c. supply was connected to the load throughout the whole of each cycle, and the mean load voltage would be zero. The mean output voltage is thus shown to be dependent upon the relative inductance and resistance of the load.

A case in which the mean output voltage is not zero, and the current is by implication not continuous, is illustrated by the waveforms of Figure 2.3. For the period when the diode conducts,

$$\hat{V}_a \sin \omega t = Ri + L\frac{di}{dt}$$

$$= Ri + \omega L \frac{di}{d\omega t}$$

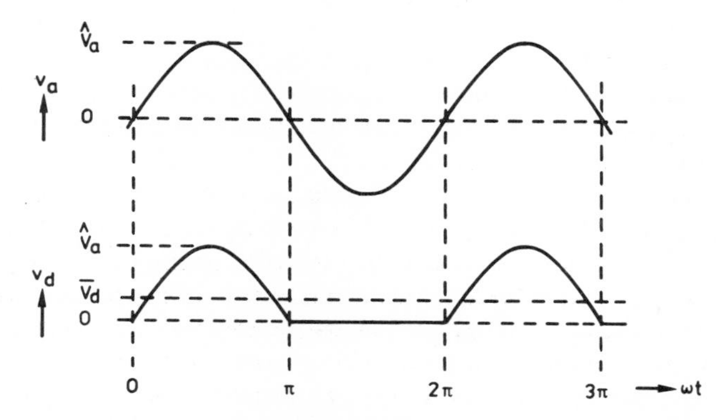

Figure 2.2 Output voltage of a single-phase half-wave rectifier with resistive load.

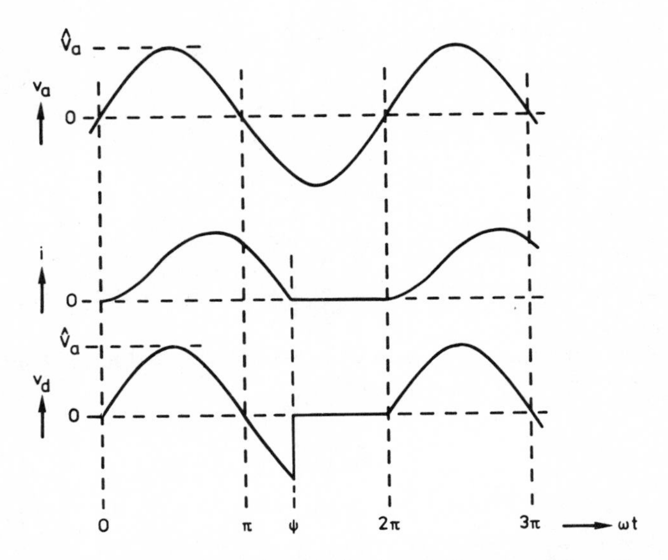

Figure 2.3 Waveforms in a single-phase half-wave rectifier with finite load inductance.

The solution of this is

$$i = \frac{\hat{V}_a}{\sqrt{(R^2 + \omega^2 L^2)}} \left\{ \sin(\omega t - \phi) + \sin\phi \, e^{-\omega t/\tan\phi} \right\} \tag{2.2}$$

where $\tan\phi = \omega L/R$.

From (2.2), the value of $\omega t = \psi$ when i falls to zero can be found, and thence the mean output voltage

$$\overline{V}_d = \frac{\omega}{2\pi} \int_0^{\psi/\omega} v_a \, dt = \frac{1}{2\pi} \int_0^{\psi} \hat{V}_a \sin\omega t \, d\omega t = \frac{V_a}{\sqrt{2}\pi}(1 - \cos\psi) \tag{2.3}$$

$\overline{V}_d/V_a$ is plotted as a function of $\omega L/R$ in Figure 2.4.

It should be clear from the above that in view of its low output voltage and the large proportion of ripple applied to the load, this simple rectifier arrangement has generally unattractive characteristics.

Two modifications of the circuit of Figure 2.1 make it of practical use at low power levels. In the circuit of Figure 2.5, a second diode is added which prevents the instantaneous voltage across the load from reversing in the negative half-cycles of the supply voltage and provides an alternative path for the load current when the first diode ceases to conduct; the load voltage waveform is thus the same as that in the basic circuit with resistive load, and the mean output voltage is similarly $V_a\sqrt{2}/\pi$. Figure 2.6 shows typical current waveforms with LR load. This circuit is sometimes referred to as the current-doubler, because with a highly inductive load the mean (whole-cycle) current drawn from the supply is very

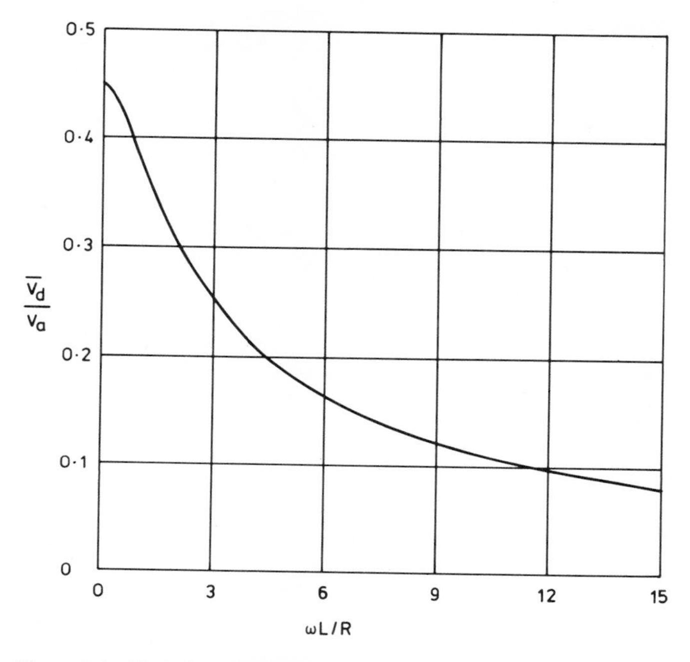

Figure 2.4 Variation of the mean output voltage of a single-phase half-wave rectifier with L/R.

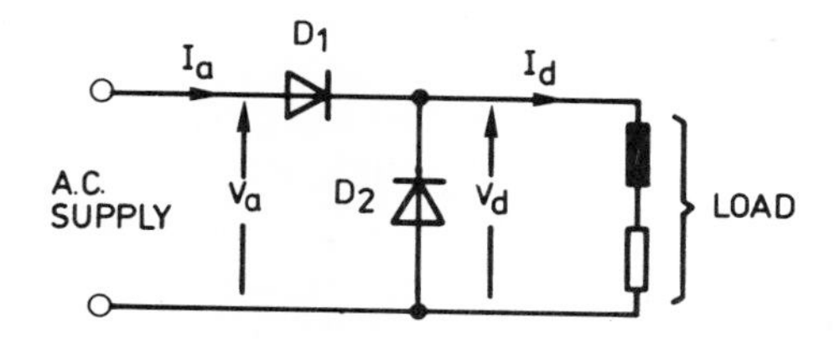

Figure 2.5 'Current-doubler' rectifier.

nearly half the d.c. load current; it has been used to some extent to supply small loads such as relays, for which the low output voltage is convenient.

The other modification of Figure 2.1 consists in adding a so-called reservoir capacitor in parallel with the load, as shown in Figure 2.7. With typical circuit relationships, that is with a large capacitor and a low a.c. circuit impedance relative to the load resistance, the capacitor voltage reaches a value close to the peak supply voltage in the positive half-cycle, and undergoes only a slow rate of decay for the greater part of the cycle until the supply voltage again approaches its positive peak (Figure 2.8). Since the diode conducts only when the supply voltage exceeds the capacitor voltage in positive half-cycles, the input current generally takes the form of short pulses, with high form factor and peak factor, and for this

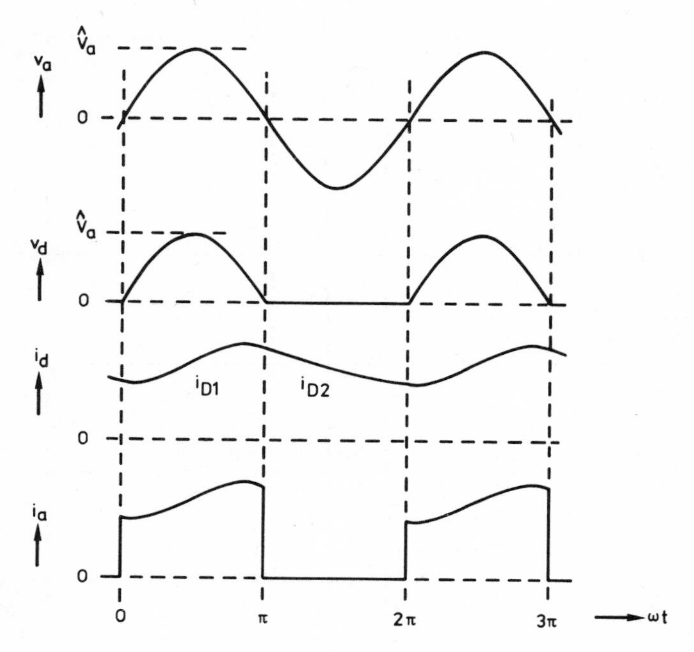

Figure 2.6 Typical waveforms in a current-doubler rectifier.

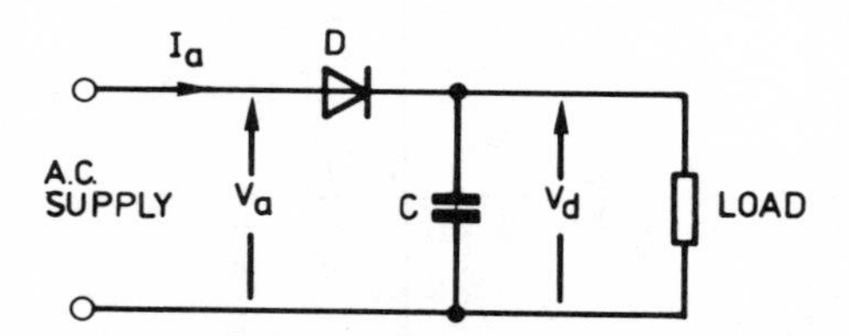

Figure 2.7 Single-phase half-wave rectifier with reservoir capacitor.

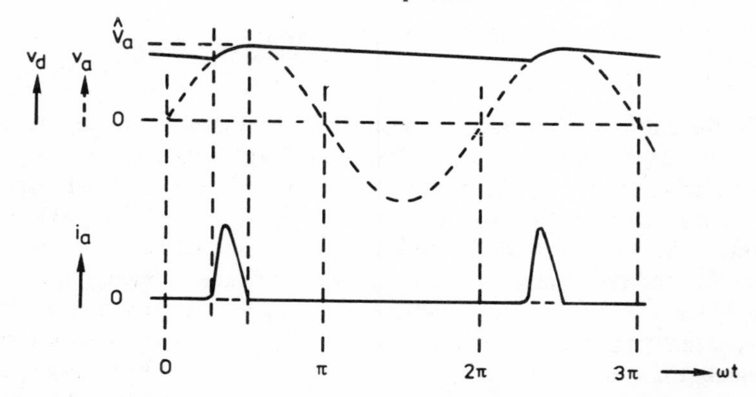

Figure 2.8 Typical waveforms in a single-phase half-wave rectifier with reservoir capacitor.

reason, among others, the circuit is not useful except for very low-power supplies, and will not be considered in detail here.

MULTI-PHASE SINGLE-WAY RECTIFIERS

Virtually all practical power rectifiers comprise one or more groups of half-wave circuits like that of Figure 2.1 connected in parallel on the output side and supplied from different phases of a multiphase supply, as illustrated in Figure 2.9. By 'phase' in this context is meant a supply line whose voltage has a distinct phase-relationship to those of the other lines, disregarding the effective identity, for some purposes, of voltages that are in antiphase.

Some care is needed to avoid confusion in view of the customary usage whereby, for example, a single-phase supply is one that can, by a simple transformer connection with a centre-tapped secondary winding, produce two phases according to the above definition, with 180° separation. To assist in this direction, rectifier connections with an even number of phases are commonly designated by classical prefixes: thus the connection shown in Figure 2.11 is termed 'bi-phase', and that in Figure 2.18 is termed 'hexaphase'. At the same time, to avoid confusion from the fact that with an even number of rectifier phases current is drawn from each phase of the a.c. supply in both half-cycles, so that a bi-phase half-wave rectifier becomes effectively a single-phase full-wave rectifier, the term 'half-wave' is preferably avoided in favour of 'single-way', which conveys the fact that there is only one current path per phase through the rectifier.

Commutation in multi-phase rectifiers

The normal output voltage waveform of a multi-phase single-way diode rectifier, as represented by Figure 2.9, is shown in Figure 2.10. All the diodes are reverse biassed, and therefore non-conducting, except the one connected to the supply terminal at the highest potential with respect to the neutral. As each supply

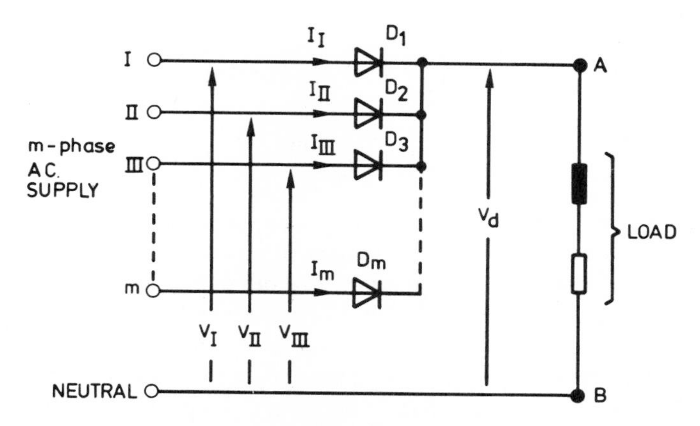

Figure 2.9 *m*-phase single-way rectifier.

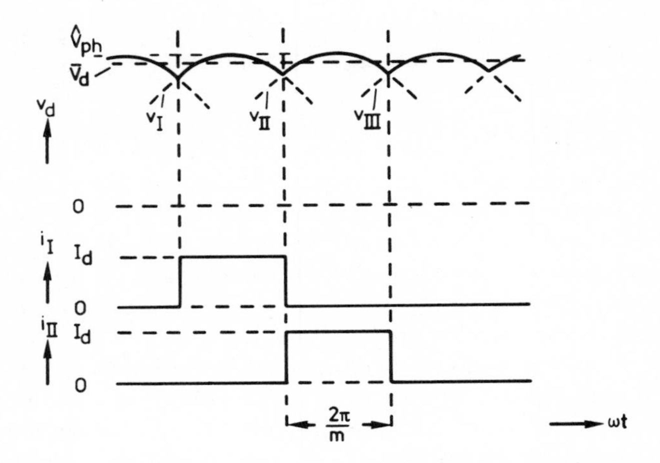

Figure 2.10 Waveforms of output voltage and input current in an m-phase single-way rectifier (Figure 2.9).

terminal in turn assumes the highest potential, the load current is transferred to the diode connected to it, and the output voltage waveform thus consists of a sequence of parts of supply phase voltages. The transference of current from one phase to another of higher potential is known as natural commutation, and the circuit is one of a class of naturally commutating or supply-commutating systems.

D.C. output voltage and A.C. input current

The output waveform of Figure 2.10, and the circuit which produces it, may be characterized by the number of voltage peaks, or 'pulses', that occur in a complete cycle of 2π electrical radians, designated by p. In the single-way circuit, $p = m$, and each diode conducts for $2\pi/m$ radians in each cycle.

The mean d.c. output voltage is given by

$$\bar{V}_d = \frac{\hat{V}_{ph}}{2\pi/m} \int_{-\pi/m}^{\pi/m} \cos \omega t \, d\omega t$$

$$= \hat{V}_{ph} \left(\frac{\sin (\pi/m)}{\pi/m} \right) \tag{2.4}$$

As m is increased, the ratio $[\sin (\pi/m)]/(\pi/m)$ approaches unity, and $\bar{V}_d \to \hat{V}_{ph}$ as $m \to \infty$. The ratio $[\sin (\pi/m)]/(\pi/m)$ may be thought of as a factor by which the output voltage is depressed from the limiting value $\hat{V}_{ph}$ by the presence of an alternating component of voltage, referred to as ripple.

Consideration of the input current waveforms is simplified by assuming that the d.c. output current is continuous and smooth, a reasonable simplification in most practical cases since most d.c. consuming circuits are inductive to some extent, and a large amount of ripple current is more often than not undesirable.

Then, since the load current flows in a particular diode only when the phase to which it is connected is at a higher potential than the others, the phase current takes the form of a rectilinear pulse of duration $2\pi/m$ radians, and amplitude I_d once per cycle, or shows in Figure 2.10.

The mean (whole-cycle) current in each diode is

$$\bar{I}_D = \frac{I_d}{m} \tag{2.5}$$

and the r.m.s. current in each diode is

$$I_D = \frac{I_d}{\sqrt{m}} \tag{2.6}$$

whence the diode current has a form factor

$$k_{fID} = \frac{I_D}{\bar{I}_D} = \sqrt{m} \tag{2.7}$$

Practical forms of single-way rectifier

In practice the bi-phase and hexaphase rectifiers depend for their existence upon associated transformers with centre-tapped secondary windings. For the moment these will be assumed to have no leakage reactance.

The bi-phase arrangement (Figure 2.11) constitutes the simplest embodiment of supply commutation, with $p = m = 2$. From equation (2.4), the d.c. output voltage is

$$\bar{V}_d = \hat{V}_{ph}\left(\frac{\sin(\pi/2)}{\pi/2}\right) = \frac{2}{\pi}\hat{V}_{ph} \quad \text{or} \quad \frac{2\sqrt{2}}{\pi}V_{ph} \tag{2.8}$$

and the waveform of current in each phase is as shown in Figure 2.12, with a pulse duration of π radians. The pulses of phase current from the two halves of the transformer secondary winding are reflected in opposite directions in the primary winding, and the current drawn from the supply is therefore of a symmetrical square waveform, with an amplitude I_d if the turns ratio between the primary

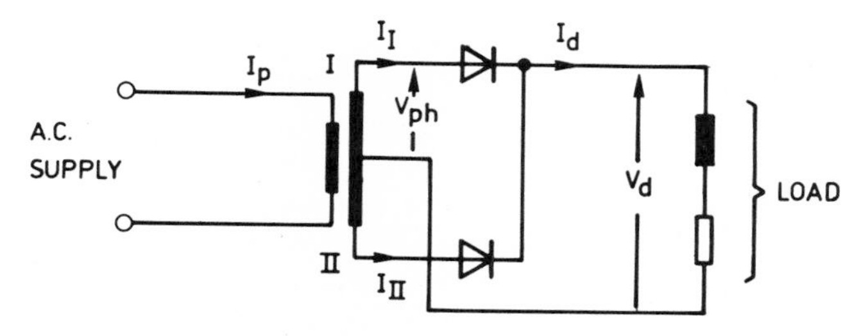

Figure 2.11 Bi-phase single-way rectifier.

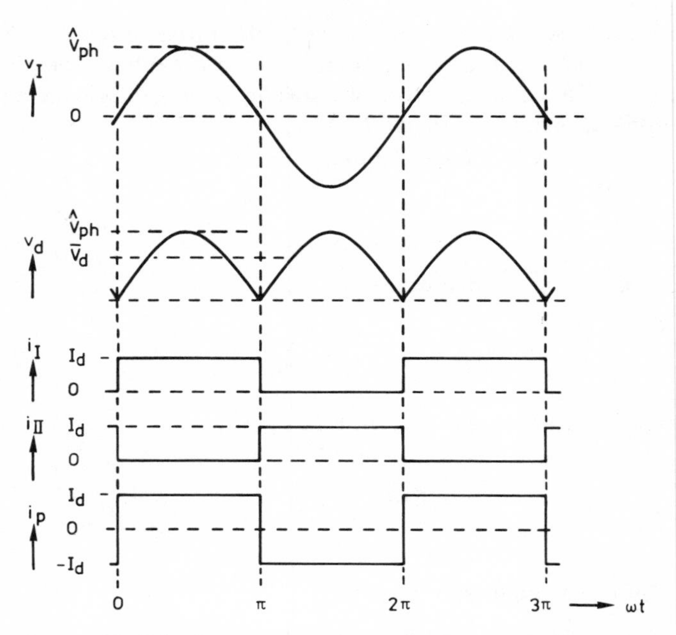

Figure 2.12 Waveforms of voltage and current in a bi-phase rectifier (Figure 2.11).

winding and each half of the secondary is unity. From equations (2.5)–(2.7),

$$\left.\begin{aligned} \bar{I}_D &= \frac{I_d}{2} \\ I_D &= \frac{I_d}{\sqrt{2}} \\ k_{fID} &= \sqrt{2} \end{aligned}\right\} \tag{2.9}$$

In the three-phase single-way rectifier, shown in Figure 2.13, $p = m = 3$. From equation (2.4),

$$\bar{V}_d = \hat{V}_{ph}\left(\frac{\sin(\pi/3)}{\pi/3}\right) = \frac{3\sqrt{3}}{2\pi}\hat{V}_{ph} \quad \text{or} \quad \frac{3\sqrt{3}}{\sqrt{2}\pi}V_{ph} \tag{2.10}$$

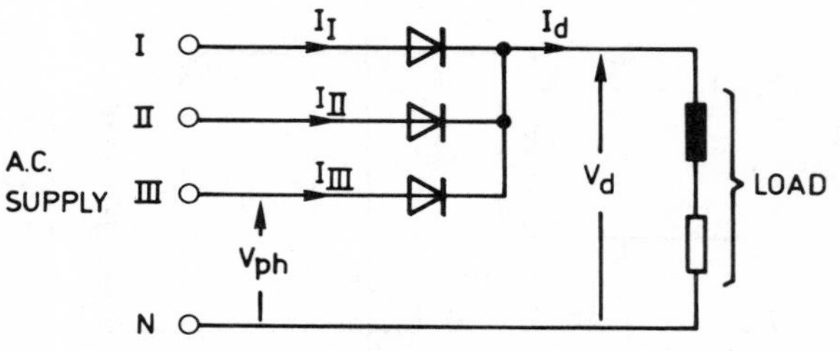

Figure 2.13 Three-phase single-way rectifier.

From equations (2.5)–(2.7),

$$\left.\begin{aligned} \overline{I}_{\mathrm{D}} &= \frac{I_{\mathrm{d}}}{3} \\ I_{\mathrm{D}} &= \frac{I_{\mathrm{d}}}{\sqrt{3}} \\ k_{\mathrm{fID}} &= \sqrt{3} \end{aligned}\right\} \quad (2.11)$$

Current is drawn from the supply in pulses of $2\pi/3$ radians, as illustrated in the waveforms of Figure 2.14. A transformer is not essential to the operation of this rectifier arrangement, if a neutral supply connection is available, but in practice it is not normally used without one, since the direct current in the supply lines—particularly the neutral line—which results from a direct connection is not usually considered acceptable at any appreciable power level, even if the desired output voltage is obtained without transformation. However, a normal transformer connection—delta–star, for example—while keeping direct current out of the supply, only transfers the problem to the transformer, which will exhibit increased magnetizing current and iron losses if subjected to d.c. magnetization.

This problem is avoided in most cases by employing a special transformer connection, known as the zig-zag connection, illustrated in Figure 2.15. Each

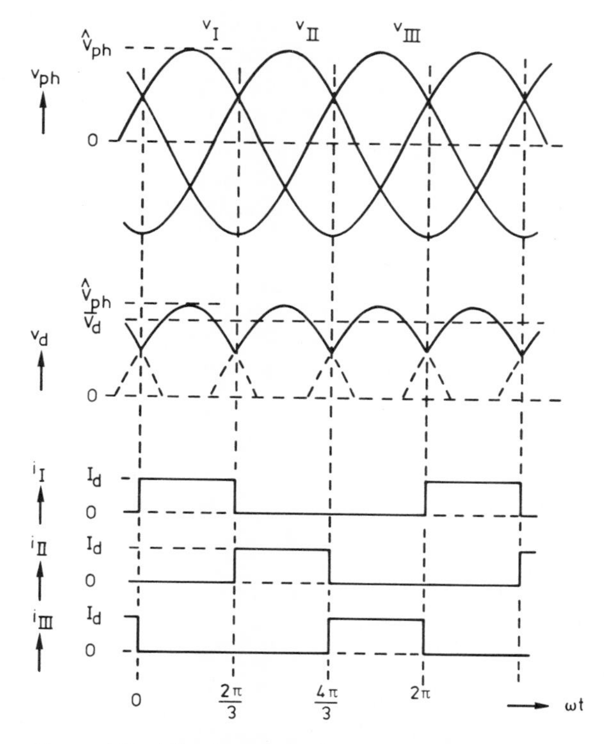

Figure 2.14 Waveforms of voltage and current in a three-phase single-way rectifier (Figure 2.13).

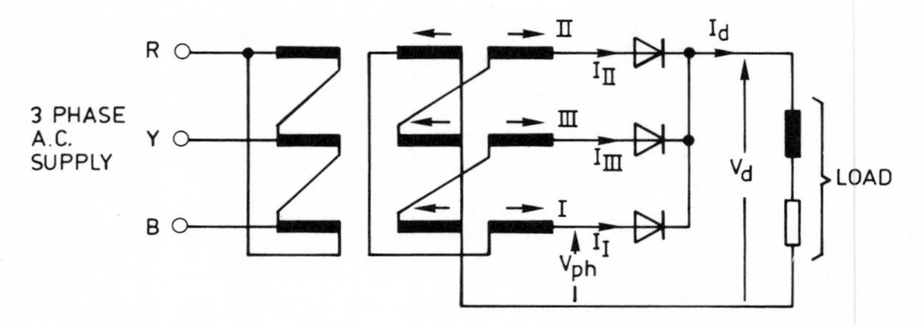

Figure 2.15 Three-phase single-way rectifier with zig-zag transformer connection.

secondary phase voltage is obtained from two equal-voltage secondary windings connected in series so that the d.c. magnetizing forces due to the two secondary windings on any one limb are in opposite directions, and balance out. The combination of secondary voltages, with a phase displacement of 60°, is illustrated by the phasor diagram of Figure 2.16.

Figure 2.17 shows the current waveforms in the circuit of Figure 2.15. The transformer primary windings have been assumed here to be connected in delta, and it will be observed that the effect of this is to reproduce in the a.c. supply lines the waveforms of the diode currents, but without the d.c. component. However, it will also be observed that the three transformer primary currents sum to zero at all times; it follows from this that the primary windings could alternatively be connected in star, without a neutral connection, with no change in the behaviour of the circuit apart from the fact that the supply current waveforms would then be simply those of the transformer primary windings.

For the purposes of Figure 2.17 all the windings are assumed to have the same number of turns, and thus

$$i_{RY} = i_I - i_{II} \quad \text{etc.} \quad \text{and}$$

$$i_R = i_{RY} - i_{BR} = 2i_I - i_{II} - i_{III} \quad \text{etc.}$$

The hexaphase rectifier shown in Figure 2.18 (Figure 2.19 is the relative phasor diagram) embodies the greatest number of phases normally (if rarely) used in a simple single-way rectifier. Here $p = m = 6$, and the duration of each diode current pulse is $2\pi/6$ radians, as shown in the waveforms of Figure 2.20.

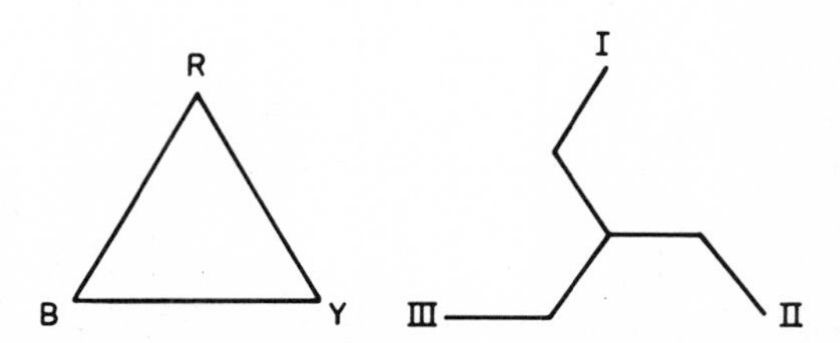

Figure 2.16 Phasor diagram of transformer voltages in Figure 2.15.

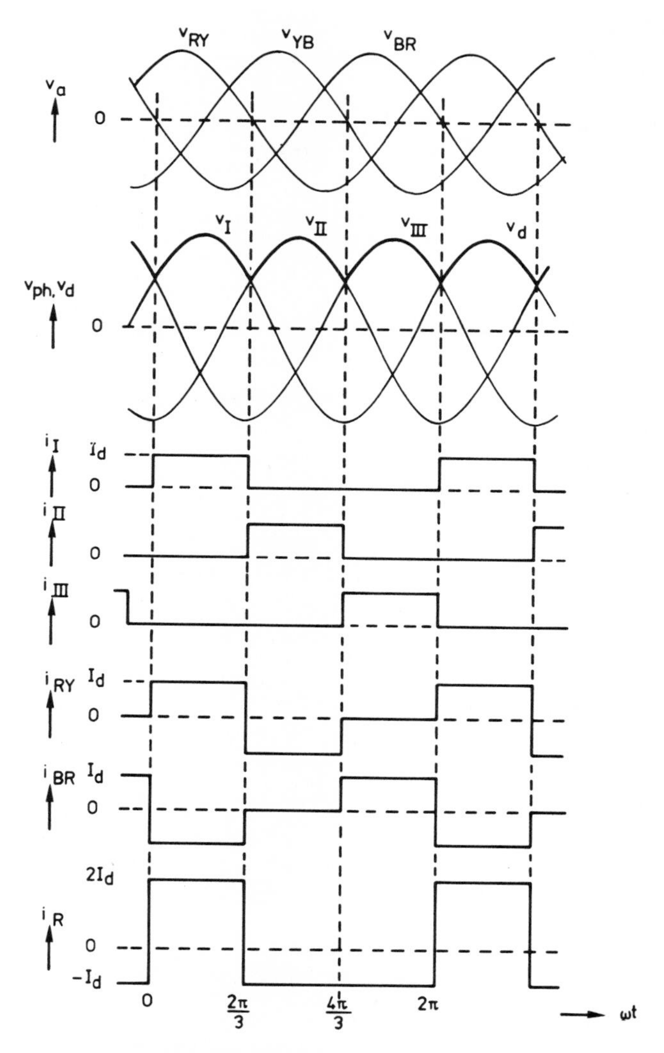

Figure 2.17 Waveforms in a three-phase single-way rectifier with delta/zig-zag transformer connection (Figure 2.15).

From equations (2.4)–(2.7),

$$\left.\begin{aligned}
\overline{V}_d &= \hat{V}_{ph}\left(\frac{\sin(\pi/6)}{\pi/6}\right) = \frac{3}{\pi}\hat{V}_{ph} \quad \text{or} \quad \frac{3\sqrt{2}}{\pi}V_{ph} \\
\overline{I}_D &= \frac{I_d}{6} \\
I_D &= \frac{I_d}{\sqrt{6}} \\
k_{fID} &= \sqrt{6}
\end{aligned}\right\} \quad (2.12)$$

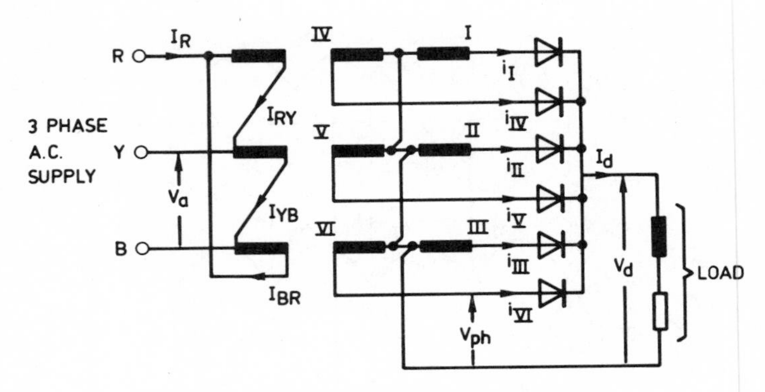

Figure 2.18 Hexaphase single-way rectifier.

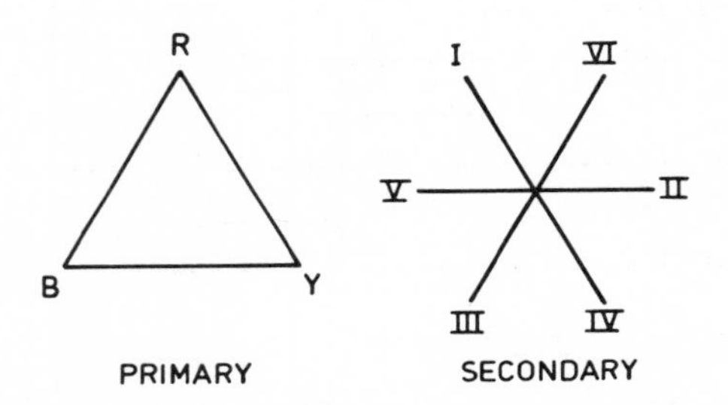

Figure 2.19 Phasor diagram of transformer voltages in Figure 2.18.

The primary currents do not sum to zero, and operation as described is therefore not possible with star-connected primary windings.

The utilization of components in simple single-way rectifiers

The expressions (2.5)–(2.7) relating to the diode currents in the simple single-way rectifiers so far discussed emphasize what is apparent from the waveforms, that with an increasing number of phases the utilization of the diodes and the transformer windings becomes progressively poorer. The losses and the current ratings of the diodes are adversely affected as the conduction angle decreases (see Chapter 1), and the transformer rating also suffers as a result of the tendency to increased copper loss as the form factor increases.

A criterion of the transformer utilization is the ratio of the d.c. output power from the rectifier to the effective volt-ampere rating of the transformer; by 'effective volt-ampere rating' is meant half the sum of the primary and secondary winding ratings, which, in the case of the single-way rectifier, are not equal. The utilization factor, as the ratio may be called, would ideally be unity, but is normally less.

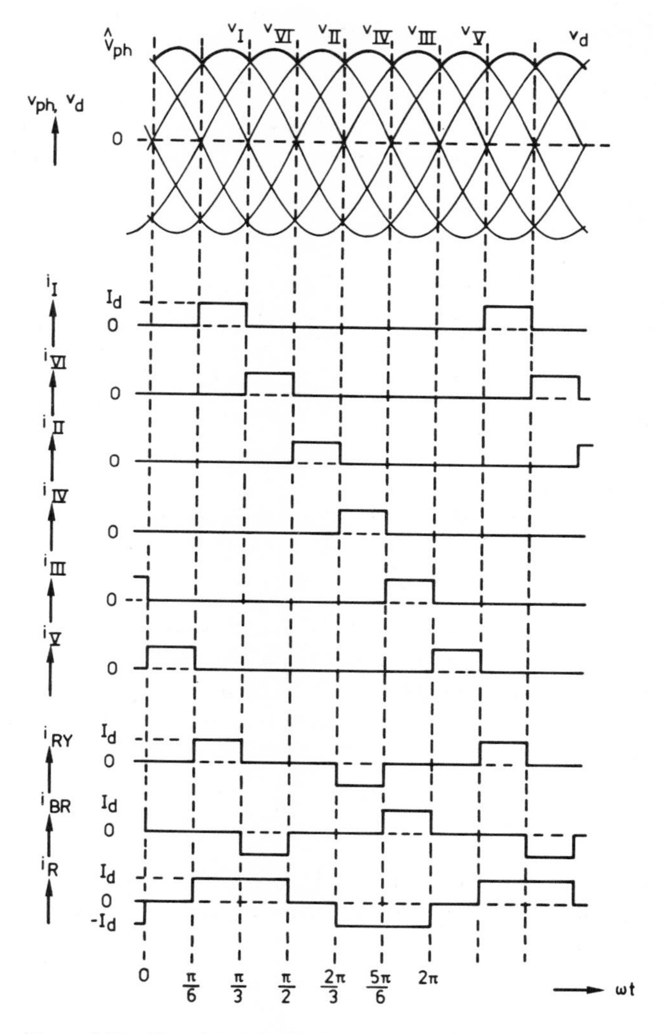

Figure 2.20 Waveforms in the hexaphase rectifier of Figure 2.18.

In the bi-phase circuit, from (2.8) and (2.9), the total secondary VA is

$$S_s = 2\,\frac{\pi}{2\sqrt{2}}\,\overline{V}_d\,\frac{I_d}{\sqrt{2}} = \frac{\pi}{2}\,P_d \tag{2.13}$$

In the primary winding, assuming unity turns ratio as before, the r.m.s. current is equal to I_d and the primary VA is

$$S_p = \frac{\pi}{2\sqrt{2}}\,\overline{V}_d I_d = \frac{\pi}{2\sqrt{2}}\,P_d \tag{2.14}$$

The transformer utilization factor is thus

$$k_{uT} = \frac{P_d}{\frac{1}{2}\left(\frac{\pi}{2} P_d + \frac{\pi}{2\sqrt{2}} P_d\right)} = 0.746 \tag{2.15}$$

Similarly, for the hexaphase rectifier of Figure 2.18,

$$S_s = 6 \frac{\pi}{3\sqrt{2}} \overline{V}_d \frac{I_d}{\sqrt{6}} = \frac{\pi}{\sqrt{3}} P_d \tag{2.16}$$

and, the primary current being again $\sqrt{2}$ times the secondary current,

$$S_p = 3 \frac{\pi}{3\sqrt{2}} \overline{V}_d \frac{I_d}{\sqrt{3}} = \frac{\pi}{\sqrt{6}} P_d \tag{2.17}$$

Therefore

$$k_{uT} = \frac{P_d}{\frac{1}{2}\left(\frac{\pi}{\sqrt{3}} P_d + \frac{\pi}{\sqrt{6}} P_d\right)} = 0.646 \tag{2.18}$$

In the case of the three-phase single-way rectifier fed from a zig-zag-connected transformer winding (Figure 2.15) the 60° phase displacement between the two secondary windings in each phase means that each winding voltage is not half, but $1/\sqrt{3}$ times the rectifier phase voltage. Therefore, from (2.10) and (2.11), the total secondary VA is

$$S_s = \frac{6}{\sqrt{3}} \frac{\sqrt{2}\pi}{3\sqrt{3}} \overline{V}_d \frac{I_d}{\sqrt{3}} = \frac{2\sqrt{2}\pi}{3\sqrt{3}} P_d \tag{2.19}$$

From inspection of the waveforms of Figure 2.17, the r.m.s. primary current, assuming that all the windings have the same number of turns, is $\sqrt{(2/3)}\, I_d$, and the total primary VA is therefore

$$S_p = \frac{3}{\sqrt{3}} \frac{\sqrt{2}\pi}{3\sqrt{3}} \overline{V}_d \frac{\sqrt{2}}{\sqrt{3}} I_d = \frac{2\pi}{3\sqrt{3}} P_d \tag{2.20}$$

The transformer utilization factor is then

$$k_{uT} = \frac{P_d}{\frac{1}{2}\left(\frac{2\sqrt{2}\pi}{3\sqrt{3}} P_d + \frac{2\pi}{3\sqrt{3}} P_d\right)} = 0.685 \tag{2.21}$$

Input current harmonics

The harmonics present in the current drawn by a rectifier from its a.c. supply often assume considerable significance. The input current to the bi-phase rectifier, as illustrated in Figure 2.12, is of square waveform, and therefore contains all the odd harmonics with amplitudes in inverse proportion to their orders:

$$\left.\begin{aligned} I_1 &= \frac{2\sqrt{2}}{\pi} I_d \\ I_n &= \frac{I_1}{n} = \frac{2\sqrt{2}}{n\pi} I_d \end{aligned}\right\} \tag{2.22}$$

(I_d here signifies the amplitude of the current square-wave, on the assumption of equal transformer windings.)

In the case of the three-phase single-way rectifier, the diode current, with the waveform shown in Figure 2.14, can be shown by Fourier analysis to contain harmonics of order n, in the series

$$n = 3r \pm 1 \tag{2.23}$$

where r is any positive integer, with amplitudes in inverse proportion to their orders:

$$\left.\begin{aligned} I_1 &= \frac{\sqrt{3}}{\sqrt{2}\pi} I_d \\ I_n &= \frac{I_1}{n} = \frac{\sqrt{3}}{n\sqrt{2}\pi} I_d \end{aligned}\right\} \tag{2.24}$$

The three diode currents being mutually displaced in phase by $2\pi/3$ at the fundamental frequency, any harmonic components are mutually displaced by $2n\pi/3$—that is, by $(3r \pm 1)\, 2\pi/3$, which for all values of r is effectively $2\pi/3$. Thus, all the harmonics, like the fundamental component, are three-phase currents, and their relative amplitudes in the supply lines are therefore not affected by the transformer connections, although the actual waveforms vary according to the phase relationships between fundamental and harmonic components.

The relationship of transformer utilization and current waveforms

Since if there are no losses the a.c. power supplied to, and delivered by, the transformer must be equal to the d.c. output power from the rectifier, the utilization factor of the transformer, as defined above, may alternatively be considered purely in terms of the power and volt-ampere ratings of its windings, without direct reference to the rectifier. With the assumed sinusoidal supply voltage waveform, the power handled by each winding is proportional to the fundamental-frequency component of current in phase with the voltage, while the VA is proportional to the total r.m.s. current. The ratio of these two current values thus gives the utilization factor of the winding, closely analogous to the power factor of a load.

As with power factor, the utilization factor of a winding may have two components, a distortion factor, k_h, which is the ratio of the fundamental-frequency component of current to the total r.m.s. current, and a displacement factor, which is the cosine of the phase angle (ϕ) of the fundamental-frequency component of current relative to the voltage:

$$k_u = k_h \cos\phi \tag{2.25}$$

If the utilization factor of the primary winding is k_{up} and that of the secondary k_{us}, it follows from the definitions that

$$\frac{2}{k_{uT}} = \frac{1}{k_{up}} + \frac{1}{k_{us}}$$

or

$$k_{uT} = \frac{2k_{up}k_{us}}{k_{up} + k_{us}} \tag{2.26}$$

Consideration of the case of the three-phase single-way rectifier circuit of Figure 2.15 will demonstrate that this approach yields the same result as the previous one. For the primary windings, as noted in (2.24), the fundamental-frequency component of current is (assuming an amplitude I_d as in Figure 2.17)

$$I_{1p} = \frac{3}{\sqrt{2}\pi} I_d \tag{2.27}$$

while the total r.m.s. primary current is

$$I_p = \sqrt{\tfrac{2}{3}}\, I_d \tag{2.28}$$

There is no phase displacement between I_{1p} and the supply (interphase) voltage ($\cos\phi = 1$), so

$$k_{up} = \frac{I_{1p}}{I_p} = \frac{3}{\sqrt{2}\pi} \frac{\sqrt{3}}{\sqrt{2}} = 0.827 \tag{2.29}$$

In the secondary windings, the fundamental-frequency component of current is found to be

$$I_{1s} = \frac{\sqrt{3}}{\sqrt{2}\pi} I_d \tag{2.30}$$

while

$$I_s = \frac{I_d}{\sqrt{3}} \tag{2.11}$$

In this case there is also a phase displacement of 30° between the fundamental-frequency current and the voltage, due to the zig-zag connection, and the secondary utilization factor is therefore

$$k_{us} = \frac{I_{1s}}{I_s} \cos\phi = \frac{3}{\sqrt{2}\pi} \frac{\sqrt{3}}{2} = 0.585 \tag{2.31}$$

Then from (2.26),

$$k_{uT} = \frac{2 \times 0.827 \times 0.585}{0.827 + 0.585} = 0.685 \tag{2.32}$$

(cf. (2.21)).

This method of analysis points to the great importance of the harmonic content of the current in the windings of the transformer as the factor mainly responsible for its imperfect utilization.

Input power factor

Just as the utilization factor of the transformer winding can be determined as the product of a distortion factor and a displacement factor, so the power factor presented by the rectifier to the a.c. supply is given by

$$k_p = k_h \cos\phi \tag{2.33}$$

where k_h ($= I_1/I$) and $\cos\phi$ relate to the current drawn from the supply terminals. It is obvious that where this current is actually that flowing in the primary winding of a transformer, the input power factor and the primary utilization factor are the same; thus it can be inferred from (2.14) and (2.20) or (2.29) above that for the bi-phase rectifier

$$k_p = \frac{2\sqrt{2}}{\pi} = 0.90 \tag{2.34}$$

and for the three-phase single-way rectifier with a zig-zag secondary connection and star-connected primary windings,

$$k_p = \frac{3\sqrt{3}}{2\pi} = 0.827 \tag{2.35}$$

Expression (2.35) also applies to the same rectifier connection with delta-connected primary windings, because although the input current waveforms are different in the two winding arrangements, the proportions of the input current harmonics are the same. It should be observed, however, that in some cases, such as the hexaphase rectifier of Figure 2.18, and some others to be discussed later, the relative harmonic content of the input current is less than that of the primary current in the transformer(s).

MULTIPLEX SINGLE-WAY RECTIFIERS

In order to obtain the benefits of a large pulse number in regard to input harmonics and output ripple without the previously mentioned penalties associated with short conduction angles, and at the same time to avoid the problem of d.c. magnetization without recourse to the somewhat inefficient zig-zag transformer connection of Figure 2.15, three-phase rectifiers are in most cases assembled from two or more three-pulse diode groups so arranged as to operate to some extent independently.

Figure 2.21 shows a circuit of this kind comprising a pair of three-pulse groups fed from oppositely connected secondary windings (phasor diagram, Figure 2.22) and supplying the load in parallel. Separate (large) smoothing inductors enable each three-pulse group to supply continuous output current and thus operate

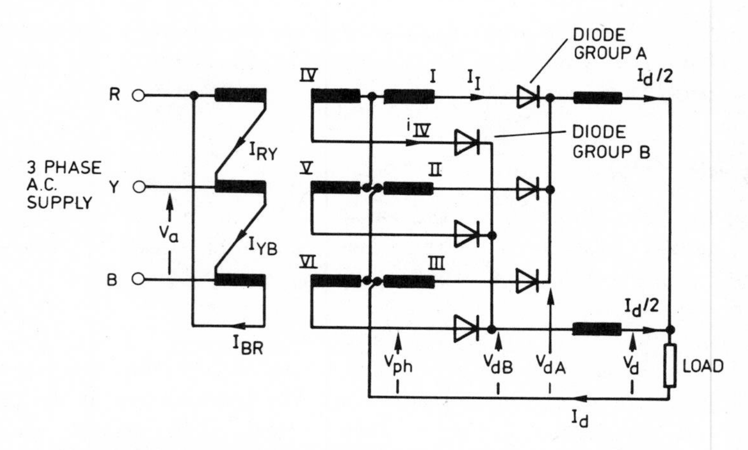

Figure 2.21 Hexaphase rectifier with separate smoothing inductors.

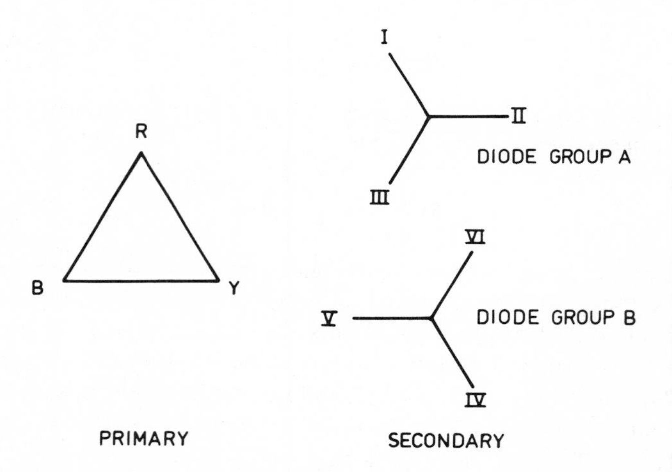

Figure 2.22 Phasor diagram of transformer voltages in Figure 2.21.

independently with 120° diode conduction in exactly the same way as the circuit of Figure 2.13, but because of the antiphase relationship of the two groups of secondary windings the combined circuit constitutes a six-pulse, or hexaphase, rectifier. A rectifier with more than one independently commutating group of diodes is sometimes referred to as a multiplex rectifier, the present example being a duplex arrangement.

Figure 2.23 shows how the diode currents of waveforms similar to those of Figure 2.14 combine in effect in the transformer primary windings and in the supply lines, producing symmetrical waveforms by virtue of the symmetry of the secondary circuits. The three primary currents add up at all times to zero, and a star primary winding with no neutral connection would therefore be equally permissible.

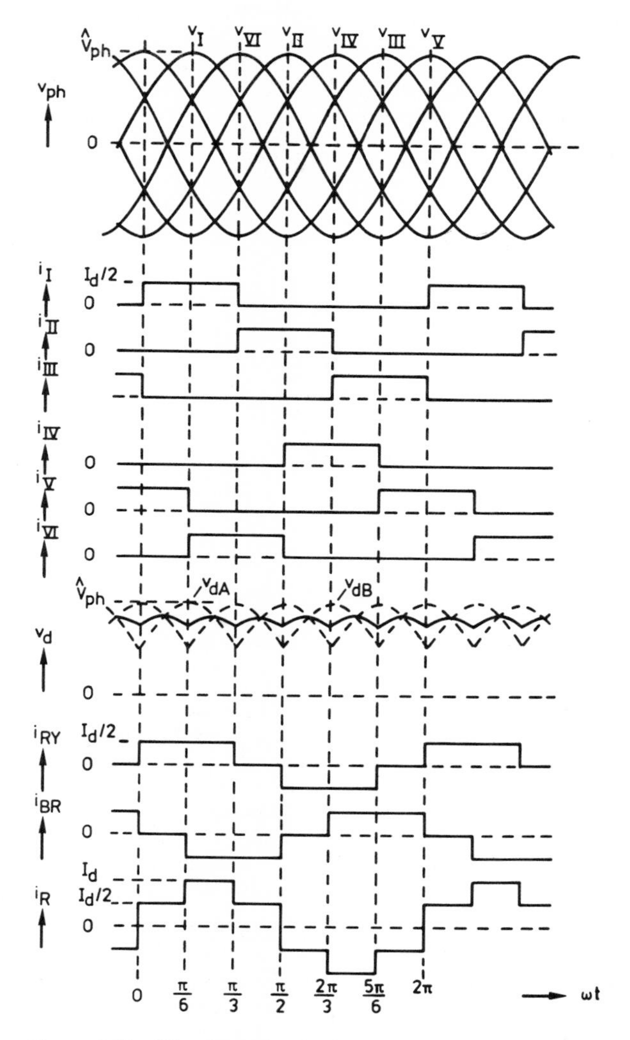

Figure 2.23 Waveforms in the hexaphase rectifier of Figure 2.21.

Since each diode group operates independently in the same way as a three-phase single-way rectifier, the relationship of d.c. to a.c. voltage is the same as for the latter, as given in (2.10). In the general expression of (2.4), m has to be taken to signify the number of phases in each commutating group.

The transformer primary current can be shown by Fourier analysis to have a fundamental-frequency component

$$I_{1p} = \frac{\sqrt{2}\sqrt{3}}{\pi} I_d \tag{2.36}$$

while
$$I_p = \frac{\sqrt{2}}{\sqrt{3}} I_d \tag{2.37}$$

The primary utilization factor is thus

$$k_{up} = \frac{I_{1p}}{I_p} = \frac{3}{\pi} = 0.955 \tag{2.38}$$

For the secondary windings,

$$I_{1s} = \frac{\sqrt{3}}{\sqrt{2}\pi} \frac{I_d}{2} \tag{2.39}$$

and
$$I_s = \frac{I_d}{2\sqrt{3}} \tag{2.40}$$

therefore
$$k_{us} = \frac{I_{1s}}{I_s} = \frac{3}{\sqrt{2}\pi} = 0.675 \tag{2.41}$$

Then from (2.26),

$$k_{uT} = \frac{2 \times 0.955 \times 0.675}{0.955 + 0.675} = 0.79 \tag{2.42}$$

This represents a considerable improvement over the corresponding figure of 0.685 for the three-phase single-way rectifier (2.32), a result of the reduction in supply harmonics and the elimination of the displacement factor in the secondary VA. The input power factor is similarly improved, being equal to k_{up} at 0.955.

The hexaphase rectifier with interphase reactor

The hexaphase rectifier of Figure 2.21 with a separate smoothing inductor for each three-pulse group can be seen, with the aid of Thevenin's theorem, to produce an output which at any instant is equal to the average of the output voltages of the two groups, subject to the smoothing effect of an inductance equal to that of the two inductors in parallel.

The fact that the average of the unsmoothed output voltages, as shown in Figure 2.23, has a six-pulse waveform and a fundamental ripple frequency six times the supply frequency is simply explained mathematically by the fact that the third harmonic and its odd multiples, i.e. those in the series $n = 3(2r-1)$, when displaced in time by $\pi/(3\omega_1)$, are displaced in phase by $n\pi/3 = (2r-1)\pi$; this is effectively π for all values of r, and the average of the two components in the series is thus zero.

Each smoothing inductor, however, has to support the whole of the output ripple voltage from its associated rectifier group: this includes the third harmonic, which in comparison with the lowest-frequency component that appears in the combined output (the sixth harmonic) is about four times as large and gives rise to about eight times the alternating flux. To avoid the unfavourable effect which this has on the design of the inductors, it is usual to employ a different arrangement,

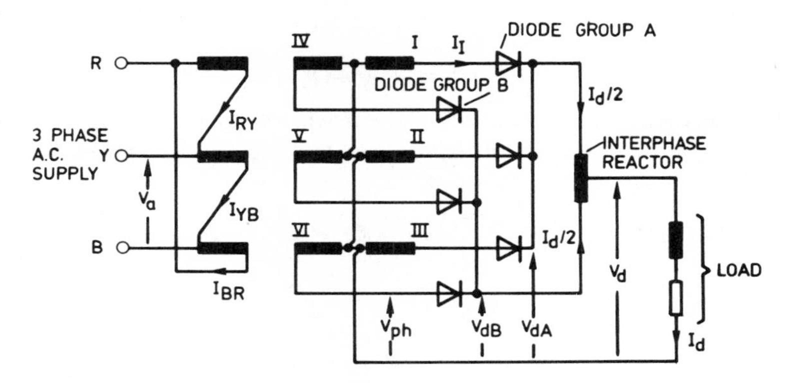

Figure 2.24 Hexaphase rectifier with interphase reactor.

shown in Figure 2.24, in which the outputs of the two rectifier groups are combined in a centre-tapped inductor, known as an interphase reactor (or interphase transformer), which plays no part in the smoothing function. If it is necessary to add inductance to the load circuit for the purpose of smoothing, only one smoothing inductor is required, designed only for the sixth and higher-order harmonics.

Assuming the rectifier to be well balanced, the output currents of the two groups flowing in opposite directions in the interphase reactor winding produce negligible d.c. magnetization of the core, and it is therefore possible to dispense with an air gap and to obtain a high inductance in a relatively small volume.

The voltage that appears across the interphase reactor is that represented by the vertically hatched area in Figure 2.25, containing all the ripple components that do not appear in the combined output—i.e. the odd harmonics. At a particular instant, say between $\omega t = 0$ and $\omega t = \pi/6$, the voltage across the reactor is

$$v_{ipr} = v_{y1} - v_{b2} = v_{y1} + v_{b1} \tag{2.43}$$

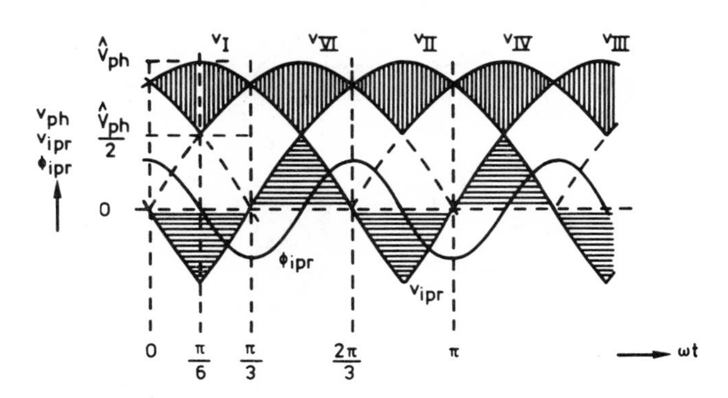

Figure 2.25 Voltage across, and flux in, the interphase reactor in a hexaphase rectifier (Figure 2.24).

Then since $$v_{y1} + v_{b1} + v_{r1} = 0,$$

$$v_{ipr} = -v_{r1} = v_{r2} \tag{2.44}$$

The roughly triangular waveform of the reactor voltage is thus composed of 60° segments of the secondary phase voltages, and the peak and mean values are

$$\hat{V}_{ipr} = \hat{V}_{ph} \sin\frac{\pi}{6} = \frac{\hat{V}_{ph}}{2} \tag{2.45}$$

and $$\bar{V}_{ipr} = \frac{6}{\pi}\int_0^{\pi/6} \hat{V}_{ph} \sin \omega t \, d\omega t = \frac{6\hat{V}_{ph}}{\pi}[-\cos \omega t]_0^{\pi/6}$$

$$= 0.256\,\hat{V}_{ph} \qquad \text{or (from (2.10))} \qquad 0.31\,\bar{V}_d \tag{2.46}$$

The current in each half-winding is $I_d/2$, and the reactor, if used as a double-wound transformer at the supply frequency (at the same peak flux), would have a rating of

$$\frac{1.1}{3} \times \frac{0.31\,V_d}{2} \times \frac{I_d}{2} = 0.028\,V_d I_d \tag{2.47}$$

The above relationships are ideal ones, and in practice the voltage across the reactor, and therefore its equivalent rating, can be about doubled by the increase in ripple due to overlap (q.v.); the reactor is nevertheless very much smaller than the transformer.

Magnetizing current in the interphase reactor

In a practical case, the triple-frequency voltage across the interphase reactor causes a magnetizing current to flow, in effect, from one rectifier group output terminal to the other, normally with the result illustrated in Figure 2.26. The magnitude of the magnetizing current is determined by the flux in the core, shown in Figure 2.25 as ϕ_{ipr}, which derives simply from the reactor voltage, and

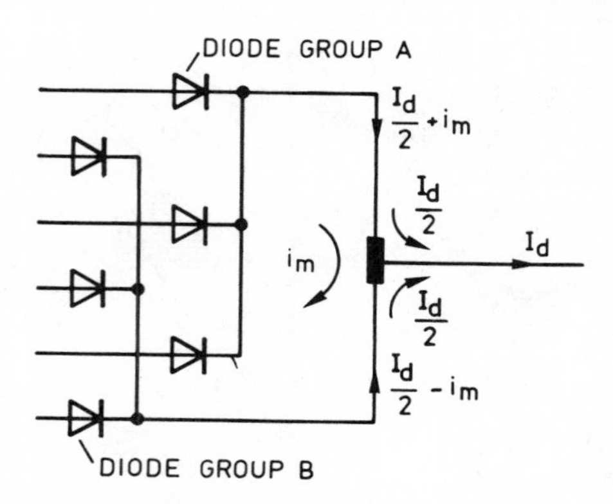

Figure 2.26 Interphase reactor magnetizing current in a hexaphase rectifier.

everything that has been said so far assumes that the peak magnetizing current is less than the load current.

If the load current is reduced sufficiently, however, i_m will rise periodically to the level of $I_d/2$, and the reactor current then becomes zero in one half of the winding. Since the diodes do not permit the current in either half to reverse, it has to remain at zero for the remainder of the reactor voltage half-cycle, and for this period the load is supplied only from the rectifier group producing the higher instantaneous voltage, with the result that the mean output voltage of the rectifier is increased.

In the extreme case when I_d is negligibly small, the magnetization of the reactor, and in consequence the voltage across it, are also negligible, and the circuit becomes indistinguishable in its operation from the simple hexaphase rectifier of Figure 2.18. The d.c. output voltage is then that of the six-pulse group given in (2.12), i.e. $3\sqrt{2}\, V_{ph}/\pi$, rather than that obtained under normal loading conditions, namely the output voltage of the three-pulse group, which from (2.10) is $3\sqrt{3}\, V_{ph}/(\sqrt{2}\pi)$. The output characteristic thus acquires an element of regulation in the region of load current below the peak magnetizing current of the reactor, rising to about 15% on open circuit. It is not usually difficult to design the reactor so that this effect is excluded from the normal range of load current.

PHASE MULTIPLICATION AND INPUT HARMONICS

Fourier analysis of the primary current waveform (or the line current waveform) in Figure 2.23 will indicate that the fundamental component is

$$I_{1p} = \frac{\sqrt{2}\sqrt{3}}{\pi} I_d \tag{2.36}$$

and that odd harmonics are present in the series

$$n = 6r \pm 1 \tag{2.48}$$

where r is any positive integer, with amplitudes in inverse proportion to their orders.

The harmonic distribution can however be deduced more easily from a general relationship which can be established for all three-phase-supplied rectifiers in which the pulse number is increased by dint of multiple commutating groups fed from phase-multiplying windings.

Since the 3-pulse rectifier presents a balanced load to a three-wire a.c. supply, it cannot draw triplen harmonic currents, and any harmonics drawn must have orders in the series

$$n = 3r \pm 1 \tag{2.23}$$

It is noted above that Fourier analysis confirms this, and the fact that all the harmonics in this series are in fact present, with amplitudes inversely proportional to their orders.

A rectifier of larger pulse number normally comprises essentially a number of 3-pulse groups supplied from phase-shifting transformer windings in such a way that the output voltage peaks occur at regular time intervals. Thus if the pulse number of the whole arrangement is p, i.e. it produces p pulses per cycle, there are $p/3$ 3-pulse groups with their supplies phase-displaced in steps of $2\pi/p$ radians. The harmonic content of the supply current then depends on the way in which the harmonic components drawn by the 3-pulse groups combine on the primary side of the transformer(s). This can be deduced from an understanding of the phase shifts that affect the individual harmonics.

Suppose that relative to some datum a particular secondary voltage is phase-shifted by an angle ϕ_{s1} at the fundamental frequency. All the components of current associated with that voltage are similarly displaced in time, and are therefore phase-shifted by an angle that is proportional to frequency, so that

$$\phi_{sn} = n\phi_{s1} \tag{2.49}$$

The phase shift in a component of current as seen from the supply terminals, ϕ_p, is a combination of the secondary phase shift ϕ_s and the phase shift which the current undergoes in its reflection into the primary circuit. At the fundamental frequency, these two phase shifts are equal and opposite, so that

$$\phi_{p1} = \phi_{s1} - \phi_{s1} = 0 \tag{2.50}$$

a necessary result, since otherwise a displacement factor less than unity would be introduced. At the nth harmonic, however, since the phase shift entailed in the reflection of the current into the primary circuit is numerically the same as for the fundamental component, the total phase shift is

$$\phi_{pn} = n\phi_{s1} \mp \phi_{s1} = (3r \pm 1)\phi_{s1} \mp \phi_{s1} \tag{2.51}$$

The $\mp$ sign here arises from the fact that a harmonic may have positive or negative phase sequence. At the nth harmonic the displacement between successive phases is

$$n\frac{2\pi}{3} = (3r \pm 1)\frac{2\pi}{3}$$

$$\equiv 2\pi/3 \text{ for } n = 3r + 1, \text{ indicating positive phase sequence}$$

or

$$-\frac{2\pi}{3} \text{ for } n = 3r - 1, \text{ indicating negative phase sequence} \tag{2.52}$$

Hence, for all values of r,

$$\phi_{pn} = 3r\phi_{s1} \tag{2.53}$$

Thus if the 3-pulse groups are phase-displaced in steps of $2\pi/p$ radians, their harmonic currents as reflected into the supply are phase-displaced in steps of $6r\pi/p$ radians, and the phasor diagram representing the addition of all the components at any one frequency comprises $p/3$ phasors at the resulting angles. The total of the phasor angles is then

$$\frac{6r\pi}{p} \times \frac{p}{3} = 2r\pi \tag{2.54}$$

which is 2π or a multiple thereof for all values of r. From this it can be deduced that the resultant is either zero or the arithmetic sum of the component magnitudes (the phasor diagram being either a regular polygon or a straight line) and that it is zero for all values of r other than those for which $6r\pi/p$ is equal to 2π or a multiple thereof, that is for values of r equal to $p/3$ or its multiples. The harmonic series for the p-pulse rectifier thus becomes

$$n = 3\left(\frac{rp}{3}\right) \pm 1 = pr \pm 1 \tag{2.55}$$

the amplitudes of all harmonics remaining inversely proportional to their orders.

12-PULSE AND 24-PULSE RECTIFIERS

By means of suitable transformer connections, the principles of phase multiplication described above may be extended to embrace any number of three-pulse groups, and in theory to reduce the input current harmonics to any desired extent. In practice, it is rarely worth going beyond twelve pulses per cycle, or at the most 24, because of the complexity of the transformer connections and the difficulty of keeping the system accurately in balance.

Figure 2.27 shows a typical twelve-pulse circuit comprising two hexaphase groups, each with its own transformer and interphase reactor, with their

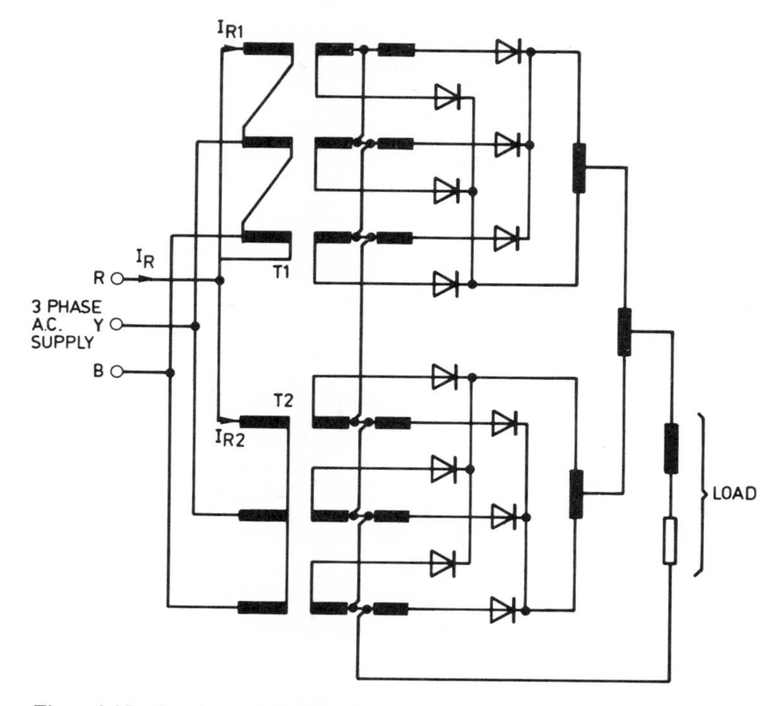

Figure 2.27 Combination of two hexaphase groups to form a twelve-pulse single-way rectifier.

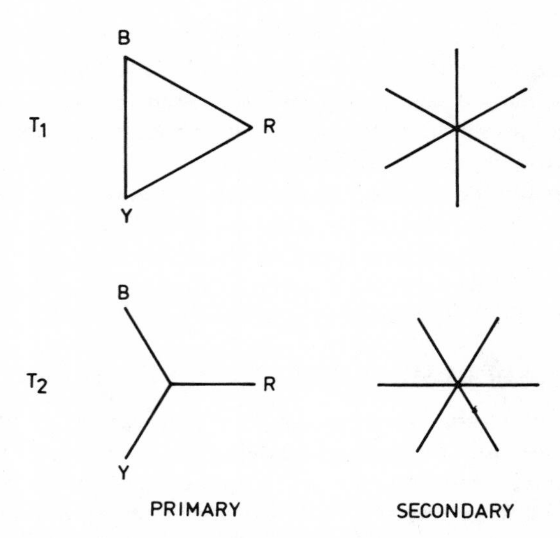

Figure 2.28 Phasor diagram of transformer voltages in Figure 2.27.

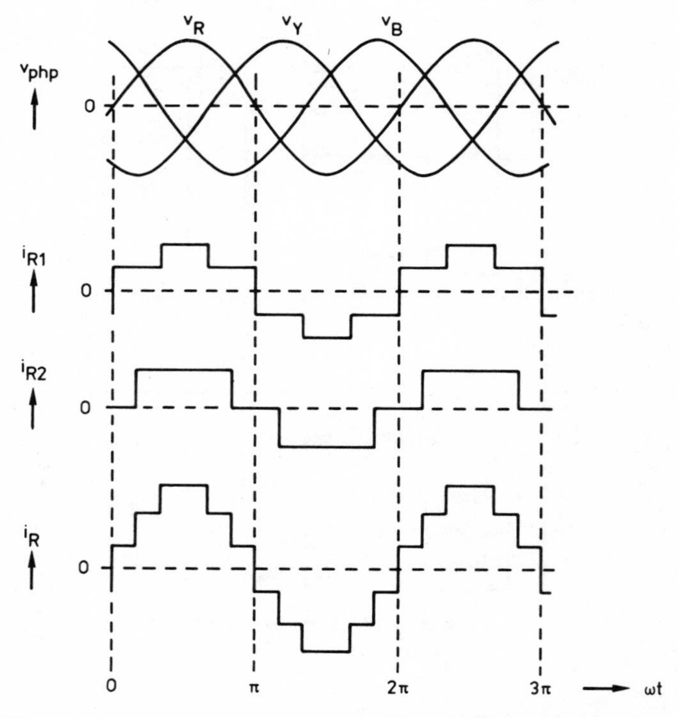

Figure 2.29 Input current waveforms in a twelve-pulse rectifier (Figure 2.27).

secondary voltages mutually displaced in phase by 30° by virtue of their different primary connections—one delta, the other star. (Other transformer connections can be used, and may be preferred from the point of view of balance.) The phasor diagram of Figure 2.28 illustrates the winding voltage relationships. A third interphase reactor is used to combine the outputs of the two hexaphase groups, supporting the six-pulse ripple voltage which appears at the hexaphase outputs but not in the combined output; the relatively low amplitude and high frequency of this ripple make this third reactor a very small component, and in fact it may often be dispensed with, given sufficient leakage reactance in the transformers.

Figure 2.29 illustrates the combination of the input current waveforms to give a resultant considerably closer than the six-pulse waveform to a sine wave, having a calculated distortion factor, or power factor, (ideally) of 0.99.

BRIDGE RECTIFIERS

Despite variations from one arrangement to another and the expedients whereby, in a polyphase rectifier, the supply current waveform can be made to

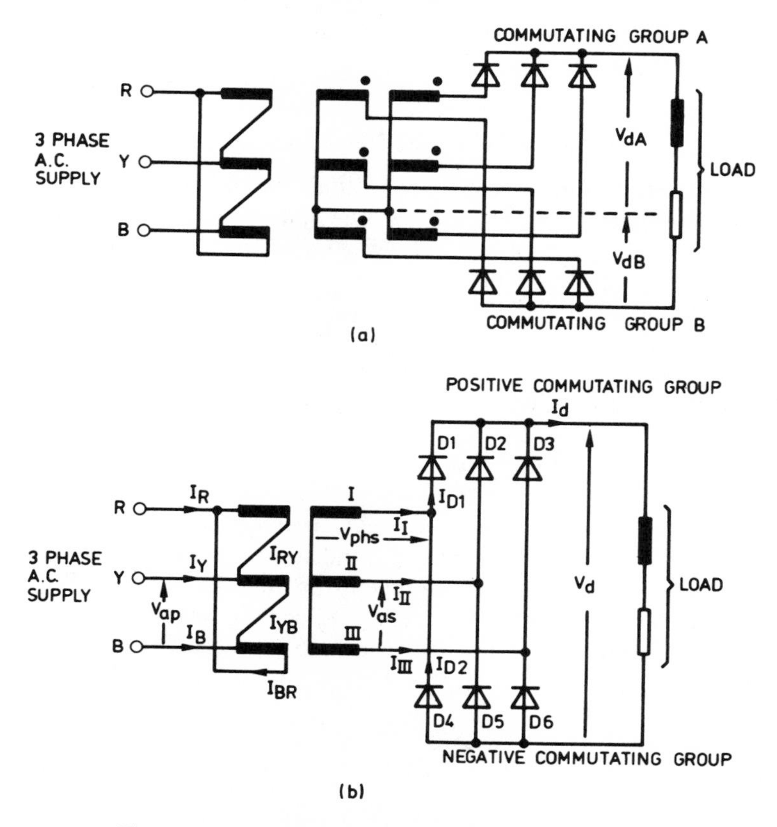

Figure 2.30 Derivation of the three-phase bridge rectifier.

approach the sinusoidal, the utilization of the transformer feeding a single-way rectifier is inevitably limited by the unidirectional nature of the secondary current; the utilization factor of 0.79 deduced above for the hexaphase rectifier (2.42) represents almost the best that can be achieved. A double-way or bridge connection offers a means of improving the utilization of the secondary windings of the transformer by enabling them to supply symmetrical alternating current.

Functionally, the bridge rectifier is most conveniently regarded as a combination of two commutating groups connected in series on the d.c. side rather than in parallel as in the hexaphase connection. The development of the bridge connection on this principle is illustrated in Figure 2.30. In Figure 2.30(a), two three-phase single-way rectifier groups, one with common cathodes and the other with common anodes, are connected in series by joining the star points of their transformer secondary windings, so that their (equal) d.c. output voltages V_{dA} and V_{dB} add in the manner shown, the supply voltages for the two groups being equal. It is immediately apparent that on each limb of the transformer the two secondary windings can be replaced by one, since they are connected at one end and at the same potential at the other. The circuit of Figure 2.30(b), which constitutes a three-phase bridge rectifier, is therefore equivalent to that of Figure 2.30(a).

The single-phase bridge is similarly derived from two bi-phase rectifiers as illustrated in Figure 2.32.

THREE-PHASE BRIDGE RECTIFIER

As in the case of the duplex single-way rectifier, the characteristics of the bridge rectifier can be simply deduced from those of its constituent commutating groups. In the three-phase bridge, as illustrated in Figure 2.31, the combination of the two alternating-current components results in primary and secondary current waveforms similar to the primary current waveforms of the hexaphase rectifier with interphase reactor. On the d.c. side, the total output voltage, instantaneous or mean, is the sum of those of the two diode groups, while each group carries the whole load current; for a given power and a given transformer secondary voltage, therefore, the bridge rectifier produces twice the output voltage of the duplex hexaphase circuit at half the output current, while the output voltage waveforms for the two circuits are identical.

It follows that in all its main external characteristics, at least, including harmonic generation, the bridge rectifier is indistinguishable from the hexaphase circuit working within the effective range of its interphase reactor.

The general expression for output voltage previously given in connection with the single-way rectifier, (2.4), is applicable to the bridge with an extra factor of two to allow for the addition of the two commutating group outputs:

$$\overline{V}_d = 2\hat{V}_{ph}\left(\frac{\sin(\pi/m)}{\pi/m}\right) \tag{2.56}$$

For practical purposes there is little to be gained by pursuing this generalization further, since normally only single-phase and three-phase bridges will be

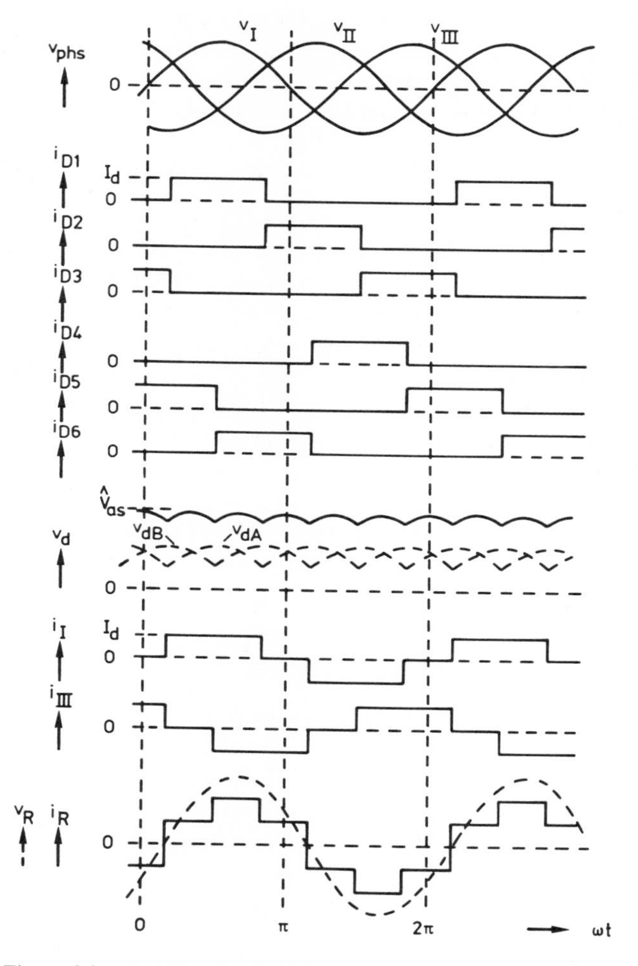

Figure 2.31 Waveforms in a three-phase bridge rectifier (Figure 2.30).

encountered. For the three-phase bridge,

$$\overline{V}_d = \frac{3\sqrt{3}}{\pi}\hat{V}_{phs} = \frac{3\sqrt{2}\sqrt{3}}{\pi}V_{phs}$$

$$= \frac{3\sqrt{2}}{\pi}V_s \tag{2.57}$$

It should be noted that in the three-phase bridge the pulse number is twice the number of phases: i.e. $p = 2m$.

The diode currents are as given for the three-phase commutating group in (2.11). The transformer secondary currents, however, are similar in harmonic content to the primary currents, which as noted above are similar to those in the hexaphase rectifier. The utilization factor of both primary and secondary windings, and therefore of the transformer, is accordingly that given by expression (2.38), as is also the input power factor:

$$k_{\text{p}} = k_{\text{u T}} = k_{\text{us}} = k_{\text{up}} = \frac{3}{\pi} = 0.955 \tag{2.58}$$

The resultant harmonic currents in the input to the bridge can be derived in a slightly different way by considering the relationship between the two sides as a time displacement of π/ω coupled with a phase-displacement of π for positive-phase-sequence components and $-\pi$ for negative-phase-sequence components. The nett phase displacement for the series of harmonics $rm \pm 1$ is then $rm\pi$, so that the harmonics corresponding to odd values of rm disappear, while the others remain. (This result is equally applicable to the single-phase bridge, in which all values of rm are even, so that the harmonic distribution is the same for the bridge as for its bi-phase components.)

SINGLE-PHASE BRIDGE RECTIFIER

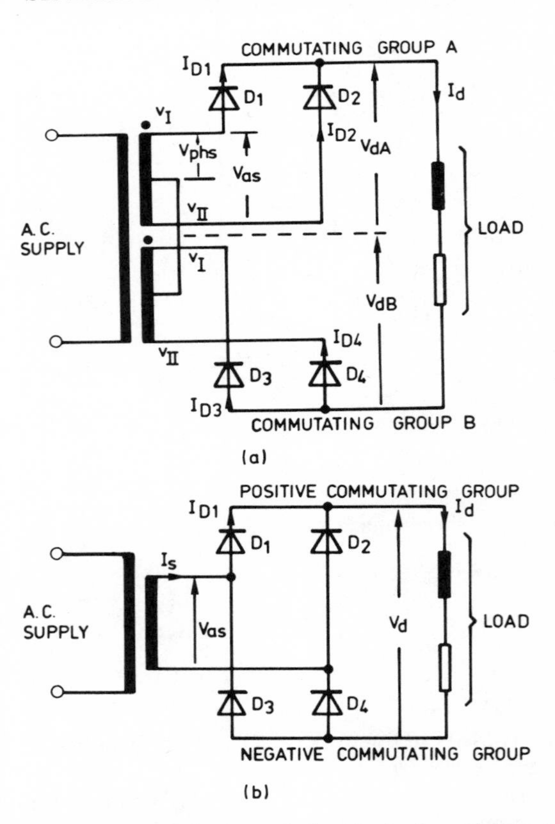

Figure 2.32 Derivation of the single-phase bridge rectifier.

In the single-phase bridge the phase displacement between the two constituent commutating groups is equivalent to the interval between successive output voltage peaks, and there is therefore no gain in pulse number in comparison with the bi-phase rectifier, the input and output waveforms of voltage and current being identical for the two circuits. The waveform of the current in the transformer secondary winding is similar to that of the primary current. Voltage and current waveforms are shown in Figure 2.33. As in the case of the three-phase bridge, the d.c. output voltage for a given transformer secondary voltage (phase voltage, as defined, or total voltage) is twice that of the single-way rectifier:

$$\bar{V}_{\mathrm{d}} = 2\hat{V}_{\mathrm{phs}}\left(\frac{\sin(\pi/2)}{\pi/2}\right) = \frac{4}{\pi}\hat{V}_{\mathrm{phs}}$$

$$= \frac{2}{\pi}\hat{V}_{\mathrm{s}} = \frac{2\sqrt{2}}{\pi}V_{\mathrm{s}} \tag{2.59}$$

Again, the utilization factors of the primary and secondary windings and of the whole transformer are the same, equal to the ratio of the fundamental component to the total r.m.s. value of the square-wave input current:

$$k_{\mathrm{p}} = k_{\mathrm{u\,T}} = k_{\mathrm{us}} = k_{\mathrm{up}} = \frac{2\sqrt{2}}{\pi} = 0.90 \tag{2.60}$$

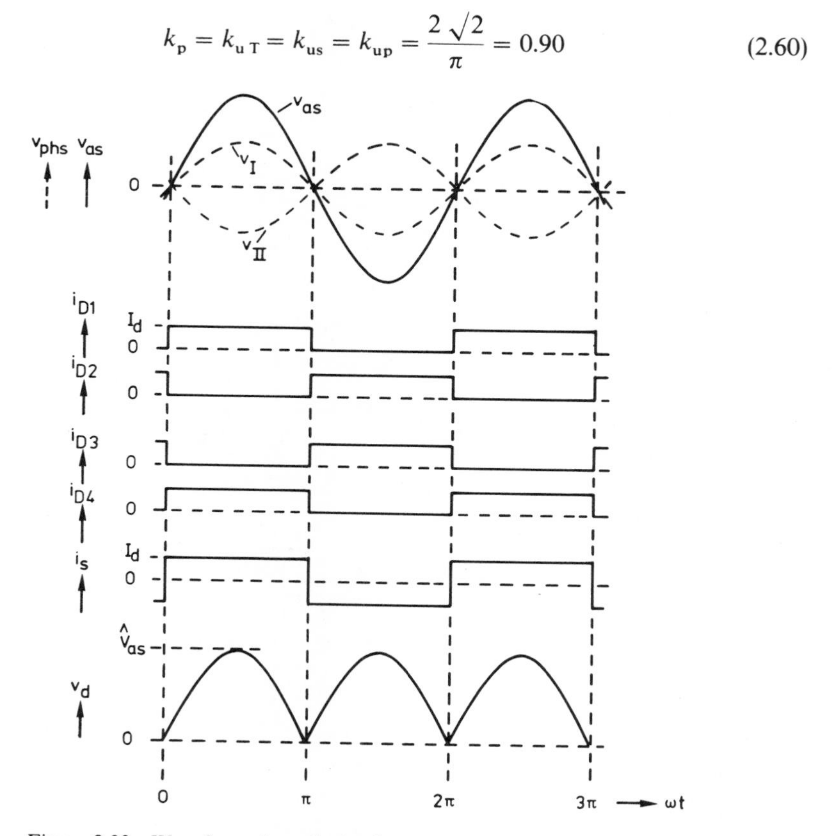

Figure 2.33 Waveforms in a single-phase bridge rectifier (Figure 2.32).

MULTIPLEX BRIDGE RECTIFIERS

The principles of phase multiplication described above may be extended to combinations of three-phase bridges with their inputs phase-displaced by half the interval between successive output voltage peaks; this technique leads to a variety of series/parallel combinations of three-phase commutating groups. Two relatively common arrangements combining two bridges to give a twelve-pulse output are illustrated in Figures 2.34 and 2.35, the first a parallel circuit with an

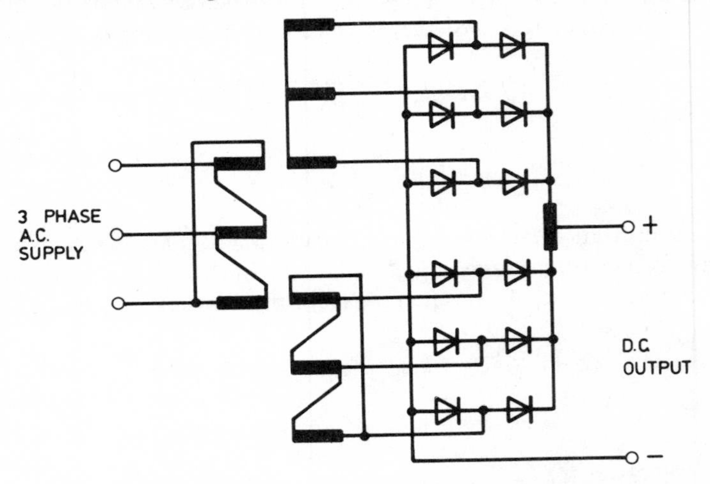

Figure 2.34 Twelve-pulse rectifier combining two three-phase bridges in parallel with interphase reactor.

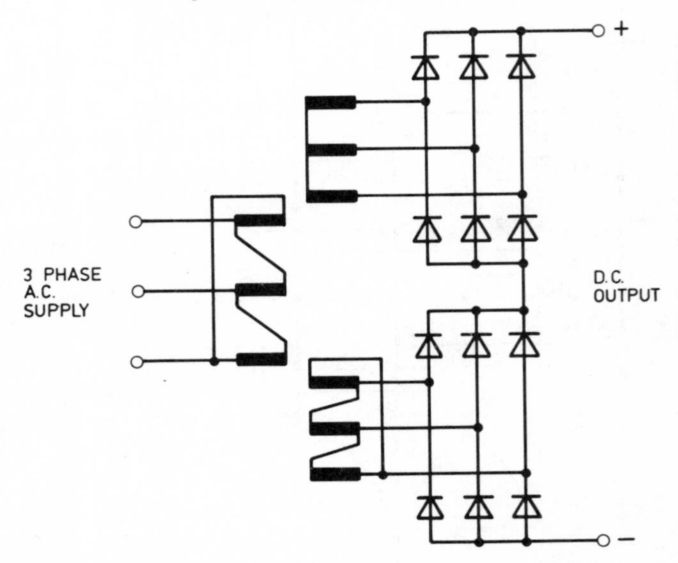

Figure 2.35 Twelve-pulse rectifier combining two three-phase bridges in series.

interphase reactor to support the six-pulse ripple components (as in the single-way circuit of Figure 2.27) while in the second the two bridges are connected in series on the d.c. side so that there are altogether four commutating groups in series. In each case the transformer connection provides the necessary 30° phase displacement between the inputs to the two bridges.

GENERAL COMPARISON OF RECTIFIER CIRCUITS

While the discussion has so far been restricted to idealized diode rectifier circuits, it will serve to indicate the general bases of practical choice between the various configurations. The principal factors relative to three-phase rectifiers are

(i) The greater the pulse number, the better is the input power factor and the lower the output ripple voltage. A pulse number greater than six, however, more often than not entails a cost penalty.
(ii) Bridge rectifiers achieve a better transformer utilization than single-way rectifiers, and can function without a transformer if the supply voltage is appropriate for the required output voltage; they are therefore more widely used than single-way rectifiers.
(iii) In comparison with a bridge, for a given output and pulse number, a single-way rectifier needs diodes of only half the mean current rating, but of twice the voltage rating, and incurs only half the conduction losses. This often makes the single-way rectifier attractive for low voltages, particularly at high currents.

CONTROLLED CONVERTERS

Discussion so far in this chapter has been confined to uncontrolled rectifiers, using only diodes as rectifying elements. If thyristors are substituted for the diodes in a rectifier circuit, an element of control is introduced whereby the direct output voltage can be varied by controlling the triggering inputs to the thyristor gates in a suitable manner.

This will be illustrated in particular terms by reference to the three-phase single-way rectifier illustrated in Figure 2.36, which is the controlled equivalent of the circuit of Figure 2.13. If each thyristor is triggered at the instant when the supply makes its anode potential positive with respect to its cathode—that is, at the instant when it would start to conduct by the process of commutation described earlier if it were a diode—the circuit behaves in the same way as the uncontrolled rectifier. If however the triggering pulses are witheld until some instant in the ensuing half-cycle, commutation is delayed, and the output voltage waveform is consequently modified.

This process is illustrated in the waveforms of Figure 2.37 for an angle of delay α, signifying a delay of α/ω between the triggering instant and the earliest possible point of commutation (the point of 'diode' commutation). Integration of the output voltage between the limits $\omega t = \alpha$ and $\omega t = 2\pi/3 + \alpha$ will show that the

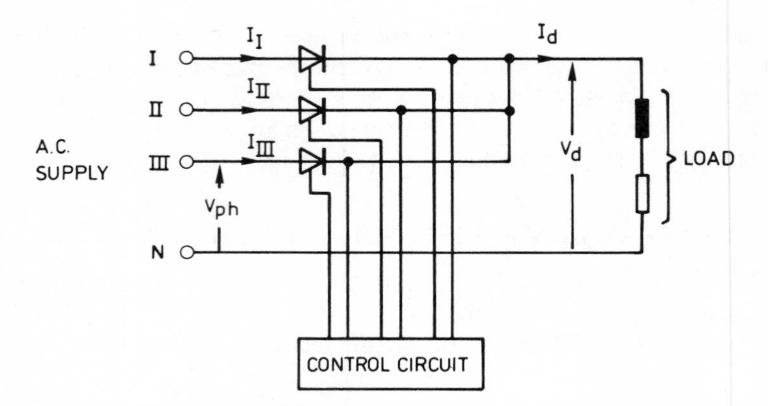

Figure 2.36 Three-phase single-way controlled rectifier.

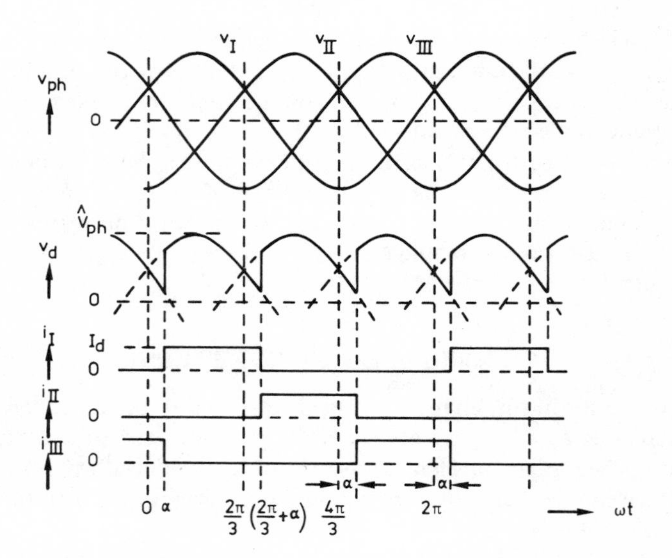

Figure 2.37 Waveforms in a three-phase single-way controlled rectifier (Figure 2.36) with delay angle α.

mean d.c. output voltage is reduced by the delayed triggering from $\hat{V}_{ph}[\sin(\pi/3)]/(\pi/3)$ to $\hat{V}_{ph}\{[\sin(\pi/3)]/(\pi/3)\}\cos\alpha$.

Obviously this expression becomes zero when $\alpha = \pi/2$ and negative* with larger angles of delay, and on the assumption made earlier that the direct current is continuous, the reversal of mean output voltage does in fact occur with values of α between $\pi/2$ and π, the waveform taking the form illustrated in Figure 2.38.

* For convenience, the direct or instantaneous output voltage from a converter is referred to in this chapter as positive when it is of the same polarity as the maximum rectified output voltage—i.e. with $\alpha = 0$.

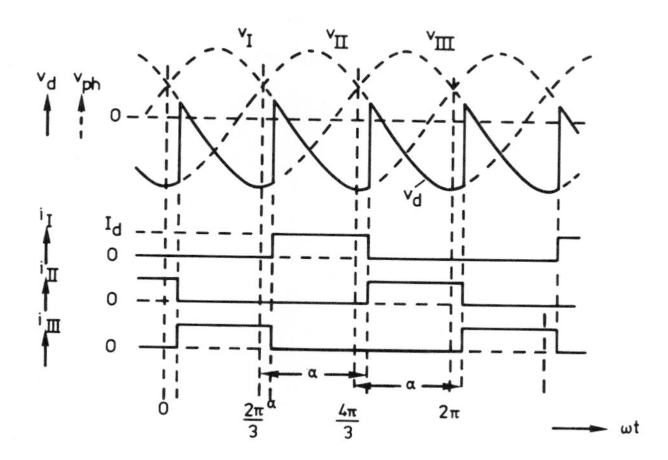

Figure 2.38 Waveforms in a three-phase single-way converter (Figure 2.36) with a delay angle of 135°.

The reversal of direct voltage, with the direction of current flow unchanged, implies a reversed power flow, and under these conditions the circuit becomes a naturally commutating inverter, feeding energy from the d.c. circuit into the a.c. supply system.

The voltage that effects commutation of the direct current from one conducting phase to the next is the voltage between the two phases—i.e. the line-to-line (interphase) voltage. Since this first assumes the necessary polarity at what has been termed zero delay angle, it falls to zero at a delay angle of π and thereafter reverses; if α is made greater than π, therefore, commutation ceases and the converter becomes inoperative with the current flowing continuously in one thyristor.

The output voltage as a function of α may be deduced in a more general form for any m-phase single-way converter by reference to the output voltage waveform of Figure 2.39 (cf. Figure 2.10 and equation (2.4)).

$$\overline{V}_{\mathrm{d}} = \frac{\hat{V}_{\mathrm{ph}}}{2\pi/m}\int_{\alpha-\pi/m}^{\alpha+\pi/m} \cos\omega t \, \mathrm{d}\omega t = \hat{V}_{\mathrm{ph}}\frac{\sin(\pi/m)}{\pi/m}\cos\alpha \tag{2.61}$$

Thus the cosine law, together with the (ideal) delay-angle range of π, applies to any single-way converter, provided that there is no uncontrolled ('free-wheeling') current path for the direct current.

With a passive load, as in Figure 2.36, as distinct from a circuit incorporating a continuous source of energy such as a mechanically driven machine, the assumption of continuous current is clearly untenable, other than transiently, if the direct voltage is reversed. In such a case, therefore, the steady-state mean output voltage follows the cosine law, as shown in Figure 2.40, only up to a point where the current becomes discontinuous, and thereafter falls gradually to zero as the delay angle reaches the limit of its range (not 180° in three-phase rectifiers, but

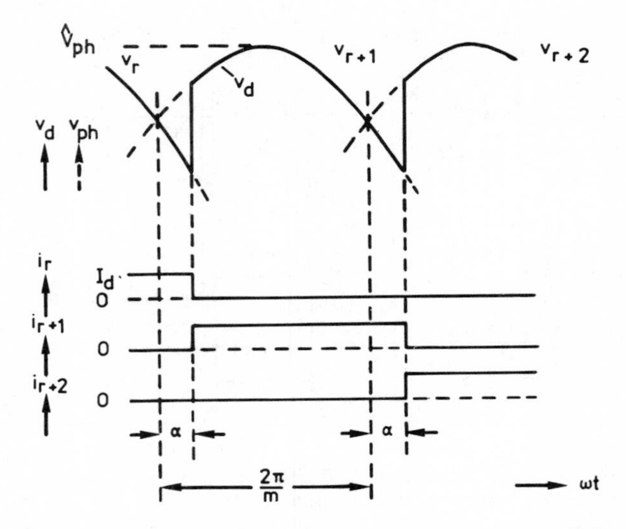

Figure 2.39 Waveforms in an m-phase single-way controlled converter.

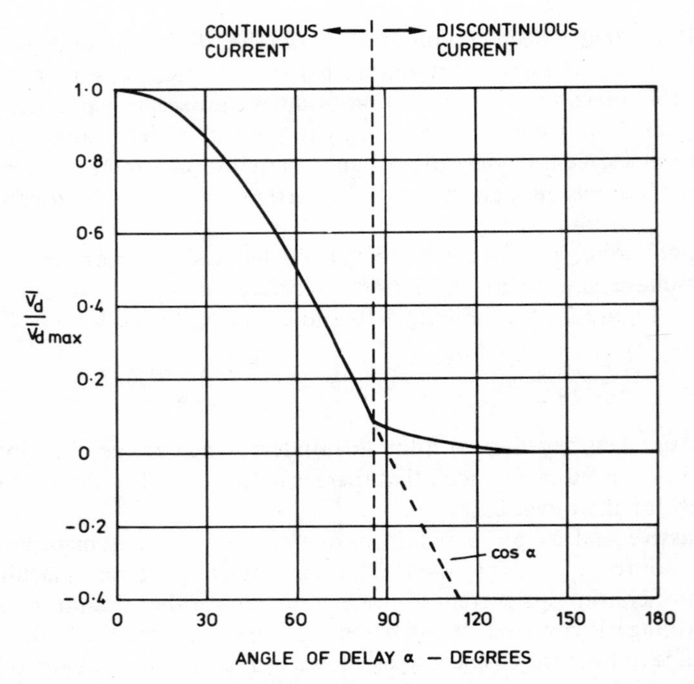

Figure 2.40 Typical control characteristic of a fully controlled converter with LR load.

the limit of the range that applies when free-wheel diodes are employed—see below).

While the behaviour of converters with discontinuous output current is important from some points of view—in the design of control systems, for example—it is outside the scope of this book and will not be considered here in more detail.

Input harmonics, power factor and transformer utilization

Reference to the waveforms illustrated in Figures 2.37 and 2.38 shows that in comparison with the corresponding uncontrolled rectifier (Figure 2.14) the only effect of delayed firing on the input current waveforms is to retard the pulses of current by a phase angle α. The input current harmonics are therefore unchanged, but the input power factor is reduced by the introduction of a displacement factor $\cos\alpha$. Since the magnitude of the input current, relative to the output current,* is not affected by α, this reduction in power factor matches the reduction in output voltage in the ratio of $\cos\alpha$ deduced above, and thus maintains the necessary power balance between input and output. The transformer utilization is reduced, at reduced output voltage, in the same ratio as the power factor.

MULTIPLEX AND BRIDGE CONVERTERS

Since it has been established that the control characteristics of all single-way controlled converters are similar, and that their input currents (relative to their output currents) are affected by delay-angle control only to the extent of a phase delay α, it follows that the principles followed earlier in explaining the combination of commutating groups of diodes into multiplex circuits and bridges can be applied equally to groups of thyristors, and a range of controlled converters corresponding to the whole range of rectifiers previously mentioned, each with the same control characteristic $\overline{V}_d = \overline{V}_{d\,max}\cos\alpha$, can be assumed without further elaboration. It should be stressed that this applies specifically to converters which consist entirely of thyristor groups (as distinct from combinations of thyristors and diodes, referred to later) and are described as fully controlled.

By way of illustration, waveforms are shown in Figure 2.42 for a fully controlled three-phase bridge (Figure 2.41) operating with an angle of delay of 75° (cf. Figure 2.31). Note that α is measured, as always, from the earliest possible instant of commutation for each thyristor, and that this coincides with one of the interphase voltage zeros. Since the instantaneous output voltage is always the difference between two phase voltages, the output voltage waveform is composed, as shown, of segments of the interphase voltage waveforms. A single-phase fully controlled bridge is illustrated in Figure 2.43, and typical waveforms in it are illustrated in Figure 2.44.

* The terms 'input' and 'output' are retained here for the a.c. and d.c. circuits respectively, as for the uncontrolled rectifier, even though the power flow may actually be reversed.

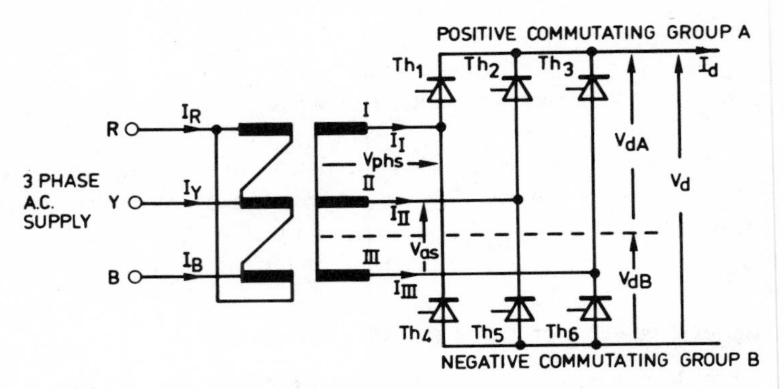

Figure 2.41 Three-phase fully controlled bridge converter.

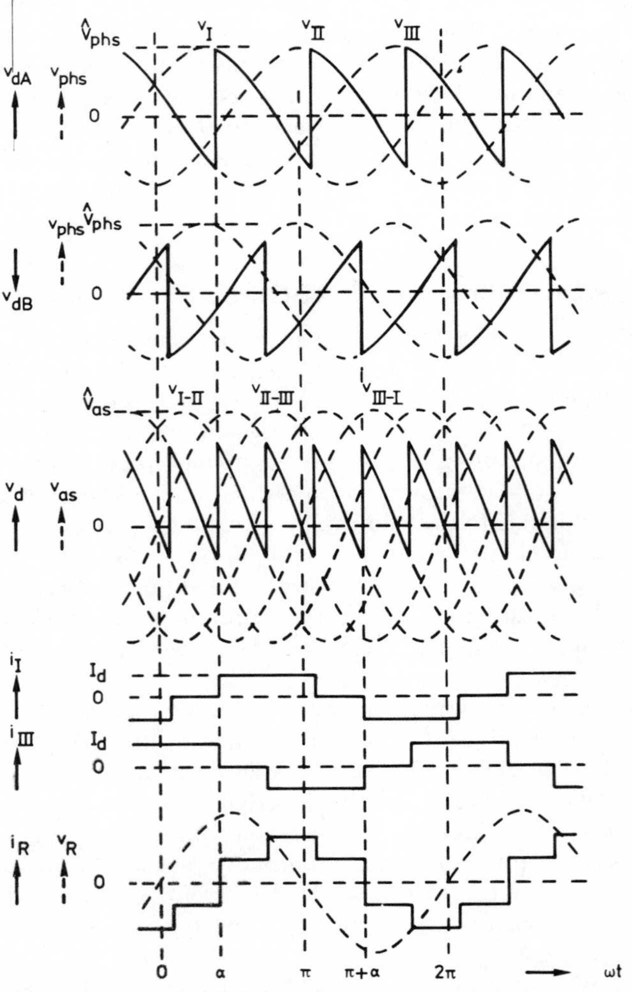

Figure 2.42 Waveforms in a three-phase fully controlled bridge converter (Figure 2.41).

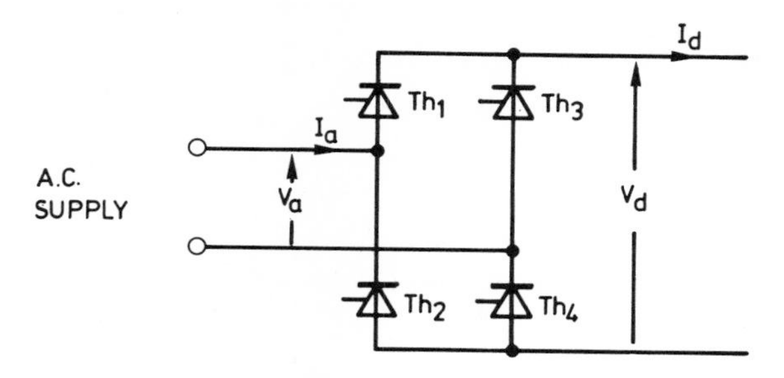

Figure 2.43 Single-phase fully controlled bridge converter.

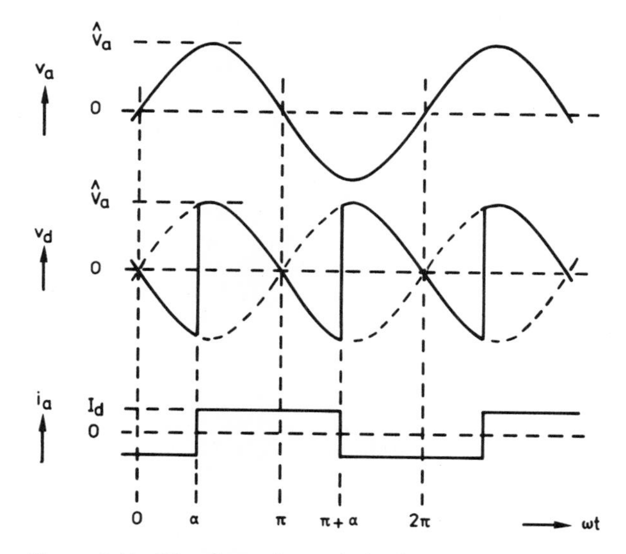

Figure 2.44 Waveforms in a single-phase fully controlled bridge converter (Figure 2.43).

THE HALF-CONTROLLED BRIDGE

When two single-way converters are combined to make a controlled bridge rectifier, it is not always necessary that both should be controlled. A bridge comprising a group of diodes in combination with a group of thyristors is known as a half-controlled bridge, and is illustrated in Figure 2.45 in three-phase and single-phase forms.

The uncontrolled half of the bridge produces a fixed d.c. output voltage, relative to the neutral point of the supply, equal to $\hat{V}_{ph} \sin(\pi/m)/(\pi/m)$; the controlled half produces a d.c. voltage depending on the delay angle, equal to

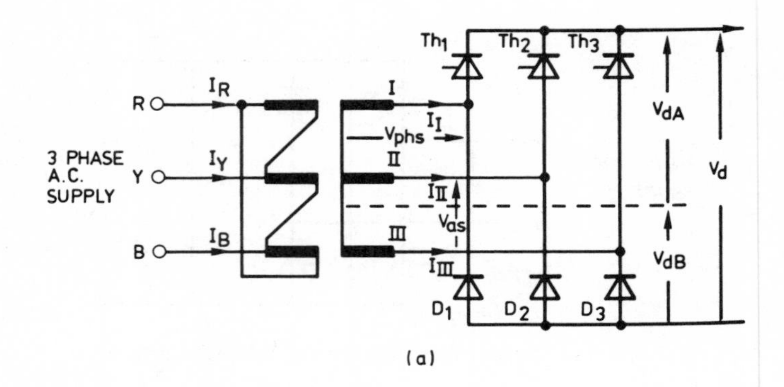

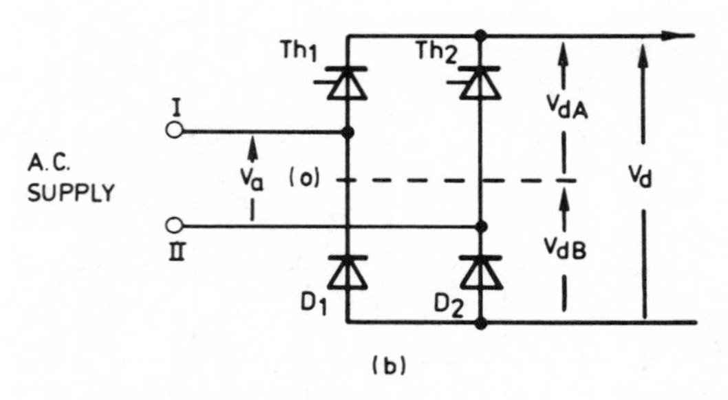

Figure 2.45 Half-controlled bridge rectifiers: (a) three-phase; (b) single-phase.

$\hat{V}_{ph} \cos \alpha \sin (\pi/m)/(\pi/m)$. The total mean output voltage is therefore

$$\bar{V}_d = \hat{V}_{ph} \frac{\sin (\pi/m)}{\pi/m} (1 + \cos \alpha)$$

which can be expressed as (2.62)

$$\bar{V}_d = \bar{V}_{d\,max} \left(\frac{1 + \cos \alpha}{2} \right)$$

This expression cannot have a negative value, and there is thus no possibility of inversion. The control characteristic of any half-controlled bridge is of the form illustrated in Figure 2.46, with an α-range of π, as in the case of the fully controlled converter.

Input current

Since the current drawn by the controlled half of the bridge is shifted in phase according to the angle of delay, while that drawn by the diode group is unaffected, the phase displacement between the input currents to the two halves is variable,

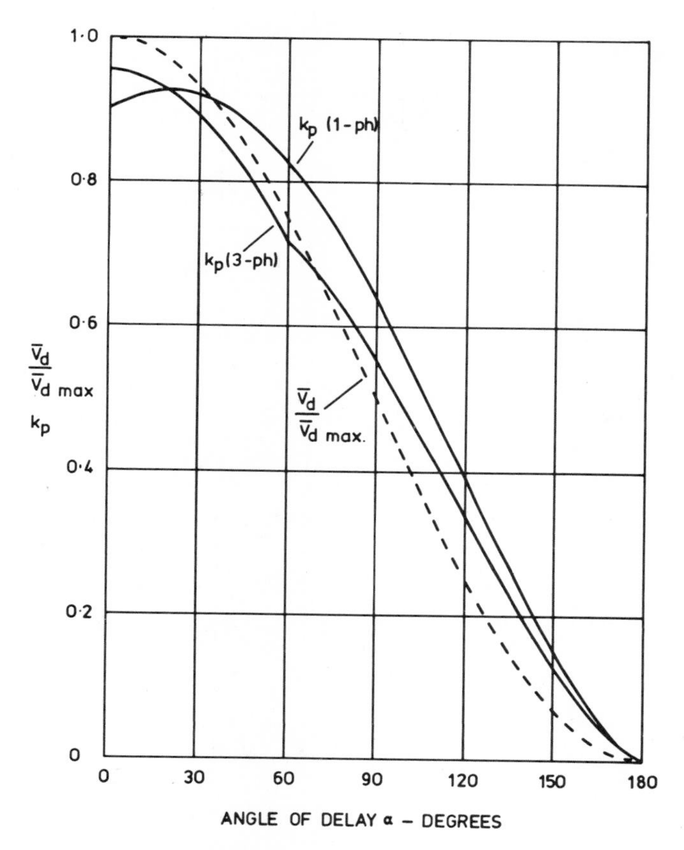

Figure 2.46 Variation with delay angle of the input power factor of half-controlled rectifiers.

and for the nth harmonic ($n = rm \pm 1$) is given by

$$\left.\begin{aligned} \phi_n = rm\pi + (rm \pm 1)\alpha, &\quad \text{equivalent to} \\ \pi + (rm \pm 1)\alpha &\quad \text{for odd values of } rm\text{, and} \\ (rm \pm 1)\alpha &\quad \text{for even values of } rm \end{aligned}\right\} \qquad (2.63)$$

In terms of the maximum value of the fundamental component, the harmonic currents are therefore

$$\left.\begin{aligned} I_n &= \frac{I_{1\,\max}}{n}\sin\frac{n\alpha}{2} \qquad \text{when } rm \text{ is odd, and} \\ I_n &= \frac{I_{1\,\max}}{n}\cos\frac{n\alpha}{2} \qquad \text{when } rm \text{ is even} \end{aligned}\right\} \qquad (2.64)$$

In the three-phase bridge, the harmonic currents corresponding to odd values of r, i.e. the even harmonics, which are eliminated in the uncontrolled and fully

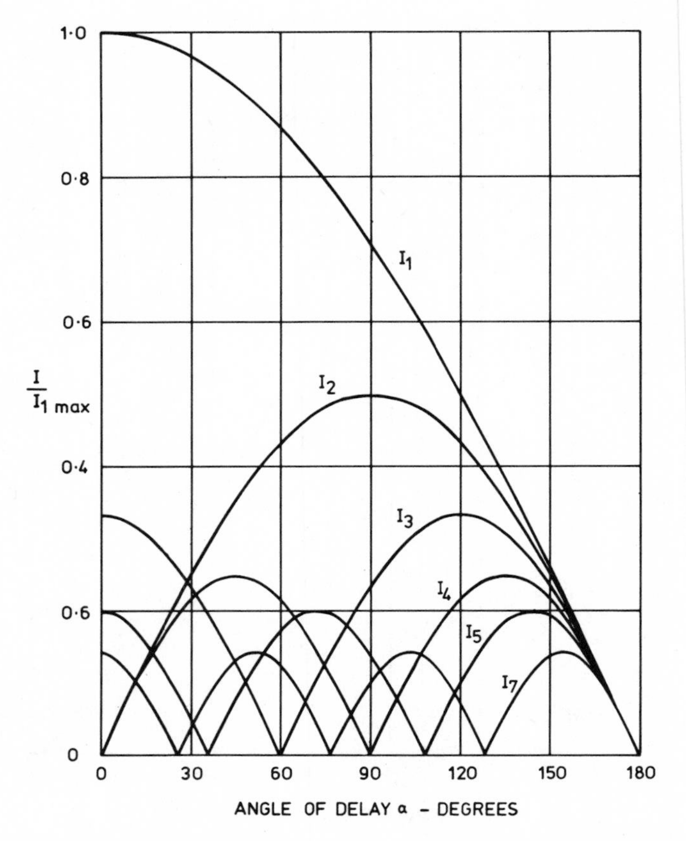

Figure 2.47 Input current harmonics in a half-controlled bridge rectifier with constant output current: single-phase: 1, 3, 5, 7, etc; three-phase: 1, 2, 4, 5, 7 . . . .

controlled bridges, re-appear as α is increased from zero, and reach their full three-pulse values when $\alpha = \pi/n$. In the single-phase case, for which rm is always even, the harmonic currents are never increased by firing delay. The variation of the lower-order harmonic currents with delay angle is illustrated for the two cases in Figure 2.47.

Typical waveforms of input current and output voltage are shown in Figure 2.48 and 2.49 for the three-phase and in Figure 2.50 for the single-phase bridge.

Input power factor of half-controlled bridges

It is apparent from inspection of the waveforms that for a given load current the r.m.s. input current of the single-phase half-controlled bridge decreases with increasing firing delay; the reduction in power factor is therefore less than in the

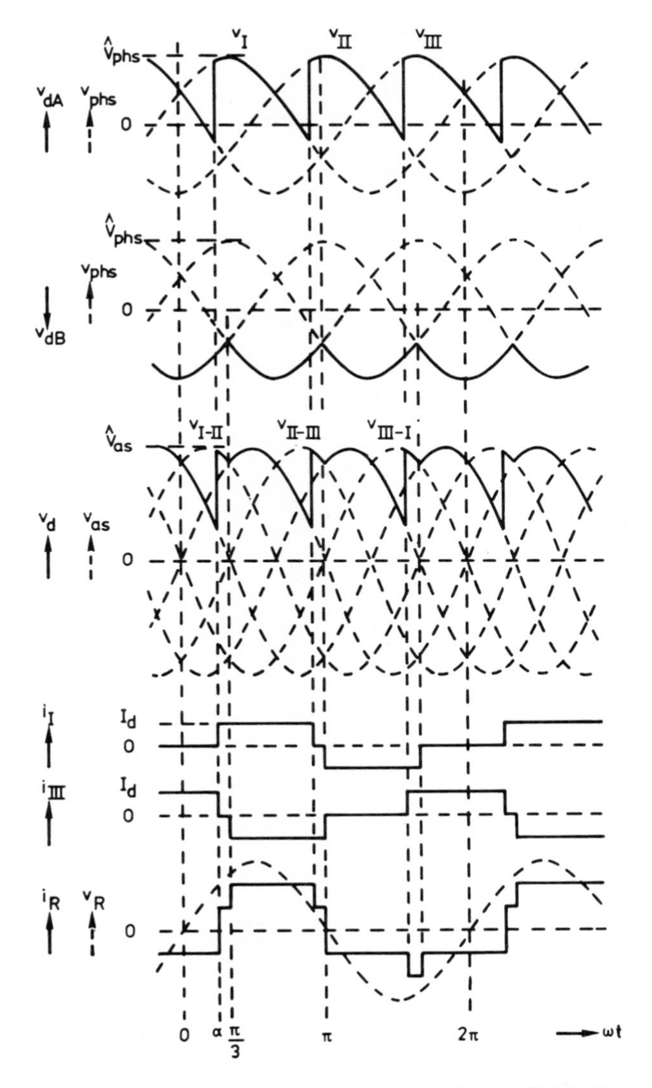

Figure 2.48 Waveforms in a three-phase half-controlled bridge rectifier (Figure 2.45(a)) ($\alpha < \pi/3$).

fully controlled bridge. With an angle of delay α, the output and input powers are

$$P = \overline{V}_d I_d = \frac{2\sqrt{2}}{\pi} V_a I_d \left(\frac{1 + \cos\alpha}{2}\right)$$

while the input VA is

$$S = V_a I_a = V_a I_d \sqrt{\left(\frac{\pi - \alpha}{\pi}\right)}$$

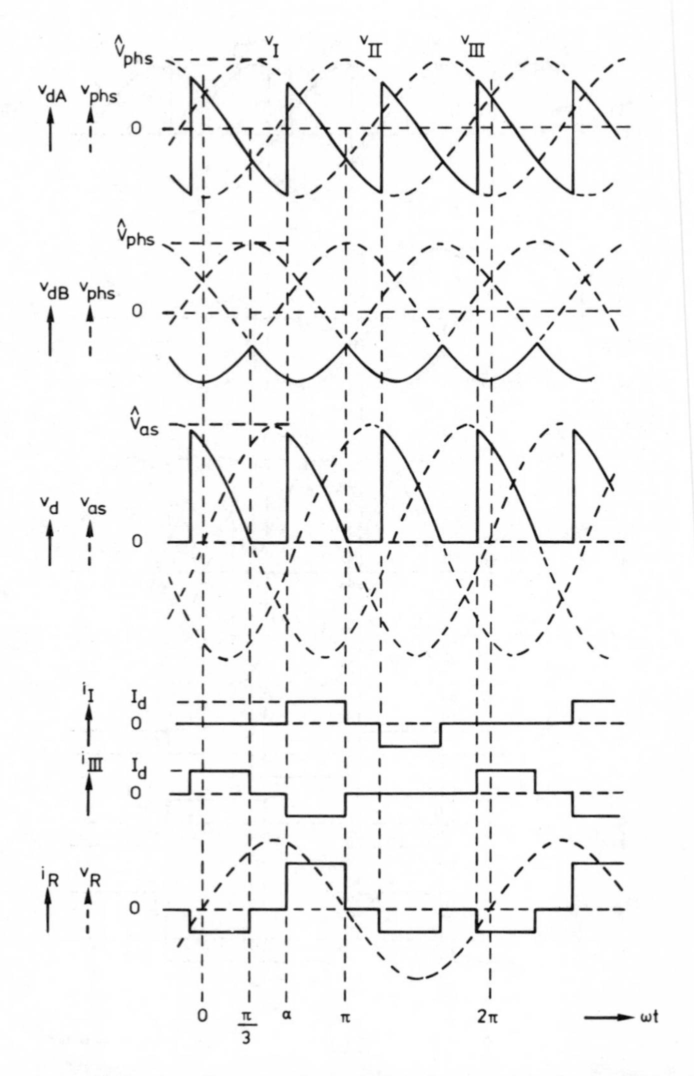

Figure 2.49 Waveforms in a three-phase half-controlled bridge rectifier (Figure 2.45(a)) ($\alpha > \pi/3$).

The power factor is therefore

$$k_p = \frac{P}{S} = \frac{\sqrt{2}\,(1+\cos\alpha)}{\sqrt{[\pi(\pi-\alpha)]}} \tag{2.65}$$

In the three-phase case, there is no reduction in r.m.s. input current so long as $\alpha < \pi/3$, and the power factor falls in proportion to $(1+\cos\alpha)/2$. When α exceeds $\pi/3$, however, the input conduction angle decreases to $(\pi-\alpha)$ (Figure 2.49), and the input VA is then

$$S = \sqrt{3}\,V_a I_d \sqrt{\left(\frac{\pi-\alpha}{\pi}\right)}$$

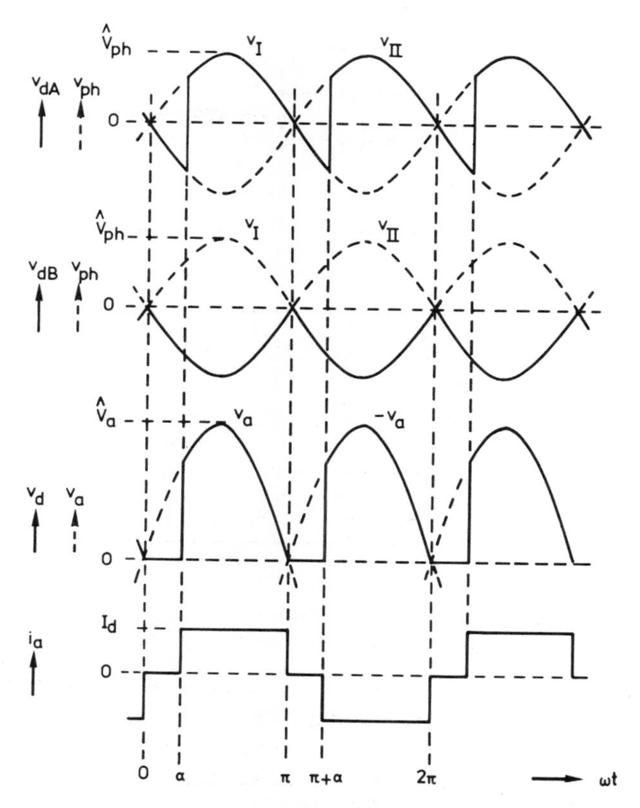

Figure 2.50 Waveforms in a single-phase half-controlled bridge rectifier (Figure 2.45(b)).

The input power is

$$P = \frac{3\sqrt{2}}{\pi} V_a I_d \left(\frac{1+\cos\alpha}{2}\right)$$

and the power factor in this region is therefore

$$k_p = \frac{P}{S} = \frac{\sqrt{3}(1+\cos\alpha)}{\sqrt{[2\pi(\pi-\alpha)]}} \tag{2.66}$$

Figure 2.46 shows the variation in power factor over the whole range of α in the two cases.

CONTROLLED RECTIFIERS WITH FREE-WHEEL DIODES

A fully controlled converter is sometimes furnished with a free-wheel diode—that is, a diode connected across its output to provide an alternative path for load current that excludes the converter thyristors, as shown in Figure 2.51, with effects which depend on the angle of delay. If α is so small that, without the diode,

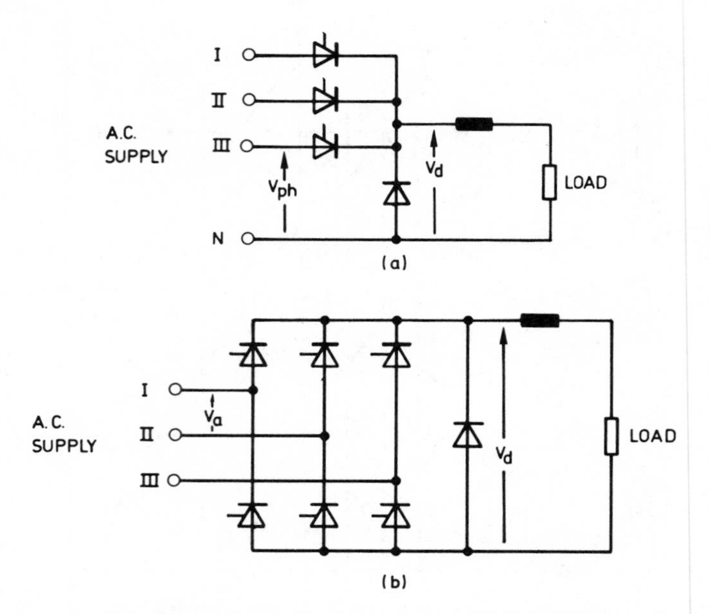

Figure 2.51 Fully controlled rectifiers with free-wheel diodes: (a) three-phase single-way; (b) three-phase bridge.

the instantaneous output voltage does not at any time become negative, the diode is always reverse-biassed, and has no effect. With a larger value of α, however, such that the output voltage tends to become negative during parts of the cycle, the diode prevents the negative excursions, periodically carrying the load current which would otherwise have to flow continuously in the converter. Since a negative output voltage cannot be produced, inversion is not possible.

Single-way rectifier

In an m-phase single-way controlled rectifier, the earliest point of commutation ($\alpha = 0$), relative to the phase voltage, is at an angle $(\pi - 2\pi/m)/2$, while the conduction angle in each phase is $2\pi/m$. The instantaneous output voltage will just reach zero, therefore, as illustrated in Figure 2.52, when $\alpha = \pi/2 - \pi/m$, and the mean output voltage will fall to zero when $\alpha = \pi/2 + \pi/m$, which defines the α range. The composite control characteristic is then

$$\left.\begin{aligned} 0 < \alpha < \left(\frac{\pi}{2} - \frac{\pi}{m}\right):\quad & \overline{V}_{\mathrm{d}} = \hat{V}_{\mathrm{ph}} \frac{\sin(\pi/m)}{\pi/m} \cos\alpha \\ \left(\frac{\pi}{2} - \frac{\pi}{m}\right) < \alpha < \left(\frac{\pi}{2} + \frac{\pi}{m}\right):\quad & \overline{V}_{\mathrm{d}} = \frac{\hat{V}_{\mathrm{ph}}}{2\pi/m} \int_{(\frac{\pi}{2} - \frac{\pi}{m} + \alpha)}^{\pi} \sin\omega t \,\mathrm{d}\omega t \\ & = \hat{V}_{\mathrm{ph}} \left\{ \frac{1 + \cos(\alpha + \frac{\pi}{2} - \frac{\pi}{m})}{2\pi/m} \right\} \end{aligned}\right\} \tag{2.67}$$

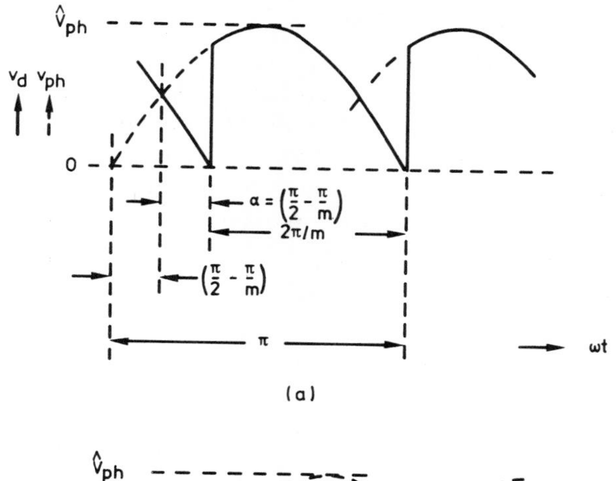

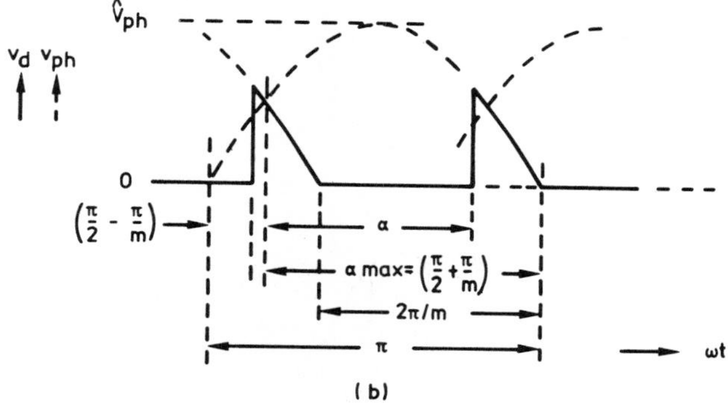

Figure 2.52 Output voltage waveforms in an m-phase single-way controlled rectifier with free-wheel diode, with two values of α.

In the bi-phase circuit ($m = 2$) $\pi/2 - \pi/m = 0$, so the control characteristic is entirely of the second form, equating to

$$\overline{V}_{\mathrm{d}} = \hat{V}_{\mathrm{ph}}\left(\frac{1+\cos\alpha}{\pi}\right) = \overline{V}_{\mathrm{d\,max}}\left(\frac{1+\cos\alpha}{2}\right) \tag{2.68}$$

i.e. similar to that of the half-controlled bridge, while the characteristic of the three-phase circuit is

$$\left.\begin{aligned} 0 < \alpha < \frac{\pi}{6}:\quad \overline{V}_{\mathrm{d}} &= \frac{3\sqrt{3}}{2\pi}\hat{V}_{\mathrm{ph}}\cos\alpha = \overline{V}_{\mathrm{d\,max}}\cos\alpha \\ \frac{\pi}{6} < \alpha < \frac{5\pi}{6}:\quad \overline{V}_{\mathrm{d}} &= \frac{3}{2\pi}\hat{V}_{\mathrm{ph}}\left\{1+\cos\left(\alpha+\frac{\pi}{6}\right)\right\} \\ &= \overline{V}_{\mathrm{d\,max}}\left\{\frac{1+\cos(\alpha+\pi/6)}{\sqrt{3}}\right\} \end{aligned}\right\} \tag{2.69}$$

These characteristics are illustrated in Figure 2.54.

Bridge rectifier with free-wheel diode

Bridge rectifiers can be considered in the same way, but the single-phase bridge with a free-wheel diode has characteristics similar to those of the more economical half-controlled bridge, and only the three-phase bridge is of practical interest. The waveforms of Figure 2.53 (cf. Figure 2.42) show that the instantaneous output voltage falls periodically to zero when $\alpha = \pi/3$ and that the mean output voltage falls to zero when $\alpha = 2\pi/3$. The control characteristic is thus

$$\left.\begin{aligned}
0 < \alpha < \frac{\pi}{3}: \quad \overline{V}_{\mathrm{d}} &= \frac{3\sqrt{3}}{\pi}\hat{V}_{\mathrm{ph}}\cos\alpha = \overline{V}_{\mathrm{d\,max}}\cos\alpha \\
\frac{\pi}{3} < \alpha < \frac{2\pi}{3}: \quad \overline{V}_{\mathrm{d}} &= \frac{3\sqrt{3}}{\pi}\hat{V}_{\mathrm{ph}}\int_{\pi/3+\alpha}^{\pi}\sin\omega t\,\mathrm{d}\omega t \\
&= \frac{3\sqrt{3}}{\pi}\hat{V}_{\mathrm{ph}}\left\{1+\cos\left(\frac{\pi}{3}+\alpha\right)\right\} \\
&= \overline{V}_{\mathrm{d\,max}}\left\{1+\cos\left(\frac{\pi}{3}+\alpha\right)\right\}
\end{aligned}\right\} \qquad (2.70)$$

This characteristic is illustrated in Figure 2.54(a).

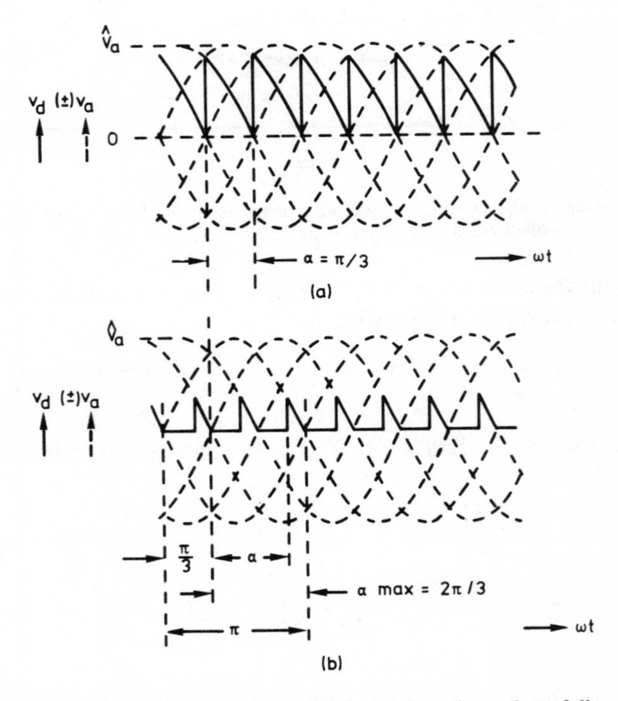

Figure 2.53 Output voltage waveforms in a three-phase fully controlled bridge rectifier with free-wheel diode, with two values of α.

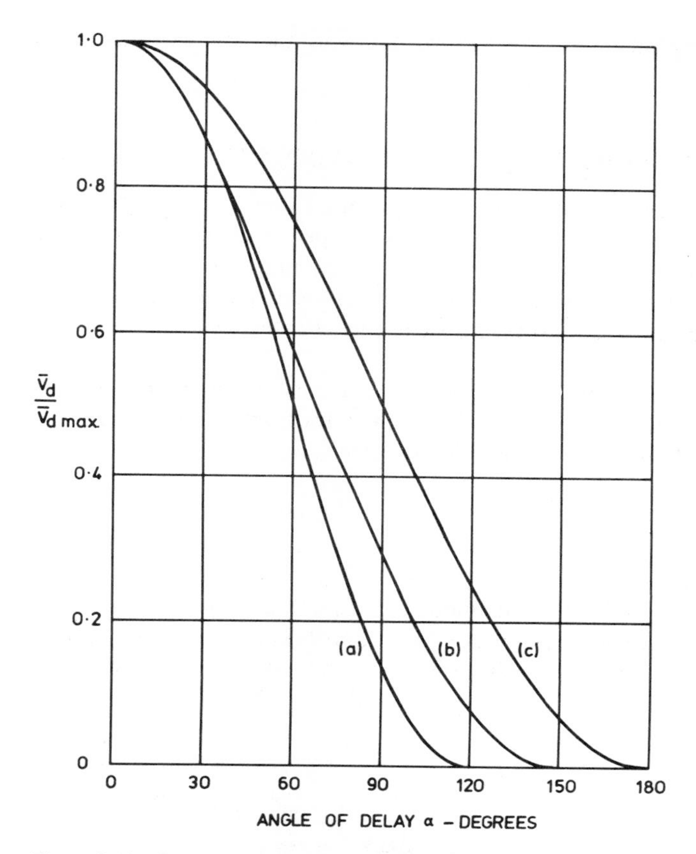

Figure 2.54 Control characteristics of fully controlled rectifiers with free-wheel diodes: (a) three-phase bridge; (b) three-phase single-way; (c) bi-phase.

Input characteristics of controlled rectifiers with free-wheel diodes

Comparison of the operation of single-way rectifiers with free-wheel diodes as exemplified by Figure 2.52 with that of half-controlled bridges illustrated in Figures 2.49 and 2.50 shows that in the region of reduced input conduction angle the only difference is that in the bridge circuits the input current pulses flow from line to line, whereas in the single-way circuits they flow from line to neutral. In terms of conduction angle, therefore, the input power factor and harmonic currents in this region are the same, and the only overall difference in input characteristics is in the three-phase circuits in the regions where the full 120° input conduction period is maintained.

The three-phase fully controlled bridge with free-wheel diode is not directly comparable to any other normal rectifier, but shows the same tendency as the other circuits in this category towards a somewhat better power factor and increased input harmonics in comparison with the same bridge without the free-wheel diode.

TWO-BRIDGE HALF-CONTROLLED RECTIFIER

The principle of connecting a controlled converter in series with an uncontrolled one, which was postulated earlier in explaining the operation of the half-controlled bridge, can be extended in suitable cases to the use of a fully controlled thyristor bridge and a diode bridge, normally three-phase, in series (Figure 2.55) with considerably greater benefit to the input power factor.

As in the half-controlled bridge, the output voltage of the uncontrolled part remains constant at $\overline{V}_{d\,max}/2$, while that of the controlled part is $(\overline{V}_{d\,max}/2)\cos\alpha$. The control characteristic is therefore the same.

$$\overline{V}_d = \overline{V}_{d\,max}\left(\frac{1+\cos\alpha}{2}\right) \tag{2.71}$$

On the input side, the fundamental-frequency load on the supply is represented by the uncontrolled rectifier drawing a constant current (relative to I_d) at unity displacement factor in parallel with the controlled converter drawing the same

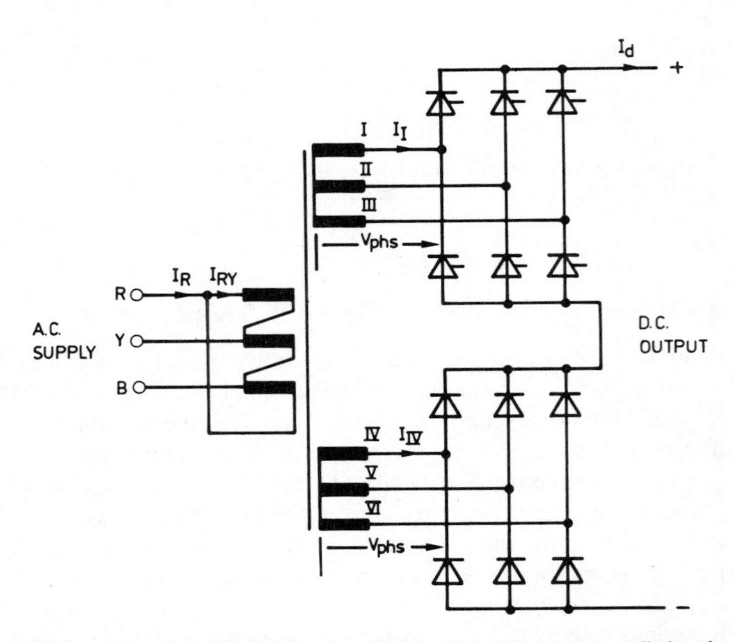

Figure 2.55 Three-phase half-controlled rectifier with fully controlled and uncontrolled bridges in series.

current but lagging the supply voltage by a phase angle α. The combined phase angle is $\alpha/2$ and the displacement factor $\cos(\alpha/2)$.

From (2.71),
$$\cos\phi = \cos\frac{\alpha}{2} = \sqrt{\left(\frac{\overline{V}_{d}}{\overline{V}_{d\,max}}\right)} \qquad (2.72)$$

The power factor does not vary in quite the same way, because the input harmonics combine in varying relationships as α varies. Simple analysis of the superimposed waveforms illustrated in Figure 2.56 lead to the power-factor characteristic shown in Figure 2.57.

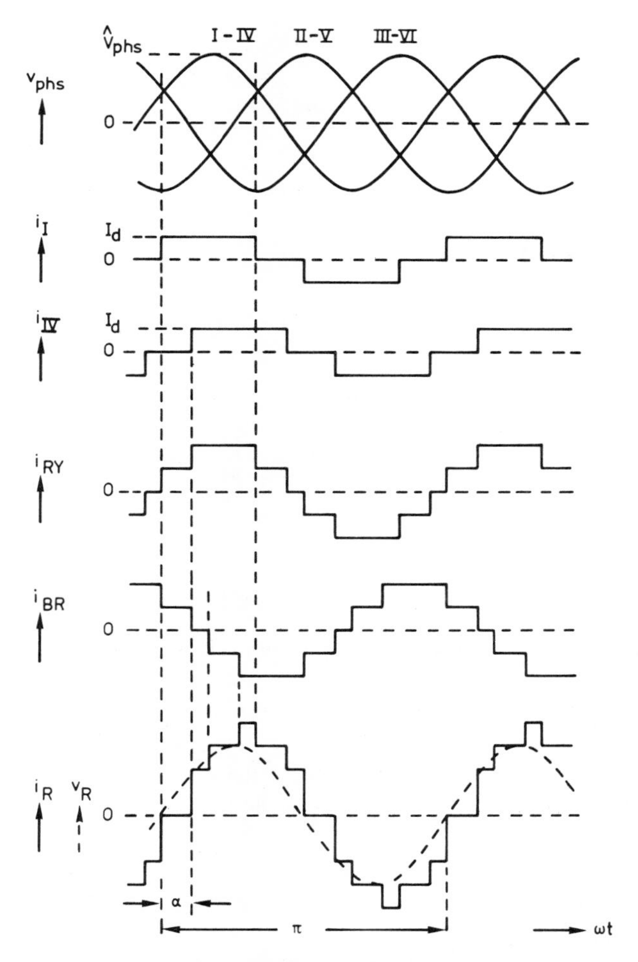

Figure 2.56 Waveforms in the two-bridge half-controlled rectifier of Figure 2.55.

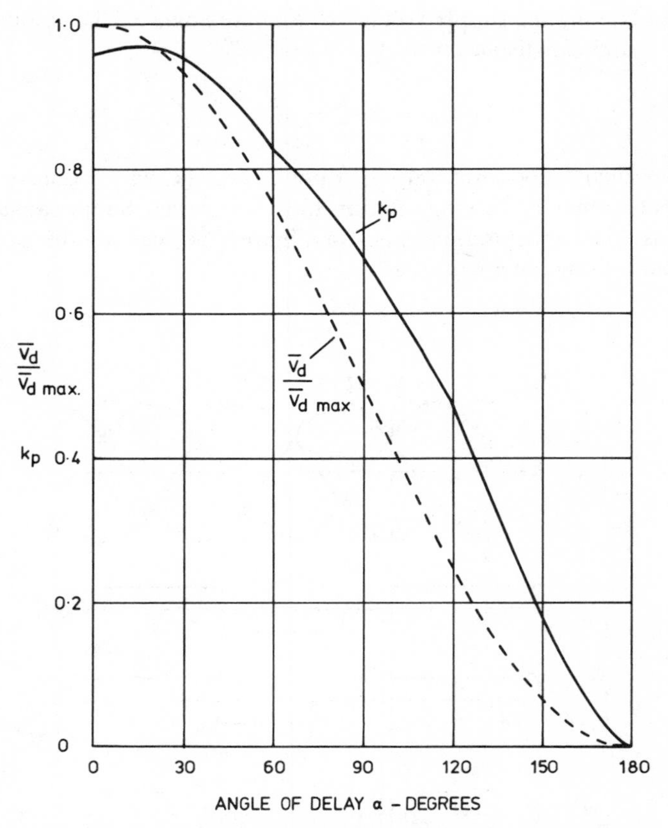

Figure 2.57 Variation with delay angle of the input power factor of the two-bridge half-controlled rectifier of Figure 2.55.

GENERALIZED OUTPUT WAVEFORMS AND RIPPLE VOLTAGE

It will have been observed that, irrespective of the particular way in which they are produced, the output voltage waveforms of all uncontrolled or fully controlled converters are ideally similar in terms of the pulse number and the peak output voltage, which may be the peak phase voltage in the case of a single-way converter or the peak interphase voltage in a bridge. The general waveform is depicted in Figure 2.58 for zero and finite angles of delay: the mean output voltage is

$$\overline{V}_{\mathrm{d}} = \overline{V}_{\mathrm{d\,max}} \cos\alpha = \hat{V}\,\frac{\sin(\pi/p)}{\pi/p}\cos\alpha \tag{2.73}$$

and the waveform repeats with a period $2\pi/p\omega$.

For the purposes of calculating the ripple content of these output waveforms it is convenient to change the ωt base to that of the fundamental ripple frequency—i.e. $p\omega$. The waveforms to be analysed are then as shown in Figure 2.59.

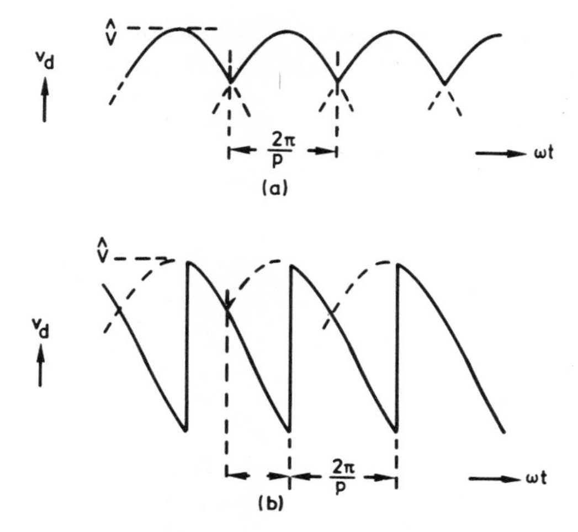

Figure 2.58 General output voltage waveform of (a) uncontrolled and (b) fully controlled converters.

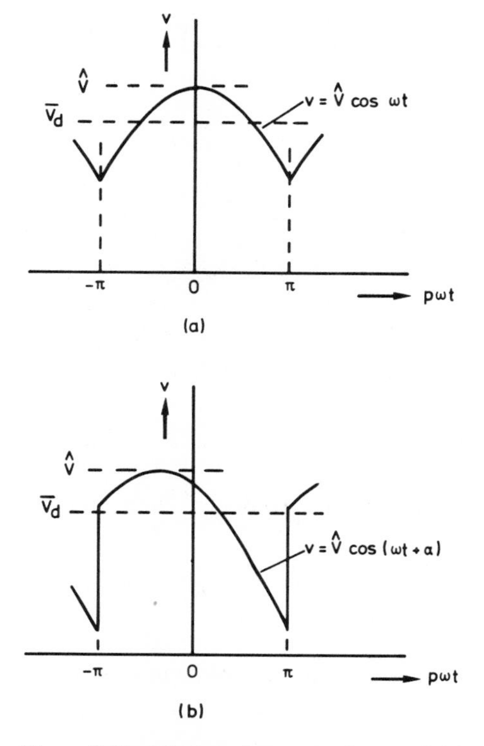

Figure 2.59 Ripple voltage waveforms of (a) uncontrolled and (b) fully controlled converters.

Considering first the uncontrolled case (Figure 2.59(a)), Fourier analysis of the waveform as drawn produces only cosine components: for the nth harmonic of the supply frequency (n is a multiple of p), $v_n = b_n \cos n\omega t$

where
$$b_n = \frac{\hat{V}}{\pi}\int_{-\pi}^{\pi} \cos \omega t \cos n\omega t \, \mathrm{d}(p\omega t)$$

$$= -\hat{V}\frac{p}{\pi}\sin\frac{\pi}{p}\cos\frac{n\pi}{p}\left(\frac{2}{n^2-1}\right) \tag{2.74}$$

Then
$$\frac{V_n}{V_\mathrm{d}} = (-)\frac{\hat{V}_n}{\sqrt{2}\,V_\mathrm{d}} = \frac{\sqrt{2}}{n^2-1} \tag{2.75}$$

The proportion of fundamental-frequency ripple, $\sqrt{2}/(p^2-1)$, is usually of more interest than that of the total r.m.s. ripple, since in an inductive d.c. circuit the higher orders are subject to a greater smoothing effect as well as being of much smaller magnitude. In fact the two values do not differ by more than 2–3 %, but if necessary the total r.m.s. ripple may be calculated by summating the series

$$\frac{V_\mathrm{r}}{V_\mathrm{d}} = \sqrt{2}\left[\left(\frac{1}{p^2-1}\right)^2 + \left(\frac{1}{4p^2-1}\right)^2 + \left(\frac{1}{9p^2-1}\right) + \cdots\right] \tag{2.76}$$

to a sufficient number of terms, or by evaluating the expression

$$V_\mathrm{r} = \sqrt{\left(\frac{1}{2\pi}\int_{-\pi}^{\pi} v^2 \, \mathrm{d}(p\omega t) - V_\mathrm{d}^2\right)} \tag{2.77}$$

It is evident from inspection of the voltage waveforms that delayed firing increases the ripple voltage at the output of a converter. The extent of the increase may be calculated by using the same technique applied above to the uncontrolled rectifier but with the voltage waveform modified to allow for the delay as in Figure 2.59(b).

Then
$$v = \hat{V}\cos(\omega t + \alpha)$$

where ω represents the fundamental ripple frequency. The nth harmonic of the supply frequency is now

$$v_n = a_n \sin n\omega t + b_n \cos n\omega t \tag{2.78}$$

$$\left.\begin{aligned} &\text{where} \quad a_n = \frac{\hat{V}}{\pi}\int_{-\pi}^{\pi} \cos(\omega t + \alpha)\sin n\omega t \, \mathrm{d}(p\omega t) \\ &\text{and} \quad b_n = \frac{\hat{V}}{\pi}\int_{-\pi}^{\pi} \cos(\omega t + \alpha)\cos n\omega t \, \mathrm{d}(p\omega t) \end{aligned}\right\} \tag{2.79}$$

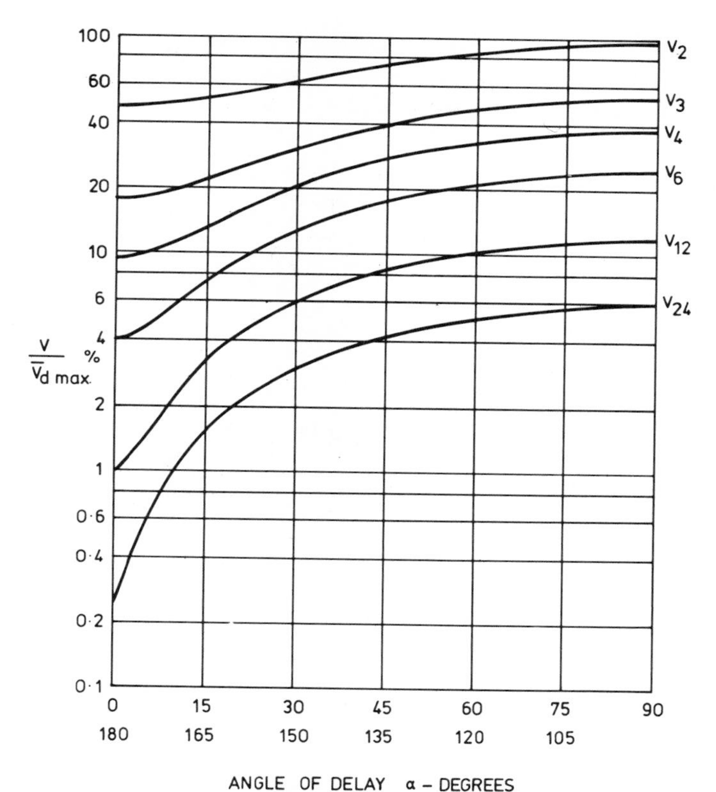

Figure 2.60 Variation with delay angle of ripple components in the output voltage of a fully controlled converter.

Evaluating these integrals gives the following expressions:

$$\left.\begin{aligned} a_n &= \overline{V}_{\mathrm{d\,max}} \sin\alpha \cos\frac{n\pi}{p}\left(\frac{2n}{n^2-1}\right) \\ b_n &= -\overline{V}_{\mathrm{d\,max}} \cos\alpha \cos\frac{n\pi}{p}\left(\frac{2}{n^2-1}\right) \end{aligned}\right\} \quad (2.80)$$

$$\frac{v_n}{\overline{V}_{\mathrm{d\,max}}} = \frac{\sqrt{2}}{n^2-1}\sqrt{(n^2\sin^2\alpha + \cos^2\alpha)} \quad (2.81)$$

This yields the family of curves shown in Figure 2.60, applicable to any fully controlled converter by selection of the appropriate series of harmonics. The maximum ripple voltage is produced when $\alpha = 90^\circ$, at which point the nth harmonic is n times greater than when $\alpha = 0$, and the fundamental ripple component from a p-pulse converter is increased by a factor p.

Output ripple from half-controlled bridge rectifiers

In a single-phase half-controlled bridge, the controlled side produces output harmonics in accordance with (2.79), (2.80) and (2.81) while the uncontrolled side

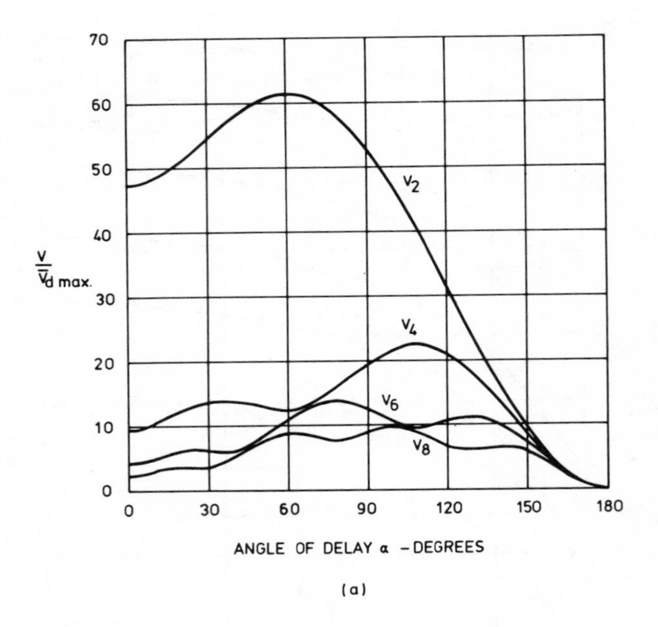

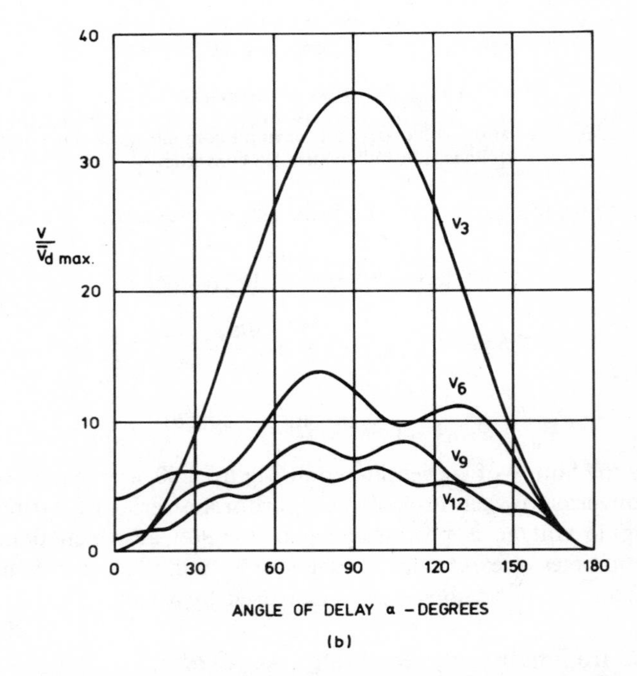

Figure 2.61 Output voltage ripple components produced by half-controlled bridge rectifiers: (a) single-phase; (b) three-phase.

produces those indicated by (2.74) and (2.75). The total of any particular (nth) harmonic results from the addition of these two components with a mutual phase displacement of $n\alpha$. Thus

$$v_n = \frac{\overline{V}_{\mathrm{dmax}}}{2}\left(\frac{2}{n^2-1}\right)\cos\frac{n\pi}{2}\,\{n\sin\alpha\sin n\omega t - \cos\alpha\cos n\omega t - \cos n(\omega t+\alpha)\} \tag{2.82}$$

from which

$$\frac{V_n}{\overline{V}_{\mathrm{dmax}}} = \frac{1}{\sqrt{2}(n^2-1)}\sqrt{[(\cos\alpha+\cos n\alpha)^2+(n\sin\alpha+\sin n\alpha)^2]} \tag{2.83}$$

For a three-phase half-controlled bridge the result is similar except that there is an additional phase displacement of $n\pi/3$ between the harmonics produced by the controlled and uncontrolled sides:

$$v_n = \frac{\overline{V}_{\mathrm{d\,max}}}{n^2-1}\cos\frac{n\pi}{3}\left\{n\sin\alpha\sin n\omega t - \cos\alpha\cos n\omega t - \cos n\left(\omega t+\frac{\pi}{3}+\alpha\right)\right\} \tag{2.84}$$

$$\left.\begin{aligned}\frac{V_n}{\overline{V}_{\mathrm{d\,max}}} &= \frac{1}{\sqrt{2}(n^2-1)}\sqrt{[(\cos\alpha+\cos n\alpha)^2+(n\sin\alpha+\sin n\alpha)^2]} \quad (n\ \text{even})\\ \frac{V_n}{\overline{V}_{\mathrm{d\,max}}} &= \frac{1}{\sqrt{2}(n^2-1)}\sqrt{[(\cos\alpha-\cos n\alpha)^2+(n\sin\alpha-\sin n\alpha)^2]} \quad (n\ \text{odd})\end{aligned}\right\} \tag{2.85}$$

The variation of individual harmonics with delay angle is illustrated by the curves of Figure 2.61. The single-phase half-controlled bridge produces substantially less ripple than its fully controlled equivalent, and is often preferred for that reason, among others: in the three-phase case, the reversion to 3-pulse characteristics with finite delay angles is frequently a serious disadvantage.

Controlled rectifiers with free-wheel diodes

A fully controlled single-phase or bi-phase rectifier with a free-wheel diode produces output voltage waveforms identical with those of the half-controlled bridge, and therefore has the same ripple characteristics, showing a considerable reduction in magnitude due to the presence of the diode. Free-wheel diodes give some reduction in the ripple from polyphase fully controlled rectifiers, but the effect is less marked, and decreases with increasing pulse number.

INTERPHASE REACTOR RATING IN CONTROLLED CONVERTERS

Since the voltage across the interphase reactor in a multiplex converter comprises the sum of those harmonic components that are produced by the converter groups but eliminated from the combined output, it would be expected that delayed firing would increase the reactor voltage by a factor roughly equal to

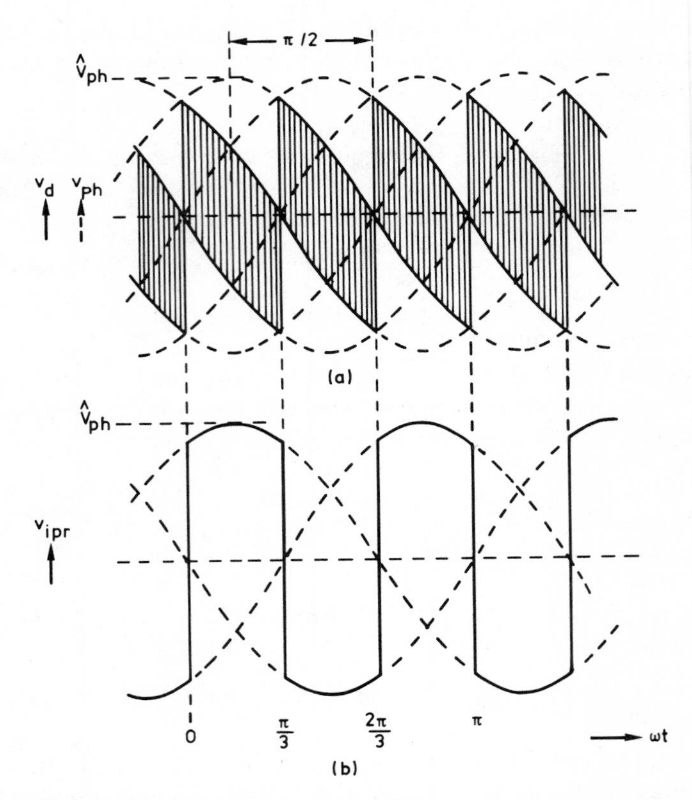

Figure 2.62 Output voltages of a hexaphase controlled converter with a delay angle of 90°: (a) outputs of the individual three-pulse groups; (b) voltage across the interphase reactor.

the relative increase in the fundamental ripple component from each group—i.e. the pulse number of the individual groups.

Figure 2.62 illustrates the common case of the hexaphase converter with $\alpha = 90°$. As in the uncontrolled rectifier (Figure 2.25) the reactor voltage waveform is composed of 60° segments of the phase voltages, but as a result of the firing delay they lie between 60° and 120° on the phase voltage waveform instead of between $-30°$ and $+30°$. The mean reactor voltage is now

$$\bar{V}_{\text{ipr}} = \frac{3}{\pi} \int_{\pi/3}^{2\pi/3} \hat{V}_{\text{ph}} \sin \omega t \, \mathrm{d}\omega t = \frac{3}{\pi} \hat{V}_{\text{ph}} = 1.15 \, \bar{V}_{\text{d max}} \tag{2.86}$$

OPERATION OF CONVERTERS WITH FINITE SOURCE INDUCTANCE

It has been assumed in the preceding discussion that the commutation of current from one supply phase to the next takes place instantaneously when the interphase voltage assumes the necessary polarity, or, in a controlled commutating group, when the thyristor connected to the 'incoming' phase is fired. In practice this is hardly possible, because of the inevitable inductance of the a.c. circuit, whether it be the inductance of the supply network, the leakage inductance of a transformer, or inductance deliberately introduced for the purposes of di/dt or dv/dt suppression. (Resistance also affects commutation but inductance usually has a dominant effect.)

Single-way converters

The basic effect of source inductance may be considered generally in terms of two consecutive phases of a polyphase commutating group (Figure 2.63). In all that follows, L_c, the commutation inductance, is the total inductance operative in the commutation process, so that, assuming the supply system to be balanced, $L_c/2$ is associated with each phase. Figure 2.64 shows the waveforms of voltage and current in the region of commutation, at (a) for the particular case of uncontrolled rectification and at (b) for the more general case of controlled conversion.

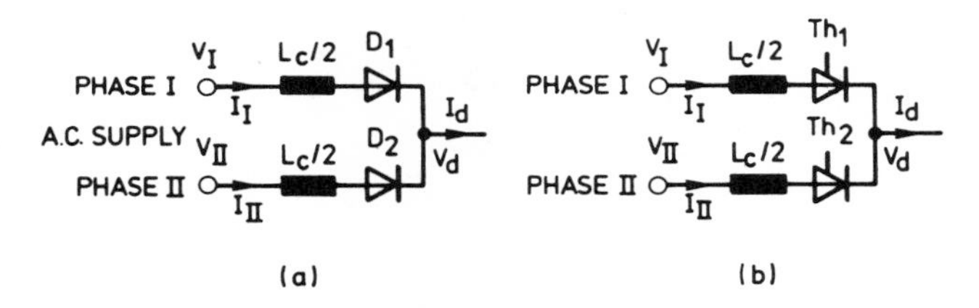

Figure 2.63 General commutation circuit with finite source inductance: (a) uncontrolled; (b) controlled.

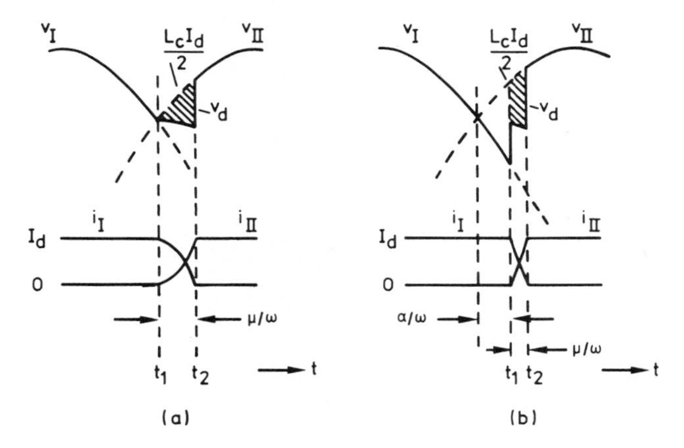

Figure 2.64 Waveforms during commutation in Figure 2.63: (a) uncontrolled; (b) controlled.

During commutation, the output current I_d is transferred from phase I to phase II at a rate which depends upon the circuit inductance and the voltage that appears across it. If I_d is assumed to be constant (at least during the commutation process) it follows that $di_I/dt = -di_{II}/dt$, and the voltages across the two equal phase inductances, $(L_c/2)\ di_I/dt$ and $(L_c/2)\ di_{II}/dt$, must therefore be equal and opposite. Hence the instantaneous output voltage v_d during the commutation period is the average of the two phase voltages. When commutation is complete, and i_I has fallen to zero, $(L_c/2)\ di_{II}/dt$ immediately becomes zero and v_d rises to equal v_{II}. The simultaneous flow of currents in the commutating phases is termed 'overlap'. Particular examples of overlap in bi-phase and three-phase converters are illustrated in Figures 2.65 and 2.66.

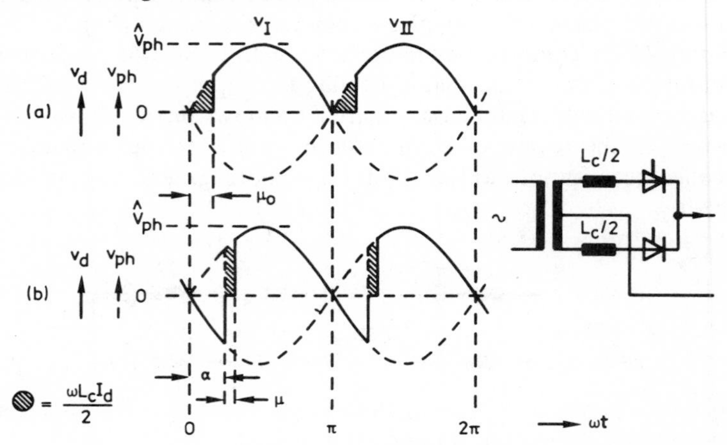

Figure 2.65 Output voltage waveforms in a bi-phase converter with overlap: (a) uncontrolled; (b) controlled.

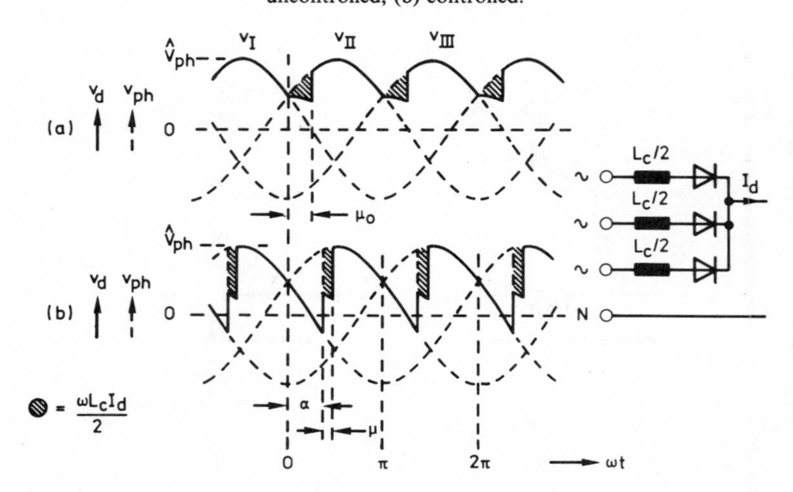

Figure 2.66 Output voltage waveforms in a three-phase single-way converter with overlap: (a) uncontrolled; (b) controlled.

Voltage–time integrals and voltage drop

The effect of overlap on the output voltage of a converter is conveniently considered in terms of voltage–time integrals. If the current i in an inductance L is changing, the induced voltage across it is

$$e = L\frac{\mathrm{d}i}{\mathrm{d}t}$$

Hence if the current changes from I_1 to I_2 in a period from t_1 to t_2, as a result of an applied voltage v (= e),

$$L(I_2 - I_1) = L\delta I = \int_{t_1}^{t_2} v\,\mathrm{d}t \tag{2.87}$$

In Figure 2.64 the area shown hatched between the phase II voltage wave (v_{II}) and the output voltage v_{d} represents the voltage–time integral across $L_{\mathrm{c}}/2$ which changes the current in it from zero to I_{d}, and irrespective of its shape is equal to $L_{\mathrm{c}}I_{\mathrm{d}}/2$. Thus the effect of overlap in a single-way converter is to subtract a voltage–time integral $L_{\mathrm{c}}I_{\mathrm{d}}/2$ from the output at every commutation, and, since there are pf commutations per second, the result is a mean voltage drop

$$\delta\bar{V}_{\mathrm{d}} = -pf\,L_{\mathrm{c}}I_{\mathrm{d}}/2 \tag{2.88}$$

(It has been assumed that I_{d} is smooth: if it is not, the value to be applied in (2.85) is the mean value during the commutation period.)

Overlap angle

The duration of the overlap period $t_2 - t_1$ (Figure 2.54) can be determined by evaluating the time integral of the interphase voltage v_{a}. If the overlap angle $\omega(t_2 - t_1)$ is μ (μ is assumed to be less than the conduction angle in each phase),

$$\int_{t_1}^{t_2} \hat{V}_{\mathrm{a}} \sin\omega t\,\mathrm{d}t = \int_{\alpha/\omega}^{(\alpha+\mu)/\omega} \hat{V}_{\mathrm{a}} \sin\omega t\,\mathrm{d}t = L_{\mathrm{c}}I_{\mathrm{d}} \tag{2.89}$$

whence

$$\cos\alpha - \cos(\alpha + \mu) = \frac{\omega L_{\mathrm{c}}I_{\mathrm{d}}}{\hat{V}_{\mathrm{a}}} \tag{2.90}$$

For the uncontrolled rectifier, in which $\alpha = 0$,

$$\cos\mu_0 = 1 - \frac{\omega L_{\mathrm{c}}I_{\mathrm{d}}}{\hat{V}_{\mathrm{a}}} \tag{2.91}$$

By referring to (2.88), the overlap angle may be related to the voltage drop:

$$\cos\alpha - \cos(\alpha+\mu) = -\frac{2\omega\delta\overline{V}_{\mathrm{d}}}{\hat{V}_{\mathrm{a}}pf} = -\frac{4\pi\delta\overline{V}_{\mathrm{d}}}{m\hat{V}_{\mathrm{a}}} \qquad (p = m)$$

$$\hat{V}_{\mathrm{a}} = 2\hat{V}_{\mathrm{ph}}\sin\frac{\pi}{m}$$

$$\therefore \cos\alpha - \cos(\alpha+\mu) = -\frac{(2\pi/m)\delta\overline{V}_{\mathrm{d}}}{\hat{V}_{\mathrm{ph}}\sin(\pi/m)} = -\frac{2\delta\overline{V}_{\mathrm{d}}}{\overline{V}_{\mathrm{d\,max}}}$$

or in terms of per-unit voltage drop $\delta_V, = \delta\overline{V}_{\mathrm{d}}/\overline{V}_{\mathrm{d\,max}}$,

$$\cos\alpha - \cos(\alpha+\mu) = -2\delta_V \tag{2.92}$$

Overlap in bridge converters

In the three-phase uncontrolled or fully controlled bridge, commutation in the two diode or thyristor groups is said to be independent, in that any commutation between two phases in one group takes place without any interaction with, or influence by, a concurrent commutation in the other group (assuming the overlap angle to be less than 60°). The effect in each group of a given source inductance is thus the same as in a single-way converter, and expressions (2.88) to (2.92) are equally applicable, with $p = 6$. Figure 2.67 illustrates the output voltage waveforms.

In the single-phase uncontrolled or fully controlled bridge (Figure 2.68), commutation is not independent, since it occurs simultaneously on the two sides in the same source inductance. The result is that in each commutating period there is on each side a loss of voltage–time integral due to the commutations on both sides. This, coupled with the fact that there are in effect four commutations per cycle for a pulse number of two, means that applying expression (2.88) would give a value of $\delta\overline{V}_{\mathrm{d}}$ only one quarter of the true value. In slightly different terms, the voltage–time integral subtracted from the output in every half-cycle is that required to change the current in the total inductance L_{c} from I_{d} to $-I_{\mathrm{d}}$:

$$\int v\,\mathrm{d}t = 2L_{\mathrm{c}}I_{\mathrm{d}}$$

hence

$$\delta\overline{V}_{\mathrm{d}} = -4fL_{\mathrm{c}}I_{\mathrm{d}} \tag{2.93}$$

Figure 2.68 shows the output voltage waveforms for the single-phase bridge. Integrating the supply voltage V_{a},

$$\int_{\alpha/\omega}^{(\alpha+\mu)/\omega} \hat{V}_{\mathrm{a}}\sin\omega t\,\mathrm{d}t = 2L_{\mathrm{c}}I_{\mathrm{d}}$$

whence

$$\cos\alpha - \cos(\alpha+\mu) = \frac{2\omega L_{\mathrm{c}}I_{\mathrm{d}}}{\hat{V}_{\mathrm{a}}} \tag{2.94}$$

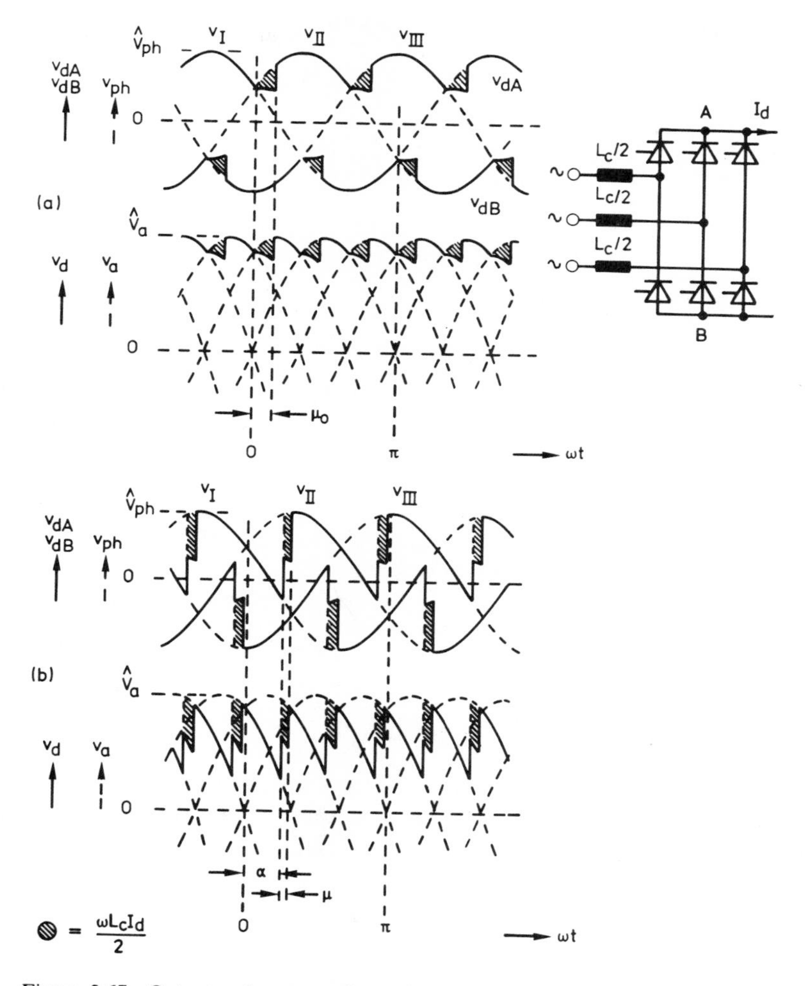

Figure 2.67 Output voltage waveforms in a three-phase bridge converter with overlap: (a) uncontrolled; (b) fully controlled.

For the uncontrolled rectifier ($\alpha = 0$),

$$\cos \mu_0 = 1 - \frac{2\omega L_c I_d}{\hat{V}_a} \tag{2.95}$$

Substituting in (2.94) from (2.93) gives the previous expression (2.92)

$$\cos \alpha - \cos (\alpha + \mu) = -\frac{4\omega\, \delta\, \overline{V}_d}{8f\, \hat{V}_a} = -\frac{\pi\, \delta\, \overline{V}_d}{\hat{V}_a}$$

$$= -\frac{2\,\delta\, \overline{V}_d}{\overline{V}_{d\,max}} = -2\delta_V \tag{2.96}$$

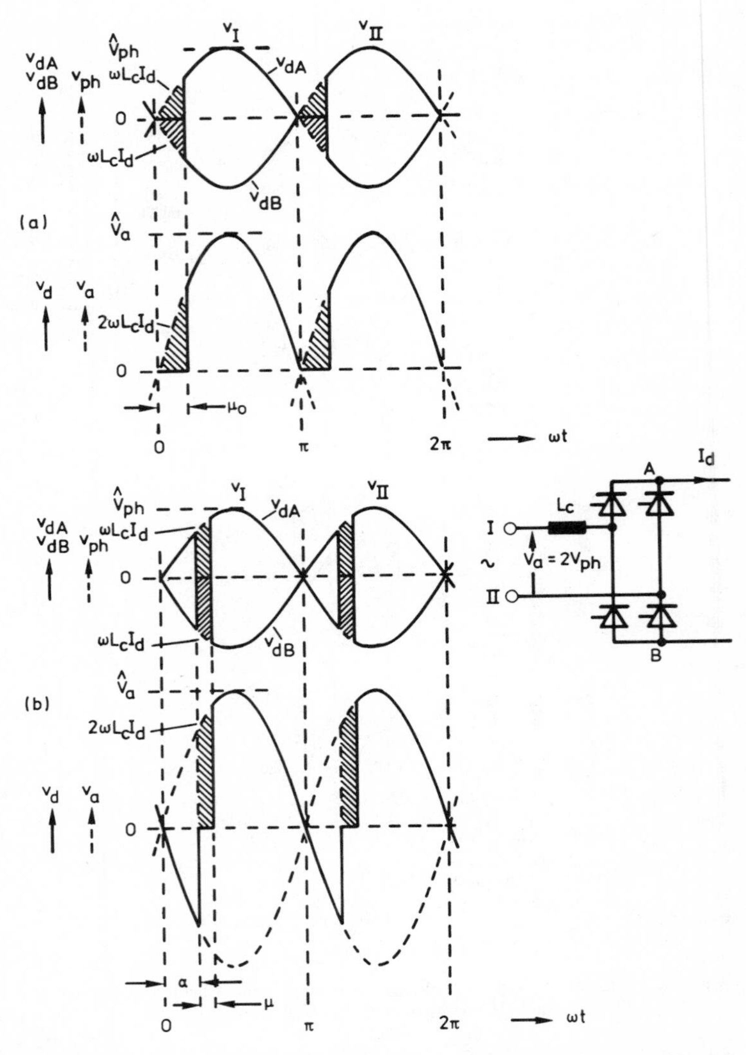

Figure 2.68 Output voltage waveforms in a single-phase bridge converter with overlap: (a) uncontrolled; (b) fully controlled.

Half-controlled bridge rectifiers with finite source inductance

Since in a half-controlled bridge the timing of the commutations on the controlled side varies according to the angle of delay, while that on the uncontrolled side remains fixed, commutation is sometimes independent and sometimes not. In the single-phase bridge, commutation is not independent with zero delay angle, but becomes independent as soon as α exceeds the overlap on the uncontrolled side commutating on its own, with a consequent halving of the voltage drop. In the three-phase case, commutation becomes non-independent when the delay angle is in the region of 60° or 120°. In general the effect is to cause slight discontinuities in the control characteristics: more detailed consideration is not warranted here.

Relationships between per-unit reactance and voltage drop

It is to be expected that in a given converter circuit the per-unit voltage-drop will be simply related to the per-unit source reactance, which by definition is the proportion of the nominal supply voltage that must be applied to the source network to cause the rated current to flow with the source short-circuited. The example of a single-phase bridge will illustrate this.

The rated source current for the purpose of the relationship—e.g. the rated secondary current of the transformer if one is provided—is the r.m.s. value of the converter input current, in this case equal to I_d (a slight approximation, ignoring the deviation of the input current waveform from the ideal square wave, due to the overlap). The p.u. source reactance is then

$$\varepsilon_x = \frac{\omega L_c I_d}{V_a} = \frac{2\pi f L_c I_d}{\overline{V}_{d\,max}} \times \frac{2\sqrt{2}}{\pi}$$

Then from (2.93)

$$\varepsilon_x = -\frac{\sqrt{2}\,\delta \overline{V}_d}{\overline{V}_{d\,max}} \quad \text{or} \quad \delta_V = -\frac{\varepsilon_x}{\sqrt{2}} \tag{2.97}$$

Applying the same process to other circuits in general use gives the following forms for δ_V:

$$\left.\begin{array}{ll} \left.\begin{array}{l}\text{Bi-phase}\\ \text{Single-phase bridge}\end{array}\right\} & \dfrac{\varepsilon_x}{\sqrt{2}} \\[2ex] \text{Three-phase single-way*} & \dfrac{\sqrt{3}}{2}\,\varepsilon_x \\[2ex] \left.\begin{array}{l}\text{Hexaphase with ipr*}\\ \text{Three-phase bridge}\end{array}\right\} & \dfrac{\varepsilon_x}{2} \end{array}\right\} \tag{2.98}$$

Overlap during inversion: loss of commutation

The increment of output voltage $\delta \overline{V}_d$ which has been referred to so far as a voltage drop becomes in effect a voltage rise when a controllable converter is inverting and the output voltage is reversed. This is illustrated by the control characteristic shown in Figure 2.69 (a), while Figure 2.69 (b) illustrates the effect on the output voltage waveform of a three-phase single-way converter.

* The secondary windings are here assumed to be closely coupled together: otherwise L_c and δ_V are somewhat higher in relation to ε_x.

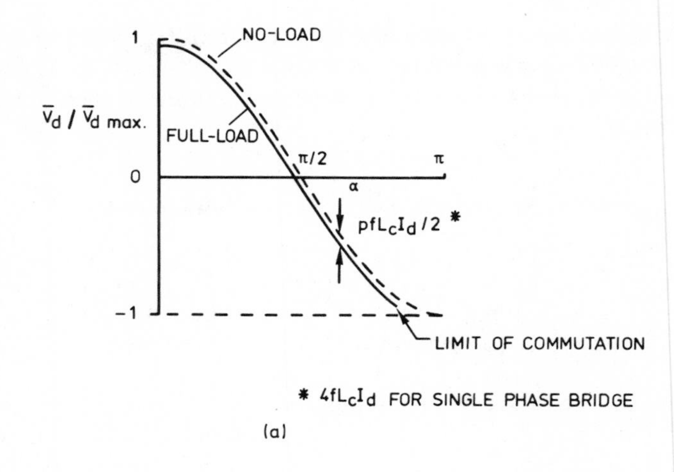

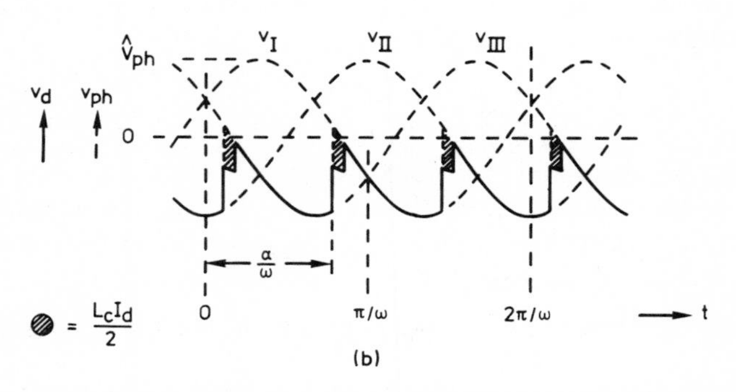

Figure 2.69 (a) Control characteristic of a fully-controlled converter with overlap; (b) output voltage waveform of a three-phase single-way converter during inversion.

The inverting characteristic reaches a limitation when α is increased to the point where the voltage–time integral that remains under the interphase voltage waveform before the voltage reverses is only just sufficient to effect commutation (Figure 2.70). A purely geometrical comparison with Figure 2.64(a) shows that

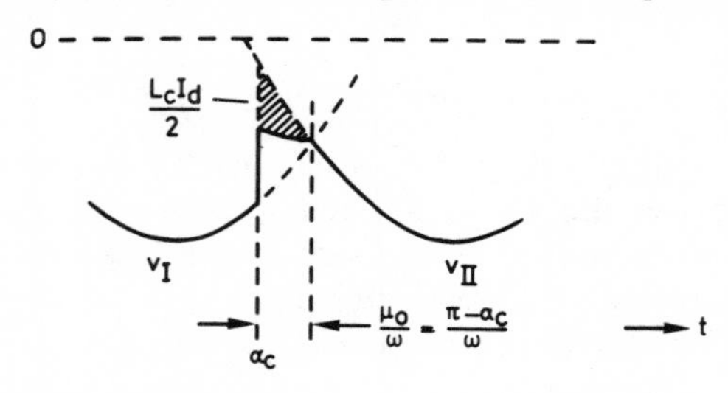

Figure 2.70 Limiting condition of overlap in a controlled converter during inversion.

$(\pi - \alpha)$ is then equal to the same overlap angle μ_0 that would be associated with rectification with zero delay angle at the same current. Thus, denoting this critical value of α by α_c, from (2.91), for a single-way converter,

$$\cos \alpha_c = -\cos(\pi - \alpha_c) = \frac{\omega L_c I_d}{\hat{V}_a} - 1 \tag{2.99}$$

alternatively,

$$\cos \alpha_c = 2\delta_V - 1 \tag{2.100}$$

If α is increased beyond α_c, the supposedly 'outgoing' thyristor is still carrying current when it becomes forward biassed again after the intersection of the phase voltage waveforms, and a failure to commutate results. Provided that the necessary conditions for commutation are subsequently re-established, the situation is not necessarily catastrophic, but while it persists the 'commutating group' affected produces no mean output voltage, the d.c. circuit being connected continuously to one phase of the a.c. supply (Figure 2.71). The maximum negative direct voltage obtainable by inversion is

$$\begin{aligned} -\bar{V}_d &= \bar{V}_{d\,max}(-\cos \alpha_c + \delta_V) \\ &= \bar{V}_{d\,max}(1 - \delta_V) \end{aligned} \tag{2.101}$$

which is equal to the maximum rectified voltage obtainable at the same current.

The above results, including (2.99), (2.100) and (2.101), are also applicable to the three-phase fully controlled bridge. For the single-phase fully controlled bridge, (2.100) and (2.101) apply, but (2.99) has to be modified to

$$\cos \alpha_c = \frac{2\omega L_c I_d}{\hat{V}_a} - 1 \tag{2.102}$$

(cf. (2.95)). The possibility of losing the direct voltage from an inverting converter through a commutation failure commonly means that considerable care must be taken to avoid increasing the firing delay excessively (or removing firing pulses altogether) so long as current is flowing continuously.

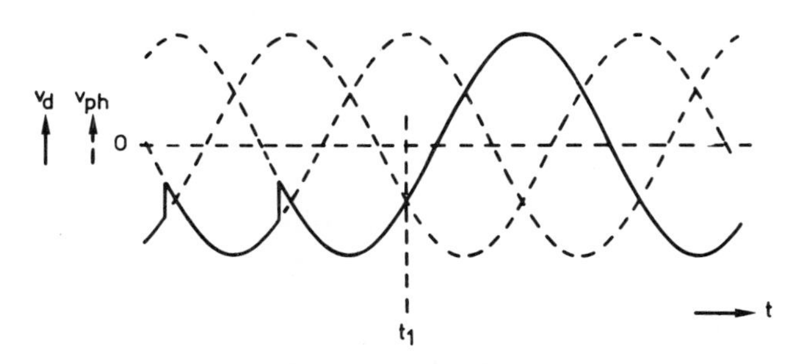

Figure 2.71 Loss of commutation in an inverting three-phase single-way converter: commutation lost at t_1.

Loss of commutation in half-controlled rectifiers

If a half-controlled rectifier is regarded as a combination of an uncontrolled rectifier and a fully controlled converter connected in series on the d.c. side, it is clear that loss of commutation on the controlled side results in a mean output voltage equal to that of the uncontrolled side alone—i.e. half the maximum output voltage—with a large ripple component at the supply frequency. Figure 2.72 illustrates this condition in a three-phase half-controlled bridge. To avoid this possibility without imposing constraints on the control system, free-wheel or 'commutating' diodes are often connected across the outputs of half-controlled rectifiers so that the load current is diverted from the conducting thyristor when the instantaneous output voltage falls to zero (given a favourable combination of diode and thyristor characteristics).

An alternative form of single-phase half-controlled bridge in which the problem of loss of commutation does not arise is shown in Figure 2.73. Otherwise

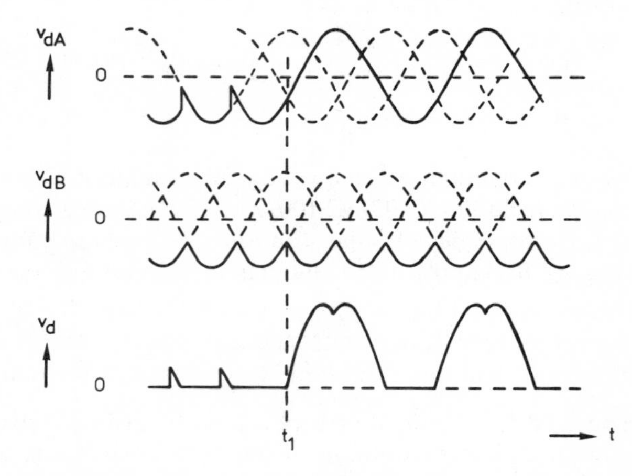

Figure 2.72 Loss of commutation in a half-controlled three-phase bridge rectifier (Figure 2.45(a)): commutation lost at t_1.

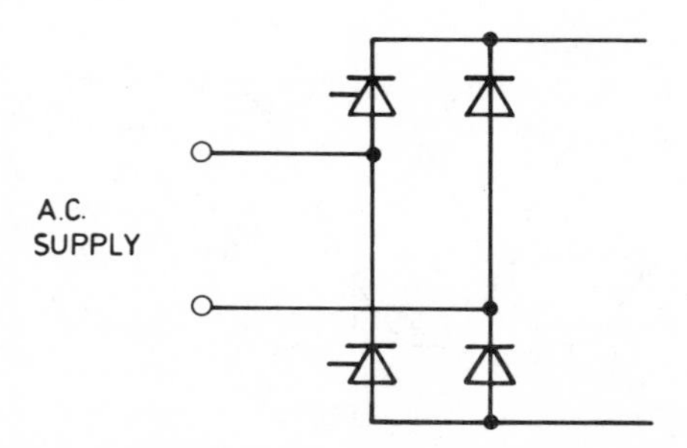

Figure 2.73 Single-phase (asymmetrical) half-controlled bridge rectifier, alternative to Figure 2.45(b).

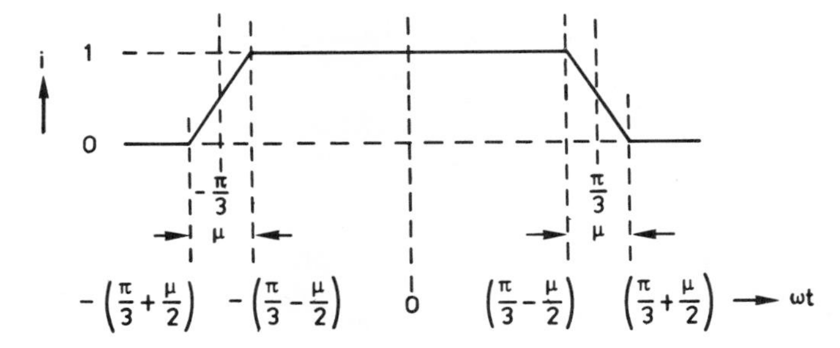

Figure 2.74 Simplified representation of the input current waveform in a three-phase commutating group.

its electrical performance is identical to that of the symmetrical circuit of Figure 2.45(b), and it is often preferred.

Reduction of input harmonics due to overlap

It might be expected, from the modified appearance of the input current waveforms of a converter with appreciable overlap due to source inductance, that the proportions of harmonics, particularly in the higher orders, would be less than those deduced from the ideal waveforms of, for example, Figure 2.17. This is confirmed by harmonic analysis, and may be shown in a simple way, with some approximation, by treating the waveforms as trapezoidal. On this assumption, Figure 2.74 represents the waveform of input current to a three-phase commutating group, which contains all the harmonics that will appear in any three-phase converter, with an amplitude of one unit, arranged for convenience so that all its harmonic components will be expressed as cosine terms. Analysis of this waveform gives the relative amplitude of the nth harmonic as

$$\begin{aligned} b_n &= \frac{2}{\pi}\int_0^{\pi/3-\mu/2} \cos n\omega t \, d\omega t + \frac{2}{\pi}\int_{\pi/3-\mu/2}^{\pi/3+\mu/2} \cos n\omega t \, d\omega t \\ &= \frac{1}{n}\,\frac{2\sin(n\pi/3)}{\pi}\,\frac{\sin(n\mu/2)}{n\mu/2} \end{aligned} \tag{2.103}$$

Thus overlap reduces any harmonic current component in a ratio of approximately $\sin(n\mu/2)/(n\mu/2)$.

Effect of overlap on output ripple

It is evident from converter output voltage waveforms such as those of Figure 2.66 that the occurrence of overlap to some extent invalidates the account of ripple voltage given earlier. To adapt the calculation of (2.78) *et seq.* to allow for overlap, it is convenient to consider the voltage waveform exemplified generally by Figure 2.64(b) as the mean of two waveforms, without overlap, with angles of delay α and $(\alpha + \mu)$, shown as A and B in Figure 2.75. Any harmonic component

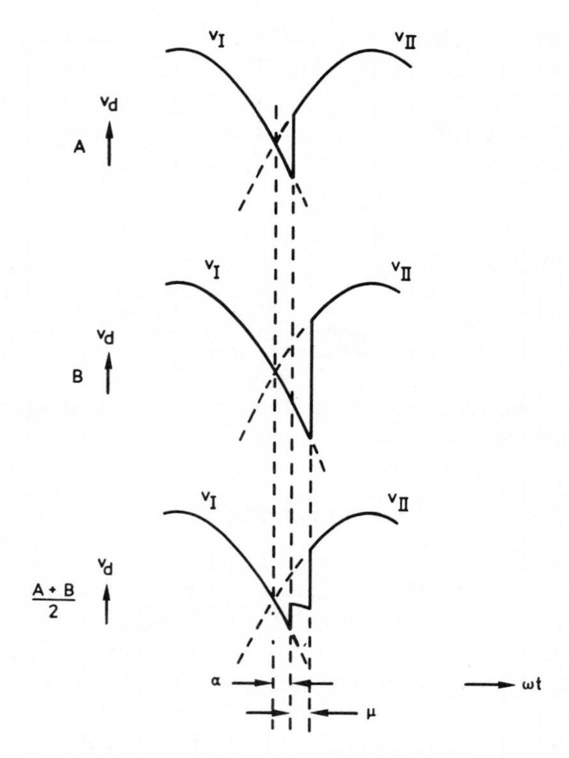

Figure 2.75 Output voltage of a converter with overlap represented as the mean of two waveforms without overlap.

of the waveform with overlap can then be calculated as the mean of the corresponding components of the two postulated waveforms A and B.

In using the expressions of (2.78)–(2.80) for this purpose, allowance must be made for the fact that, in the method of calculation adopted, increasing the angle of delay from α to $(\alpha+\mu)$ entails advancing the waveform by a phase angle μ (see Figure 2.59), implying a phase advance of $n\mu$ at the nth harmonic. Thus, making the necessary adjustment to waveform A, the nth harmonic component is given by

$$V_n = \frac{\cos(n\pi/p)}{n^2-1}\overline{V}_{\mathrm{d\,max}}\{n\sin\alpha\sin n(\omega t+\mu)-\cos\alpha\cos n(\omega t+\mu)$$

$$+n\sin(\alpha+\mu)\sin n\omega t-\cos(\alpha+\mu)\cos n\omega t\} \qquad (2.104)$$

The most significant aspect of the result is, in fact, the increase in ripple in the uncontrolled rectifier. Putting $\alpha=0$ in (2.104) enables it to be simplified to

$$V_n = \frac{\overline{V}_{\mathrm{d\,max}}}{\sqrt{2}(n^2-1)}\sqrt{[(n\sin\mu+\sin n\mu)^2+(\cos n\mu+\cos\mu)^2]} \qquad (2.105)$$

The effect is most noticeable in rectifiers which would otherwise be expected to produce a relatively smooth output. Taking as an example a three-phase bridge rectifier with a commutation angle of 26° (resulting in 0.05 p.u. voltage drop), the sixth harmonic voltage is increased approximately from 0.040 $\overline{V}_d$ to 0.064 $\overline{V}_d$ and the twelfth harmonic from 0.010 $\overline{V}_d$ to 0.025 $\overline{V}_d$.

SMOOTHING

For many purposes a rectifier is required to provide a substantially smooth direct voltage, and it is often necessary to attenuate the ripple voltage by means of a smoothing filter. Normally an LC low-pass filter is employed (Figure 2.76), and this may comprise one, two or occasionally more sections.

It is not necessary here to consider the design of smoothing filters in great detail. The attenuation of an LC section as shown in Figure 2.76(a) is normally calculated with sufficient accuracy by ignoring the load admittance and considering only the filter components themselves. Thus, at some particular harmonic frequency ω_n,

$$\left.\begin{aligned} V_{no} &= V_{ni}\frac{1/\omega_n C}{\omega_n L - 1/\omega_n C} \\ \text{or}\quad \frac{V_{ni}}{V_{no}} &= \left(\frac{\omega_n}{\omega_0}\right)^2 - 1 \end{aligned}\right\} \qquad (2.106)$$

where $\omega_0 = 1/\sqrt{(LC)}$.

Obviously, from (2.106), the attenuation decreases sharply as ω_n approaches ω_0, and if (ω_n/ω_0) is less than $\sqrt{2}$ the ripple component is actually increased. It is therefore a common constraint upon the design of filters that the resonant frequency should not nearly coincide with any harmonic frequency that may be

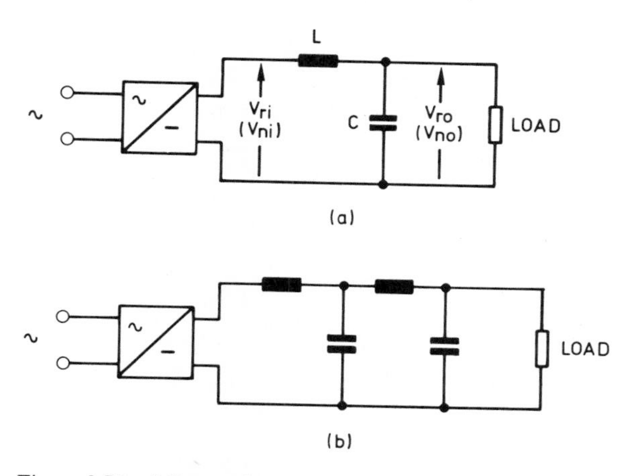

Figure 2.76 *LC* smoothing filters: (a) single-section; (b) two-section.

present in the ripple spectrum, even though attenuation may not be necessary at the frequency in question, and the component at that frequency may arise only through imperfections in the rectifier—a second-harmonic from a six-pulse rectifier, for example, which may result from an unbalanced three-phase input. Care may be necessary, too, with certain loads, which may modify the resonant frequency of the filter or introduce additional resonances at frequencies of interest.

Critical inductance

For the sake of a clear basis of exposition, it has been assumed in considering the behaviour of converters that the output current is continuous under all conditions, and in general smooth, by virtue of an adequate inductance in the d.c. circuit. In many practical situations, however, continuous current flow can be maintained by the inductance only within certain limits. This is illustrated in Figure 2.77 for the case of an uncontrolled three-phase single-way rectifier.

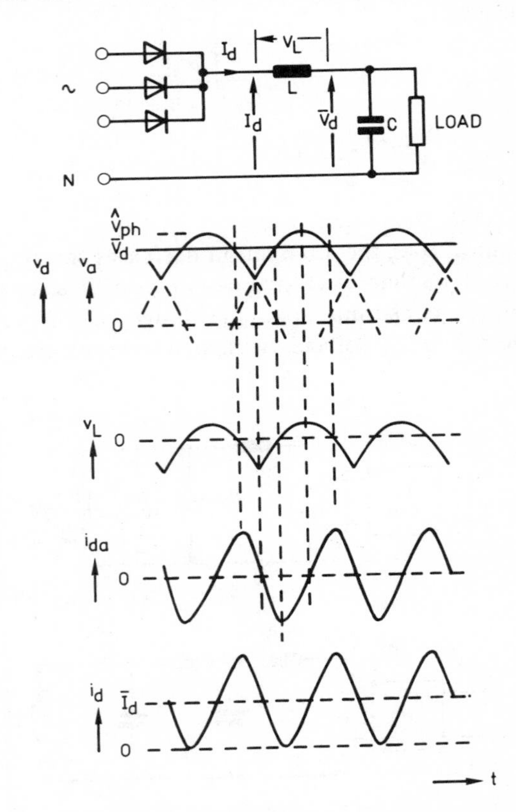

Figure 2.77 Waveforms in a three-phase single-way rectifier with critical smoothing inductance. (I_{da}: alternating component of I_d.)

Assuming that the voltage across the capacitor is perfectly smooth, the voltage across the inductor is the difference between the instantaneous output voltage of the rectifier, v_d, and its average value $\overline{V}_d$. So long as the rectifier functions as hitherto assumed and produces the output voltage waveform shown in Figure 2.77 (overlap is neglected), the result is an alternating component of current in the inductor which is not affected by the load current on which it is superimposed (except possibly through variation of L with load current).

The output current of the rectifier is thus continuous so long as the instantaneous sum of the alternating component i_{da} and the mean output current $\overline{I}_d$ remains above zero. Since I_{da} varies inversely as L, continuity depends on an adequate inductance: the value of inductance necessary just to sustain continuous current under given conditions is known as the critical inductance.

The result of operating with less than the critical inductance is illustrated in Figure 2.78. The instantaneous voltage across the inductor reverses at t_1: at t_2 a sufficient voltage–time integral has been developed across it to reduce i_d to zero, and since the inductor can support no voltage without current flowing in it v_d jumps to the level of $\overline{V}_d$. $\overline{V}_d$ is therefore increased relative to its value with continuous current. In the limit, when $\overline{I}_d = 0$, $\overline{V}_d$ rises to the peak input voltage.

The primary reason for attaching importance to the critical inductance is the undesirability of this rise in output voltage within the normal range of operating current. In many cases this is the significant factor in determining the amount of inductance that is required, rather than the optimum design of a filter from the point of view of ripple attenuation, and it is sometimes attractive deliberately to make the inductor non-linear by virtue of an unusually small gap in the core (the so-called 'swinging choke') to optimize the design over the load range.

Three particular aspects of critical inductance may usefully be considered in detail—that discussed above, the corresponding case with a controlled converter, and the case mentioned earlier of a controlled converter with an L–R load.

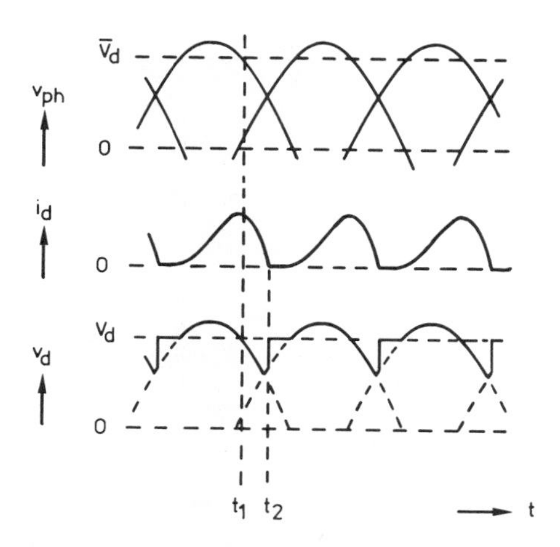

Figure 2.78 Waveforms in a three-phase single-way rectifier with less-than-critical inductance.

Uncontrolled rectifier

Given the symmetry of the voltage and current waveforms, the critical condition for the uncontrolled rectifier is easily determined as that which produces a peak-to-peak alternating component of current in the inductor equal to twice the mean load current (Figure 2.77). The general case of a *p*-pulse rectifier is illustrated by the waveforms of Figure 2.79.

The voltage–time integral applied to the inductor between t_1, when $i_d = 0$, and t_2, when $i_d = 2\overline{I}_d$, is

$$\int_{t_1}^{t_2} v_L \,\mathrm{d}t = \int_{t_1}^{t_2} (\hat{V}\sin\omega t - \overline{V}_d)\,\mathrm{d}t$$

$$= \frac{2\hat{V}\cos\omega t_1}{\omega} - \frac{2\overline{V}_d}{\omega}\left(\frac{\pi}{2} - \omega t_1\right) \tag{2.107}$$

But $\int_{t_1}^{t_2} v_L \,\mathrm{d}t = 2L_0\overline{I}_d$ (see (2.87))

Hence, if $\overline{V}_d/\overline{I}_d = R$,

$$\frac{\omega L_0}{R} = \frac{\hat{V}}{\overline{V}_d}\cos\omega t_1 + \omega t_1 - \frac{\pi}{2} \tag{2.108}$$

from which, since

$$\frac{\overline{V}_d}{\hat{V}} = \sin\omega t_1 = \frac{\sin(\pi/p)}{\pi/p}$$

$$\frac{\omega L_0}{R} = \sqrt{\left[\left(\frac{\pi/p}{\sin(\pi/p)}\right)^2 - 1\right]} + \sin^{-1}\left(\frac{\sin(\pi/p)}{\pi/p}\right) - \frac{\pi}{2} \tag{2.109}$$

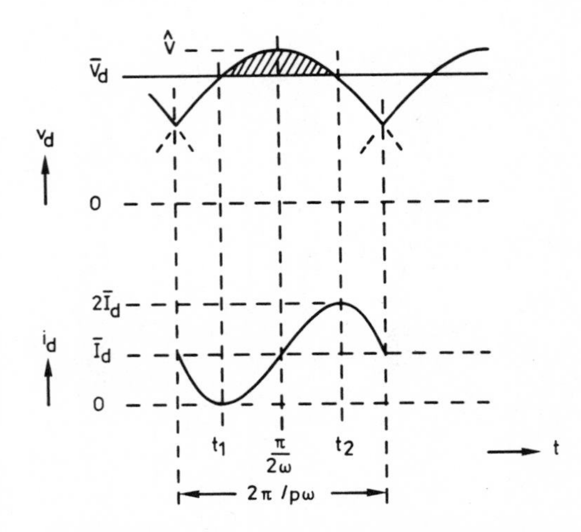

Figure 2.79 Output waveforms in a *p*-pulse rectifier with critical inductance.

This leads to the following values:

p	2	3	6	12
$\omega L_0/R$	0.33	0.083	0.0095	0.0012

Controlled converter with smooth output voltage

In the case of a controlled converter, the lack of symmetry in the output voltage waveform makes the calculation more complicated. The complication may be reduced, however, without much loss of information, by considering only that range of delay angle for which the instantaneous output at the instant of commutation exceeds the mean output voltage.

The general case within this range is illustrated by the waveforms of Figure 2.80. It can be assumed that with critical inductance $i_d = 0$ at $t = 0$ (i.e. at the instant when the step increase in v_d produces a reversal of the slope of current waveform). The mean current is therefore obtained by a straightforward integration:

$$\overline{I}_d = \frac{\omega p}{2\pi} \int_0^{2\pi/\omega p} i_d \, dt \tag{2.110}$$

i_d is itself the result of the integral

$$i_d = \frac{1}{L_0} \int_0^t v_L \, d\tau \tag{2.111}$$

$$= \frac{1}{L_0} \int_0^t \left\{ \hat{V} \sin\left(\omega\tau + \frac{\pi}{2} - \frac{\pi}{p} + \alpha\right) - \overline{V}_d \right\} d\tau \tag{2.112}$$

$$= \frac{\overline{V}_d}{L_0} \int_0^t \left\{ \frac{\pi/p}{\cos\alpha \sin(\pi/p)} \left(\omega\tau + \frac{\pi}{2} - \frac{\pi}{p} + \alpha\right) - 1 \right\} d\tau \tag{2.113}$$

The result of the double integral is

$$\frac{\omega L_0}{R} = \tan\alpha \left(1 - \frac{\pi}{p} \cot\frac{\pi}{p}\right) \tag{2.114}$$

and is valid so long as (in the rectifying case)

$$\hat{V} \sin\left(\frac{\pi}{2} - \frac{\pi}{p} + \alpha\right) > \hat{V} \frac{\sin(\pi/p)}{\pi/p} \cos\alpha$$

or, by transpositions

$$\tan\alpha \not< \frac{p}{\pi} - \cot\frac{\pi}{p} \tag{2.115}$$

The variation of $\omega L_0/R$ with α for a 3-pulse converter is illustrated by the graph of Figure 2.81.

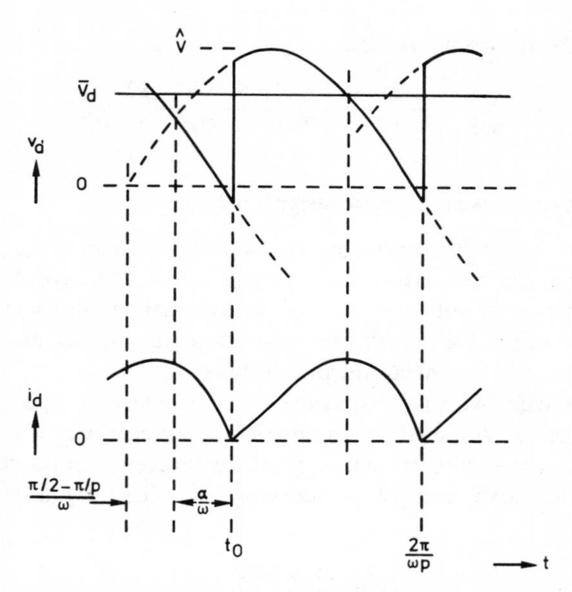

Figure 2.80 Output waveforms in a *p*-pulse controlled converter with critical inductance.

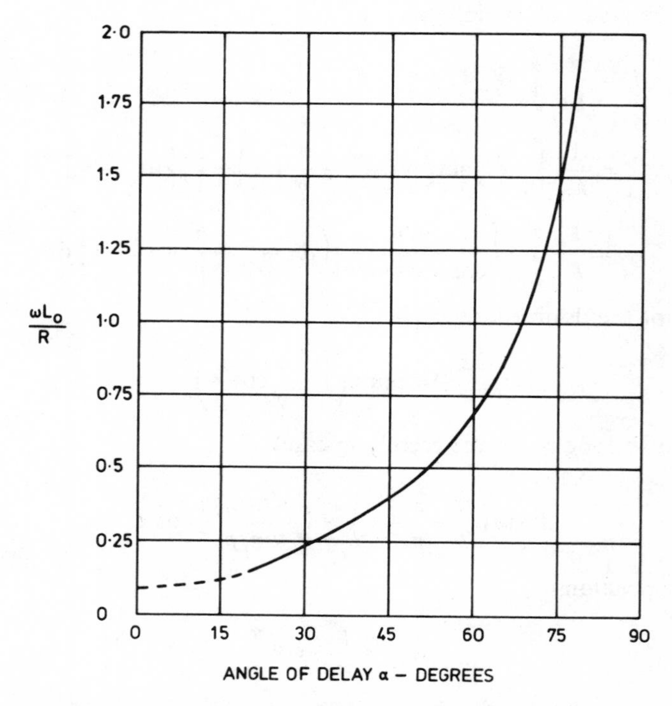

Figure 2.81 Variation of critical inductance with delay angle in a three-pulse controlled converter.

Controlled converter with *LR* load

The limiting condition for a p-pulse fully controlled converter with a load comprising inductance L in series with resistance R (see Figure 2.40) is illustrated by the waveforms of Figure 2.82. With a critical angle of delay α_c (i.e. the largest value of α with which the output current is continuous),

$$i_d = \frac{\hat{V}}{\sqrt{(R^2+\omega^2L^2)}}\left\{\sin\left(\omega t+\alpha_c+\frac{\pi}{2}-\frac{\pi}{p}-\phi\right)-\sin\left(\alpha_c+\frac{\pi}{2}-\frac{\pi}{p}-\phi\right)e^{-Rt/L}\right\}$$

$$= \frac{\hat{V}}{\sqrt{(R^2+\omega^2L^2)}}\left\{\cos\left(\omega t+\alpha_c-\frac{\pi}{p}-\phi\right)-\cos\left(\alpha_c-\frac{\pi}{p}-\phi\right)e^{-Rt/L}\right\} \quad (2.116)$$

where $\phi = \tan^{-1}(\omega L/R)$.
When $\omega t = 2\pi/p$, $i_d = 0$; therefore

$$\cos\left(\alpha_c+\frac{\pi}{p}-\phi\right) = \cos\left(\alpha_c-\frac{\pi}{p}-\phi\right)e^{-(2\pi/p)/\tan\phi} \quad (2.117)$$

from which

$$\tan\alpha_c = \frac{\cos\left(\frac{\pi}{p}-\phi\right)-e^{-(2\pi/p)/\tan\phi}\cos\left(\frac{\pi}{p}+\phi\right)}{\sin\left(\frac{\pi}{p}-\phi\right)+e^{-(2\pi/p)/\tan\phi}\sin\left(\frac{\pi}{p}+\phi\right)} \quad (2.118)$$

Without a smoothing capacitor or other source of steady load voltage, critical inductance has no significance in relation to uncontrolled or half-controlled rectifiers, or rectifiers with free-wheel diodes.

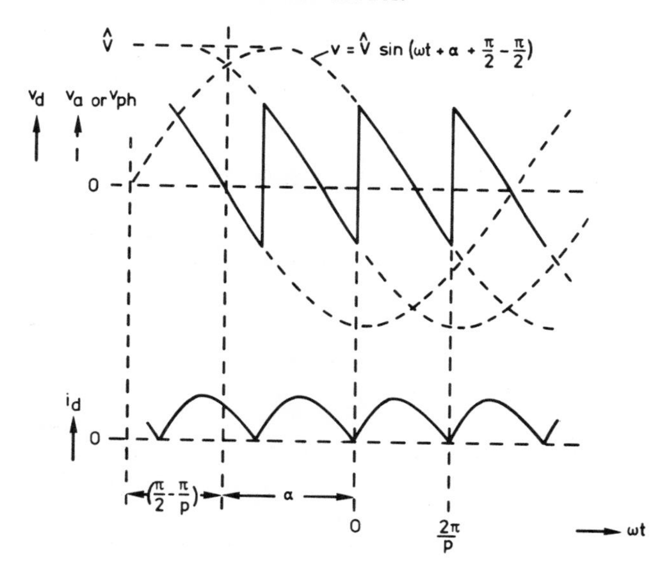

Figure 2.82 Output waveforms in a fully controlled converter with *LR* load at the critical delay angle.

APPLICATIONS OF CONVERTERS: DOUBLE CONVERTERS

The fully controlled converter, as described above, can produce a reversible direct output voltage with output current in one direction, and in terms of a conventional current/voltage diagram (Figure 2.83) is said to be capable of operation in two quadrants, the first and second. Such a range of operation is useful for certain purposes, examples being the control of a d.c. torque motor, i.e. a motor used to provide unidirectional torque with reversible rotation (Figure 2.84), and a d.c. transmission link between two a.c. systems in which power can be transmitted in either direction according to the polarity of the voltage with current flow always in one direction (Figure 2.85). Equally, a converter may be used under steady-state conditions in the first quadrant only, but transiently in the second quadrant in order to extract energy from the load quickly and thereby improve the response of the system to changing command signals.

If four-quadrant operation of a d.c. motor is required—i.e. reversible rotation and reversible torque—a single converter needs the addition of either a changeover contactor to reverse the armature connections or a means of reversing the field current in order to change the relationship between the converter voltage and the direction of rotation of the motor. Both of these are practicable in suitable circumstances, but the best performance is obtained by using two converters—

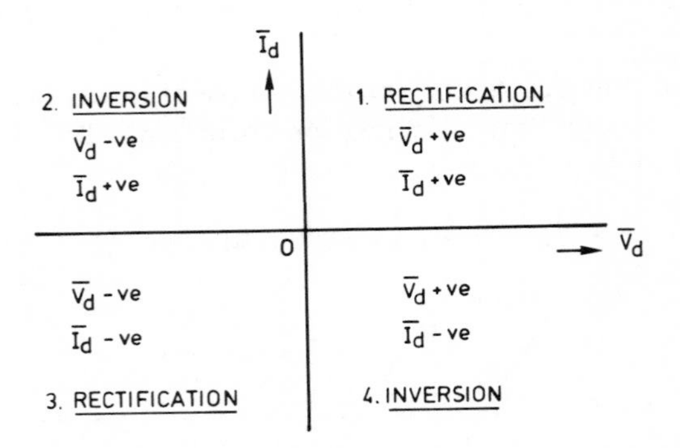

Figure 2.83 Current–voltage diagram representing possible conditions of operation of a controlled converter.

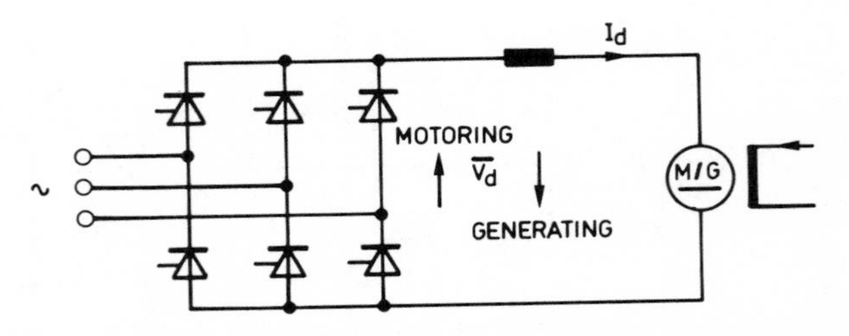

Figure 2.84 Fully controlled three-phase bridge converter controlling a torque motor in two quadrants.

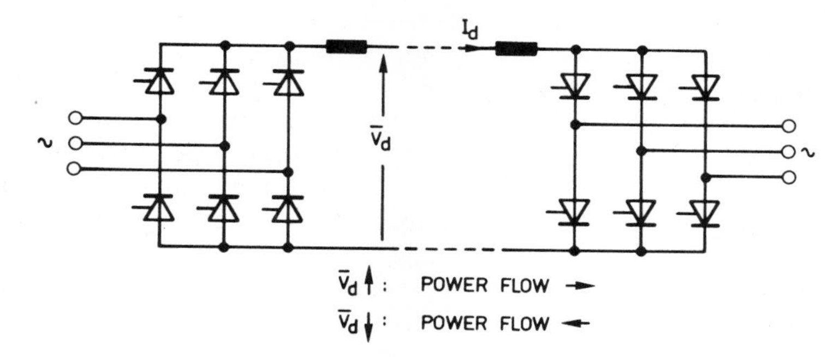

Figure 2.85 Two fully controlled two-quadrant converters in a d.c. transmission link for reversible power flow.

usually two bridges—in the inverse parallel connection shown in Figure 2.86(a). The four possible quadrants of converter operation thus resulting can be translated into four (steady-state) combinations of motor torque and rotation as illustrated in Figure 2.86(b).

Two modes of control are possible in a double converter such as that of Figure 2.86. In the circulating-current-free mode, the firing of the thyristors in the two

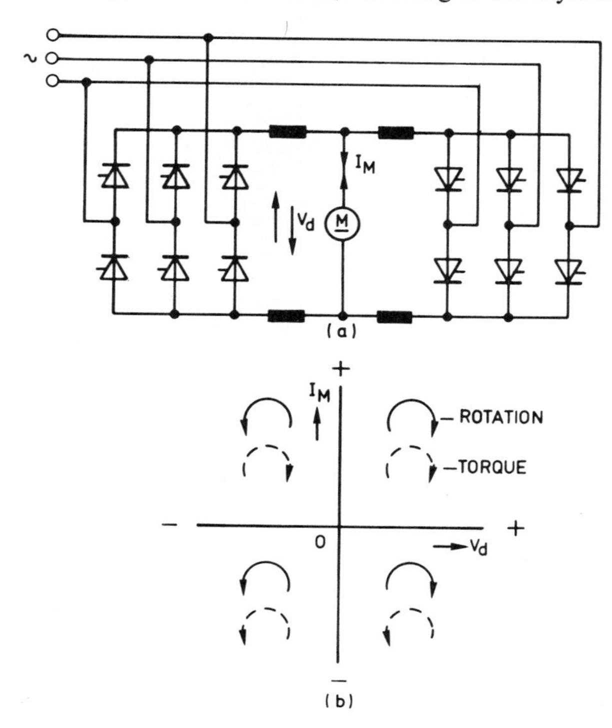

Figure 2.86 (a) Fully controlled double-bridge converter for four-quadrant control of a d.c. motor; (b) torque–rotation diagram corresponding to the converter current–voltage diagram.

converters is so constrained that current flows in only one converter at any time, while the other is in a totally blocking condition. In the circulating-current mode, a relatively small current is allowed to flow between the two converters, limited by appropriate delay-angle control and adequate inductance in the d.c. loop.

A double converter with control signals appropriately varied in a cyclic manner can produce an alternating output, at a frequency substantially less than the supply frequency, and by virtue of its ability to operate in all quadrants can generate a prescribed waveform of voltage—e.g. sinusoidal—regardless of the load characteristics. Such an arrangement is known as a cycloconverter, and has useful applications in a.c. motor drives and frequency-changers.

More detailed consideration of converter control is beyond the scope of this book.

CHAPTER 3

A. C. Regulators

The inclusion of thyristor power regulators in a book which is primarily concerned with converters could be considered anomalous, since in their normal form their essential function is simply to regulate the flow of current from a supply to a load, rather than to effect a true power conversion. They do, however, represent a sufficiently significant usage of thyristors to warrant at least an introduction.

Thyristor a.c. regulators may be considered akin to naturally commutating converters in that the cessation of conduction in the thyristors is brought about by the alternating supply voltage. Commutation, however, in the true sense of a transfer of current from one branch of a circuit to another, is not essentially part of the operation of a simple a.c. regulator, and occurs only under certain conditions.

SINGLE-PHASE REGULATOR

The simplest means of regulating the power delivered from an a.c. supply to a load is that illustrated in Figure 3.1. This is in fact identical to the rudimentary half-wave rectifier circuit of Figure 2.1, but since its only normal use is to control very small heating loads it is necessary in the present context to recognize only a purely resistive load, leaving its characteristics to be deduced in a simple manner from those of the much more useful full-wave circuit, shown with a resistive load in Figure 3.2. Output voltage, or current, waveforms are illustrated in Figure 3.3 for various output voltages.

As with a controlled rectifier, variation of the output of a thyristor a.c. regulator is achieved by triggering the appropriate thyristor—that is, the one to which the supply is presenting a forward voltage—at a variable point in each half-cycle, after

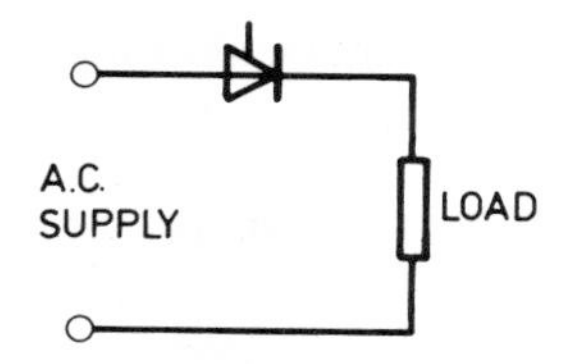

Figure 3.1 Half-wave a.c. power regulator.

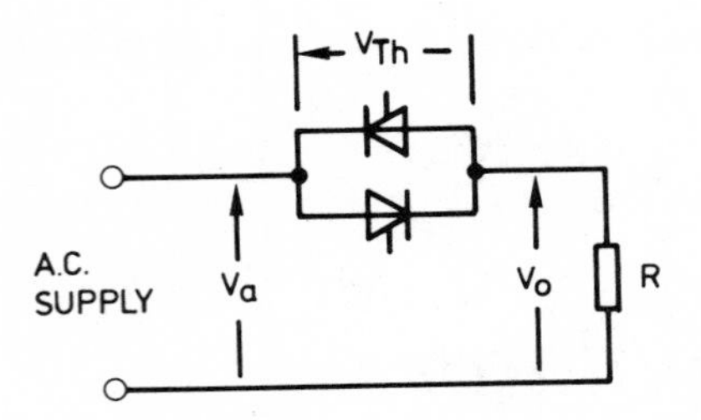

Figure 3.2 Single-phase full-wave regulator with resistive load.

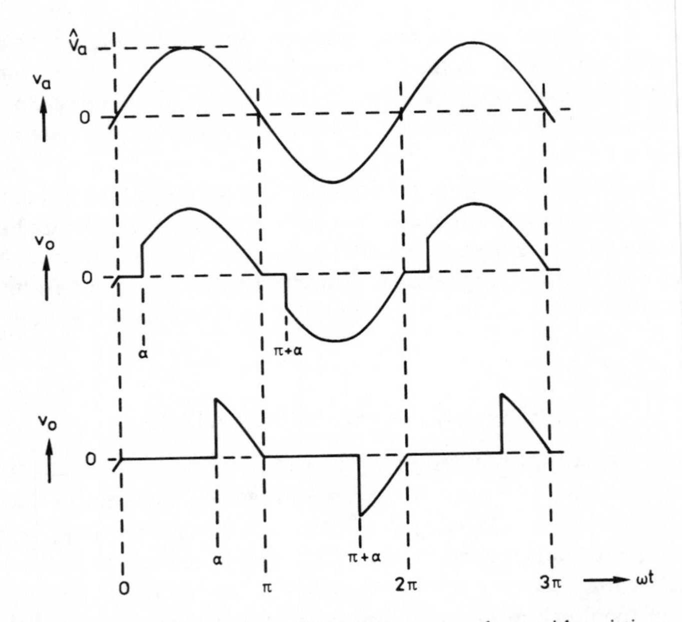

Figure 3.3 Waveforms in single-phase a.c. regulator with resistive load, for two delay angles.

a delay α/ω measured from the supply voltage zero. Conduction ceases at the end of the half-cycle when the instantaneous supply voltage, and therefore the current in the thyristor, fall to zero.

The mean (half-cycle) output voltage can be obtained by integrating the supply voltage between the limits α and π.

$$\left.\begin{aligned}\overline{V}_o = \frac{1}{\pi}\int_{\alpha}^{\pi} \hat{V}_a \sin \omega t \, d\omega t &= \frac{\hat{V}_a}{\pi}(1+\cos\alpha) \\ &= \overline{V}_a \frac{(1+\cos\alpha)}{2}\end{aligned}\right\} \quad (3.1)$$

This quantity, however, is rarely of interest. The more significant quantity, the r.m.s. output voltage, is

$$\left.\begin{aligned} V_o = \sqrt{\left(\frac{1}{\pi}\int_{\alpha}^{\pi} \hat{V}_a^2 \sin^2 \omega t \, d\omega t\right)} &= \hat{V}_a\sqrt{\left(\frac{2(\pi-\alpha)+\sin 2\alpha}{4\pi}\right)} \\ &= V_a\sqrt{\left(1-\frac{2\alpha-\sin 2\alpha}{2\pi}\right)} \end{aligned}\right\} \quad (3.2)$$

and the output power is

$$P_o = \frac{V_o^2}{R} = \frac{V_a^2}{R}\left(1-\frac{2\alpha-\sin 2\alpha}{2\pi}\right) \quad (3.3)$$

These values are plotted in Figure 3.4.

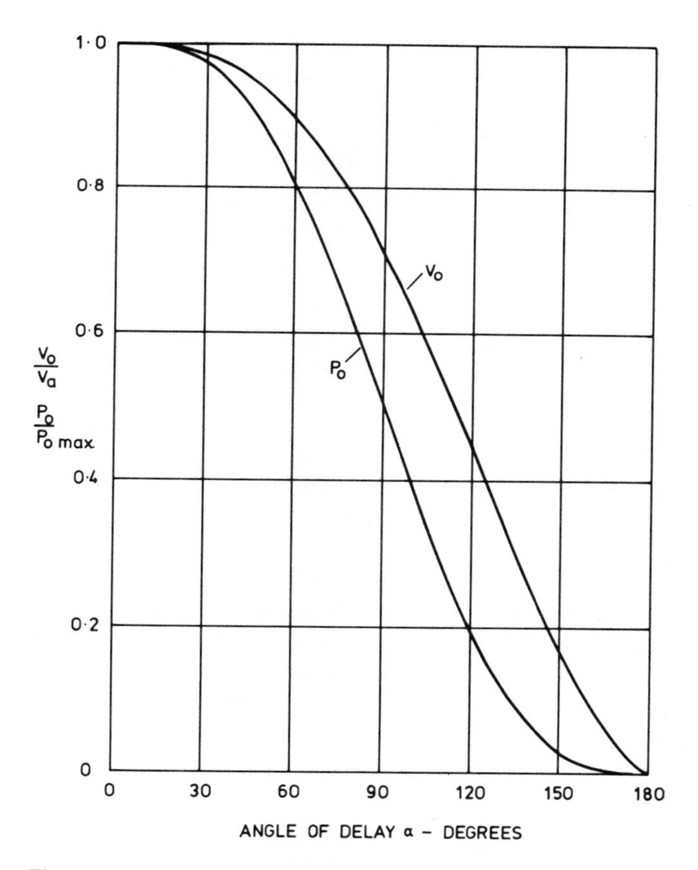

Figure 3.4 Variation with delay angle of r.m.s. output voltage and output power of a single-phase a.c. regulator with (fixed) resistive load.

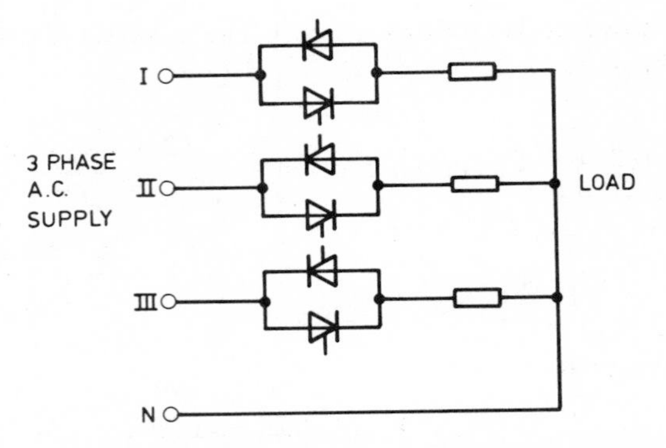

Figure 3.5 Star-connected three-phase arrangement of a.c. regulators with neutral connection.

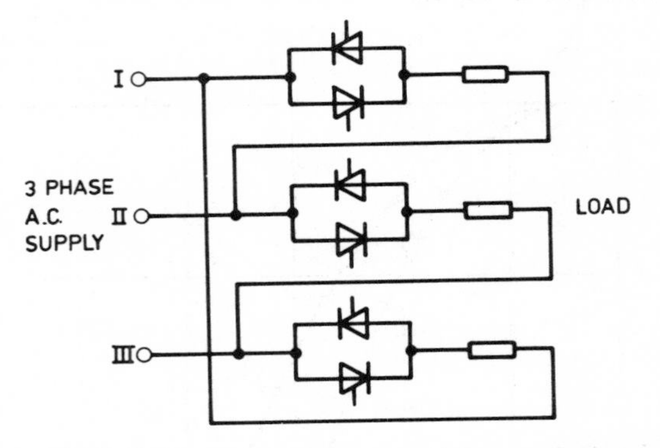

Figure 3.6 Open-delta-connected three-phase arrangement of a.c. regulators.

THREE-PHASE A.C. REGULATORS

Three-phase regulators may take a variety of forms. Figures 3.5 and 3.6 illustrate two arrangements, with divisible loads, which function simply as groups of single-phase regulators, in the first case in star formation with a neutral connection and in the second case in a delta circuit. These arrangements need little further comment beyond pointing out that in the star circuit such triplen harmonic currents as flow in the loads flow also in the supply lines (and the neutral connection), whereas this is not possible in the three-wire delta system, assuming it to be balanced, so that it generates supply harmonics only of odd orders other than triplen.

Three-phase three-wire a.c. regulators

Removing the neutral connection from a star-connected circuit such as that of Figure 3.5, as in Figure 3.7, complicates its operation somewhat, since the phase outputs no longer depend solely on switching in one line. Understanding is facilitated, however, by considering first the various constitutions of the output

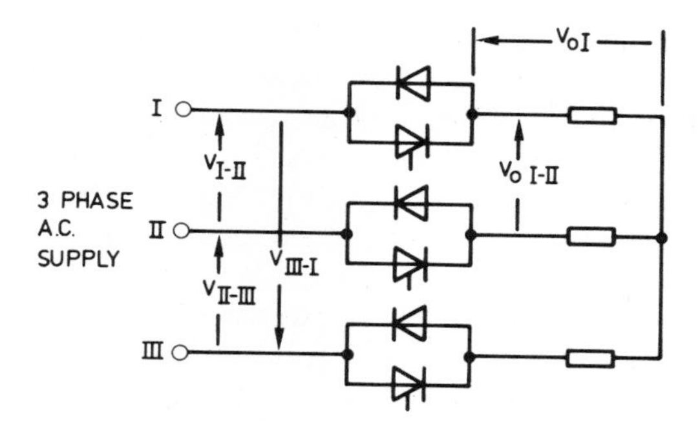

Figure 3.7 Six-thyristor three-phase a.c. regulator without neutral connection.

voltage and thyristor voltage waveforms resulting from the limited number of thyristor conduction patterns. If the direction of the line currents is ignored, there are only four such patterns: these are illustrated in Figure 3.8 and tabulated, together with the resulting forms of the output phase voltage V_{oI} and the interphase voltage V_{oI-II}, in Table 3.1.

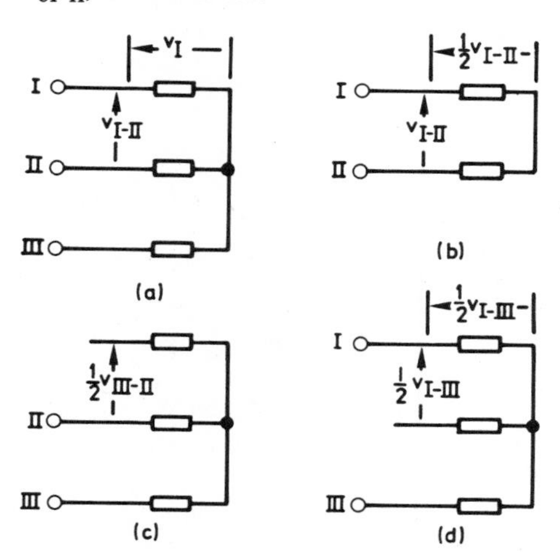

Figure 3.8 Patterns of voltages in a three-phase three-wire a.c. regulator with balanced load.

Table 3.1
Pattern of voltages in a three-phase three-wire a.c. regulator.

	Lines conducting	Lines non-conducting	Output voltage Phase v_{oI}	Interphase v_{oI-II}
(a)	All	None	v_I	v_{I-II}
(b)	I, II	III	$\frac{1}{2}v_{I-II}$	v_{I-II}
(c)	II, III	I	0	$\frac{1}{2}v_{III-II}$
(d)	III, I	II	$\frac{1}{2}v_{I-III}$	$\frac{1}{2}v_{I-III}$
(e)	None	All	0	0

The condition of full output shown in Figure 3.9(a) determines the reference $\alpha = 0$ for each thyristor at the point where current starts to flow through it into the resistive load. With a small firing delay, as in Figure 3.9(b) for $\alpha = 30°$, conduction in each phase ceases 180° after the reference point, and the condition of all lines conducting is re-established as each thyristor is fired. When α reaches 60° (Figure 3.9(c)), the firing of one thyristor causes another, previously conducting, to turn off by a process of commutation, with the result that there are never more than two lines conducting together. When α exceeds 90°, the conduction periods become less than the interval between successive firing pulses; it is then necessary to fire pairs of thyristors simultaneously to establish conducting paths, which means that each thyristor has to receive two firing pulses in each cycle, separated by 60°, as shown in Figure 3.9(d) and (e).

Under the latter conditions of reduced conduction, the current in each line falls to zero at an instant determined by the waveform of the interphase supply voltage, not the phase-to-neutral voltage, namely at 150° from the zero of the delay-angle range. 150° is thus the maximum operative angle of delay, giving zero output.

The r.m.s. output voltage may readily be computed as a function of delay angle from the waveforms shown, giving the characteristic shown in Figure 3.10 (a).

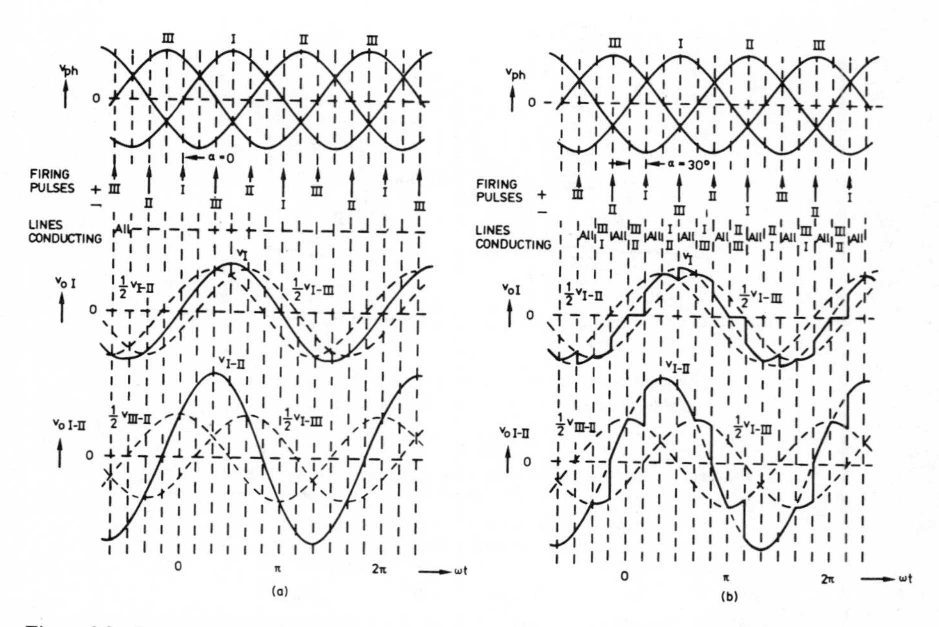

Figure 3.9 Voltage waveforms in a three-phase six-thyristor regulator with balanced resistive load. (Firing pulses to thyristors conducting current to and from the load are designated + and − respectively.) (a) $\alpha = 0$; (b) $\alpha = 30°$; (c) $\alpha = 60°$; (d) $\alpha = 90°$; (e) $\alpha = 120°$.

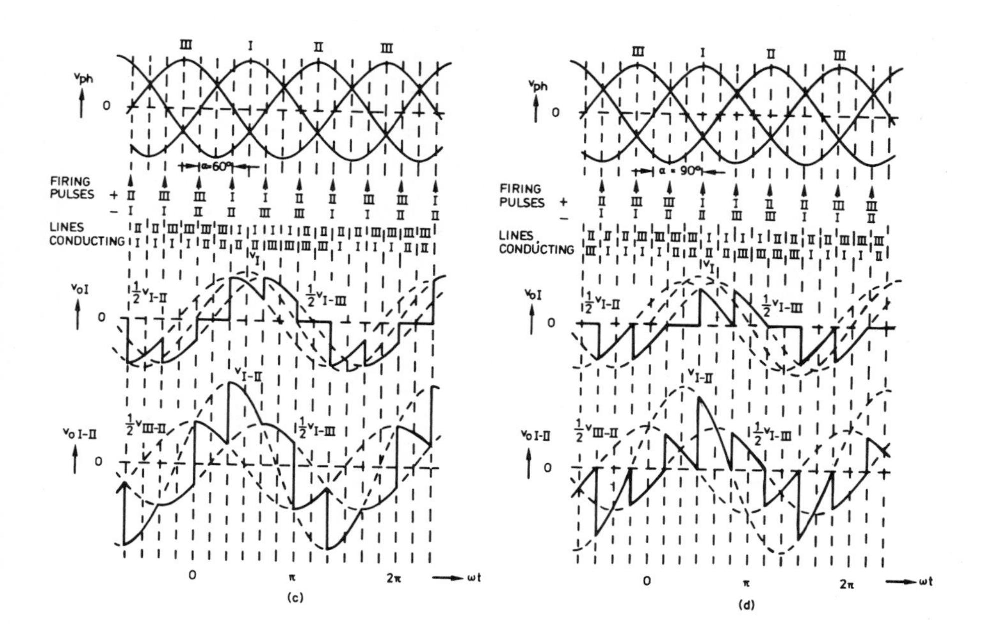
v ph
FIRING PULSES
LINES CONDUCTING
α=60°
v oI
½v I-II
½v I-III
v I
v oI-II
½v III-II
v I-II
0
π
2π
ωt
(c)
α = 90°
(d)

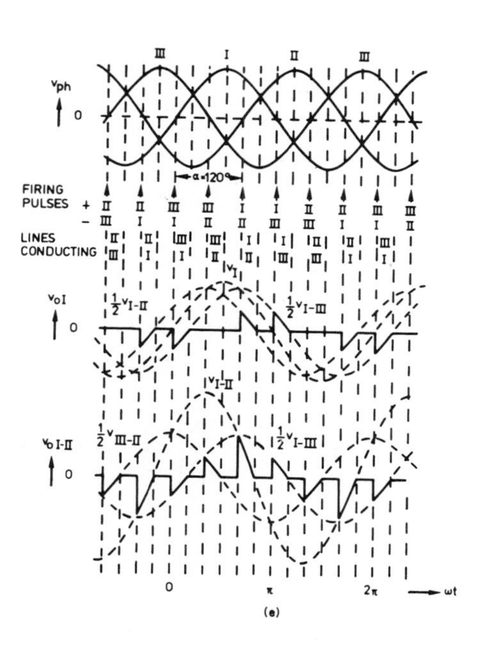
v ph
FIRING PULSES
LINES CONDUCTING
α=120°
v oI
½v I-II
½v I-III
v I
v oI-II
½v III-II
v I-II
0
π
2π
ωt
(e)

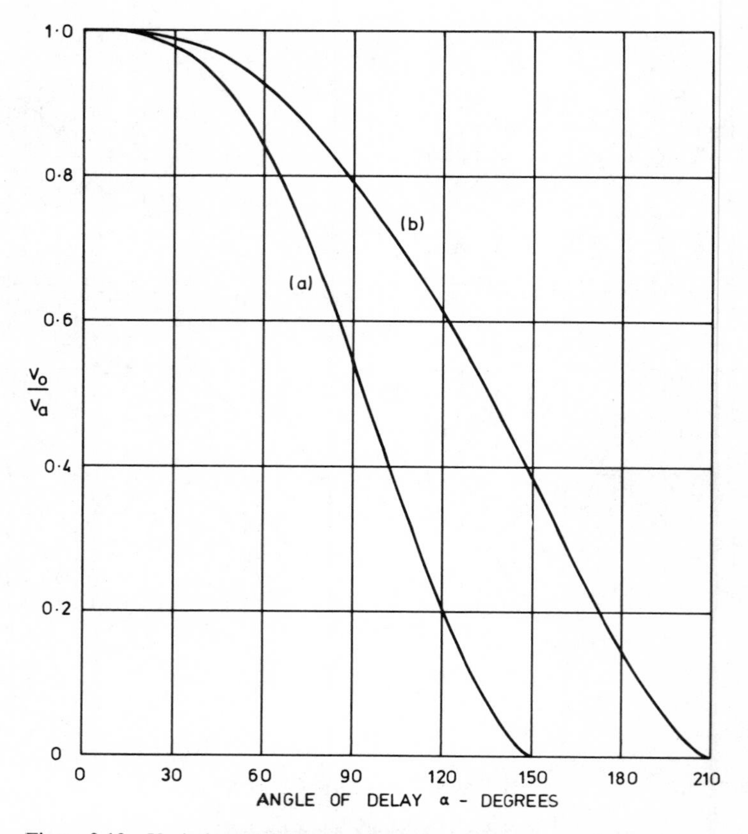

Figure 3.10 Variation with delay angle of the r.m.s. output voltage of three-phase three-wire regulators with resistive load: (a) six-thyristor; (b) thyristor–diode.

Thyristor–diode regulators

Every path for current in the three-phase six-thyristor regulator of Figure 3.7 includes two thyristors, which is more than is basically necessary to achieve control of the output voltage, and it is possible to replace three of them by diodes as in Figure 3.11. The technique previously applied to the six-thyristor regulator

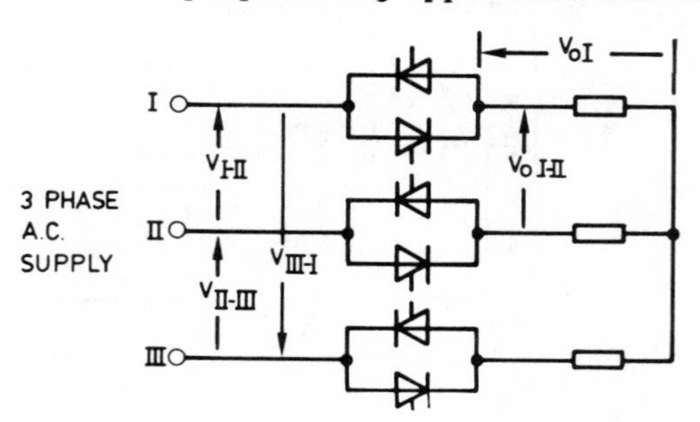

Figure 3.11 Thyristor-diode three-phase a.c. regulator.

yields the waveforms shown in Figure 3.12(a)–(c); these are asymmetrical, as might be expected from the presence of only three firing pulses per cycle, and the operative range of delay angles, with resistive load, is 210°. The control characteristic is shown in Figure 3.10(b).

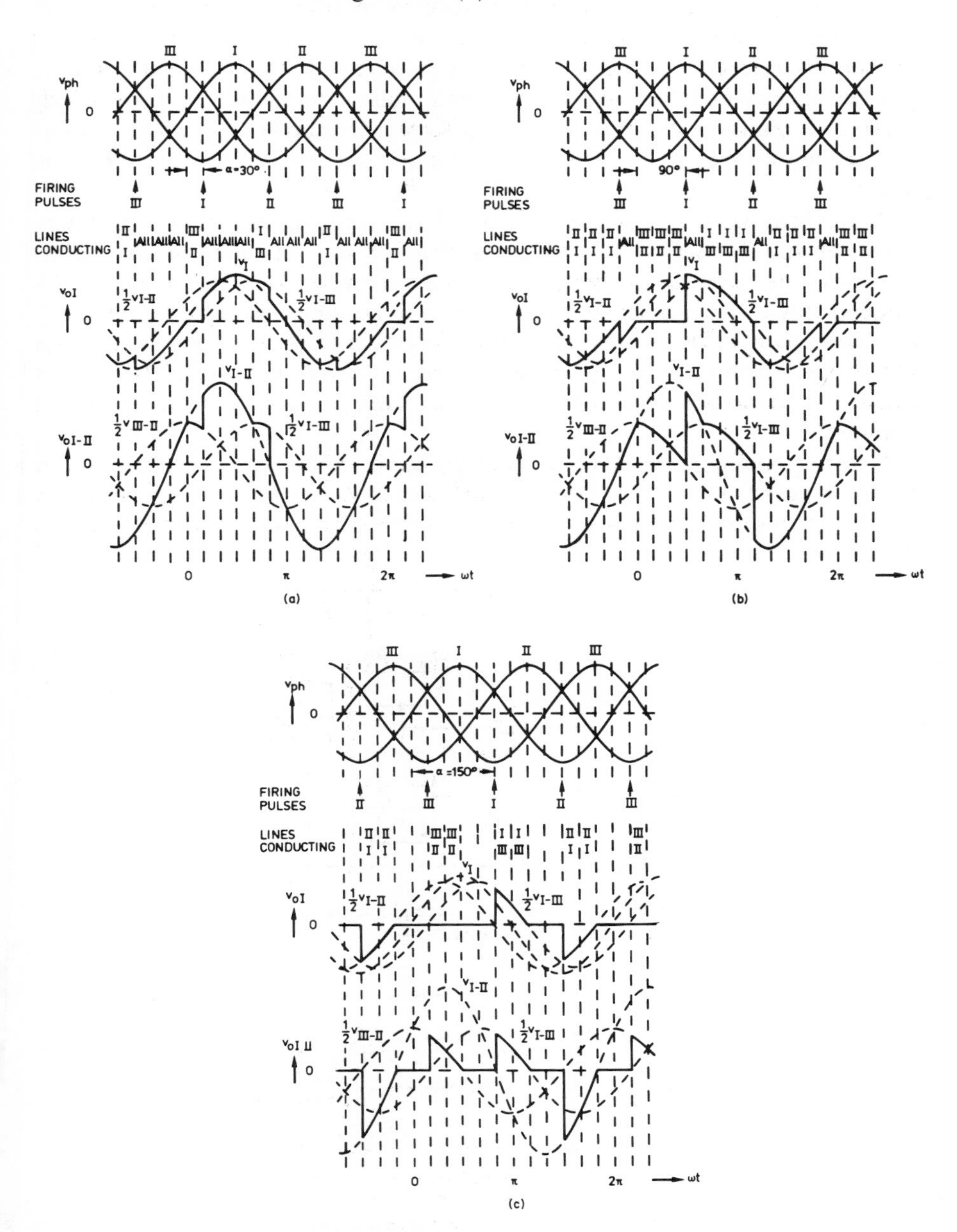

Figure 3.12 Voltage waveforms in a three-phase thyristor-diode regulator with balanced resistive load; (a) $\alpha = 30°$; (b) $\alpha = 90°$; (c) $\alpha = 150°$.

Open-star regulators

Comparison of the three-wire regulators of Figures 3.7 and 3.11 with the four-wire regulator of Figure 3.5 reveals a disadvantage in the three-wire regulators in regard to thyristor utilization, in that, while the maximum current is the same whether the neutral is connected or not, the maximum thyristor voltage without the neutral connection is higher by a factor of 3/2, and the thyristors are likely in practice to be rated for the full interphase voltage to allow for certain fault conditions. Similarly the three-wire circuits compare unfavourably with the open-delta circuit of Figure 3.6, where the thyristor currents, for a given power, are lower by a factor of $\sqrt{3}$.

A thyristor utilization close to that of the single-phase or the four-wire regulator can be achieved with a star load if the star can be opened to accommodate a delta-connected group of thyristors as in Figure 3.13. Functionally this arrangement can be shown to be similar to that of Figure 3.7, producing identical load waveforms, but, since each thyristor is part of only one current path instead of two, the mean thyristor current is halved. There is the additional advantage that the simultaneous firing of pairs of thyristors is not required.

A corresponding three-thyristor circuit (Figure 3.14) shows a disadvantage in thyristor ratings compared with the thyristor–diode circuit of Figure 3.11, to which it is functionally similar, although it has the slight merit of dispensing with the diodes.

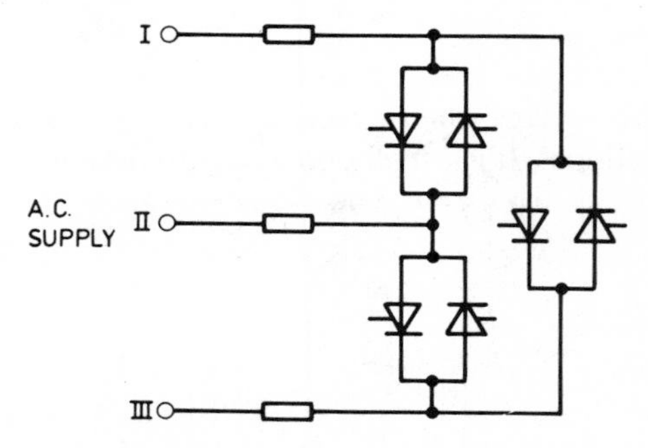

Figure 3.13 Open-star three-phase regulator with six thyristors.

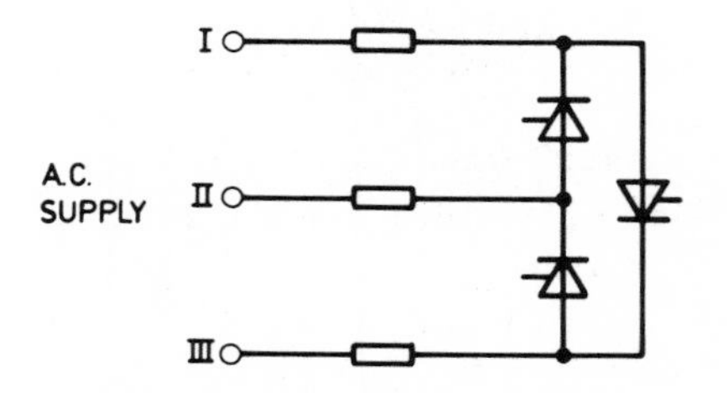

Figure 3.14 Open-star three-phase regulator with three thyristors.

Thyristor voltage ratings

Thyristors in three-phase regulator circuits without a neutral connection may in general be subjected to blocking voltages up to the peak interphase supply voltage. The six-thyristor circuit of Figure 3.7 is an exception, in that the thyristor voltage does not normally exceed $\sqrt{3}/2$ times the supply voltage, but this is true only so long as the load is balanced.

REGULATORS WITH INDUCTIVE LOAD

Reactance in a load controlled by a thyristor a.c. regulator considerably modifies its output waveforms and control characteristics. Loads that include appreciable capacitance often give rise to serious problems in regard to peak currents and harmonics, and the following discussion is concerned only with loads comprising resistance and inductance.

Figure 3.15 shows a single-phase regulator with a series LR load. At full output, the output current lags the voltage by the load phase angle $\phi = \tan^{-1} \omega L/R$ as shown in Figure 3.16. Since the thyristors carry the positive and negative half-cycles of current alternately with this phase displacement, they have to be fired at $\omega t = \phi$ and $\omega t = \pi + \phi$: in other words, relative to the supply voltage waveform, which is the only fixed reference if the load is variable, full output is obtained when the delay angle $\alpha = \phi$.

If, with short firing pulses, the delay angle is reduced below ϕ, the situation illustrated in Figure 3.17 results, wherein conduction in one thyristor lasts for more than 180° while the other does not conduct at all because it is not experiencing forward voltage when it receives its firing pulse. The output current is therefore unidirectional. It is usually necessary to design the control system so that this undesirable and possibly dangerous condition cannot arise, either by extending the firing signals at least to $\omega t = \phi$ or $(\pi + \phi)$, or by using a synchronizing or clamping circuit to prevent the application of a firing pulse to a thyristor until conditions are favourable to conduction.

With α in the operative region between ϕ and π, the waveforms with a series LR load are of the form shown in Figure 3.18, and the characteristic of r.m.s. output voltage against delay angle is typically as illustrated in Figure 3.19.

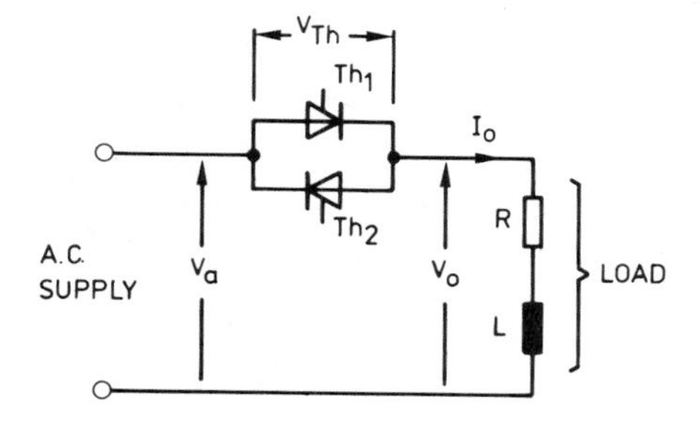

Figure 3.15 Single-phase a.c. regulator with series LR load.

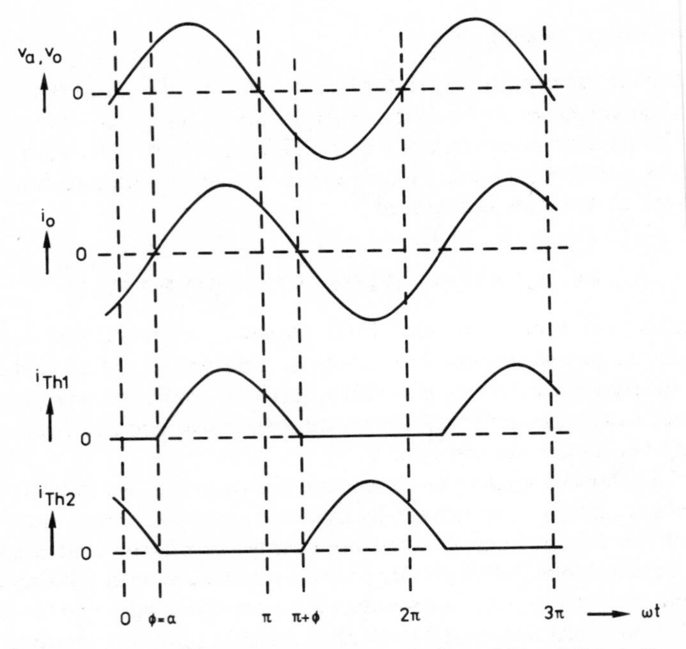

Figure 3.16 Waveforms in the circuit of Figure 3.15 at full output.

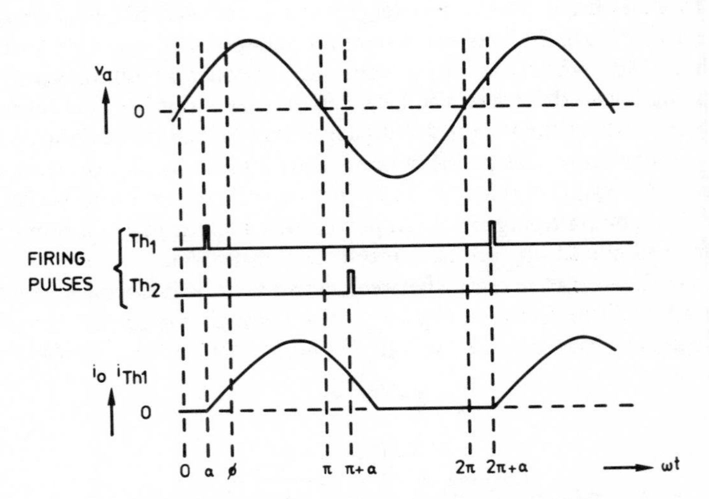

Figure 3.17 Waveforms in the circuit of Figure 3.15 with $\alpha < \phi$

If the load consists of inductance and resistance in parallel, the output waveforms differ somewhat from those for the series *LR* load as a result of the path which exists for current circulating within the load when the thyristors are not conducting, but the control characteristics and general behaviour are closely similar.

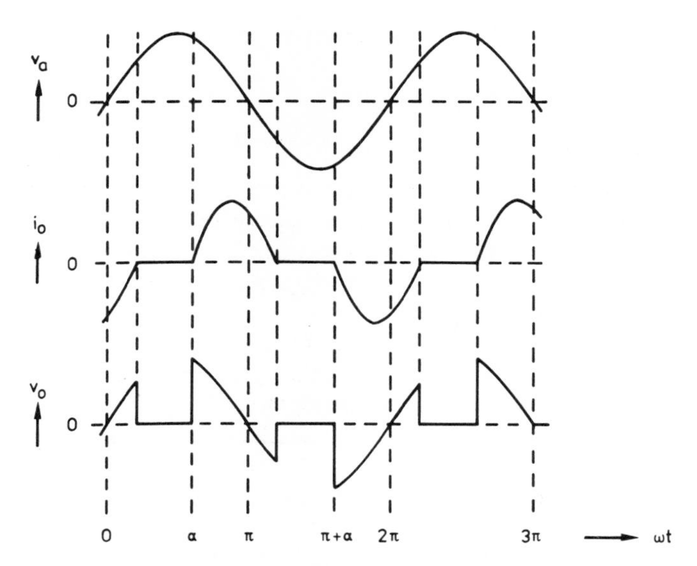

Figure 3.18 Waveforms in the circuit of Figure 3.15 with $\alpha > \phi$

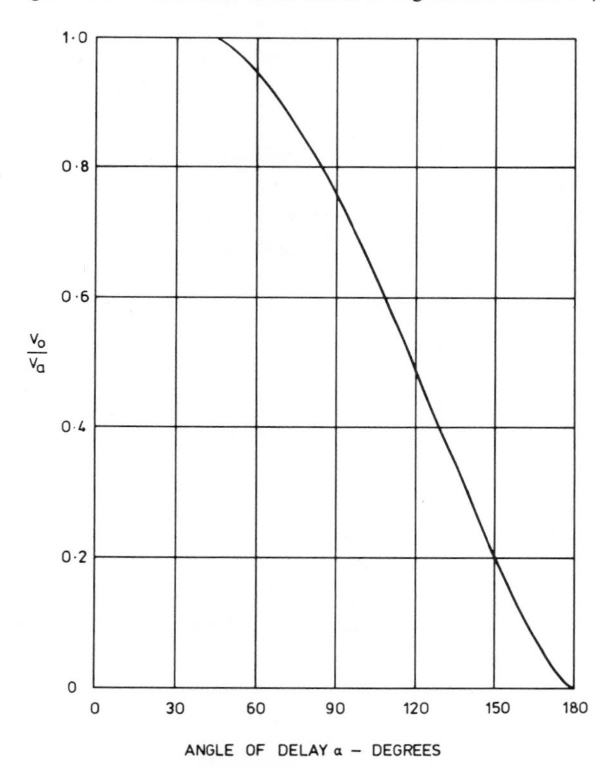

Figure 3.19 Control characteristic of single-phase regulator (Figure 3.15) with series *LR* load ($\phi = 45°$).

Similar effects are observed in three-phase regulator circuits with inductive loads.

A.C. REGULATORS CONJOINED WITH RECTIFIERS

An important application of thyristor a.c. regulators is in the control of rectifier output voltage in cases where there is a practical advantage in using the thyristors at a voltage level different from that of the rectifier, from the point of view either of conduction losses or cost. Such cases normally arise in rectifiers for high current at low voltage or for high voltage.

Disregarding the transformer, placing thyristors in the a.c. lines feeding a rectifier, as illustrated in Figure 3.20, has the same effect as incorporating them in the rectifier itself, since each direction of current flow in the a.c. circuit is associated with a particular arm of the rectifier, except that the diode rectifier always provides a path for current in the d.c. circuit. A diode rectifier controlled by a thyristor regulator of appropriate configuration, having as many thyristors as there are diodes, is thus equivalent functionally to the corresponding controlled rectifier with the addition of a free-wheel diode.

The use of such a combination has no point unless a transformer is interposed between the regulator and the rectifier. However, the equivalence demonstrated above is unaffected provided that the relationships between the rectifier input currents and the regulator currents are not changed by the transformer connections. Thus the arrangements shown in Figure 3.21 afford true equivalents, apart from secondary effects due to transformer magnetizing current etc., of the controlled bridge and hexaphase rectifiers. (This statement would, however, require some qualification if the open-delta primary circuit in Figure 3.21(b) were replaced by the three-wire regulator of Figure 3.7 with a delta primary winding.)

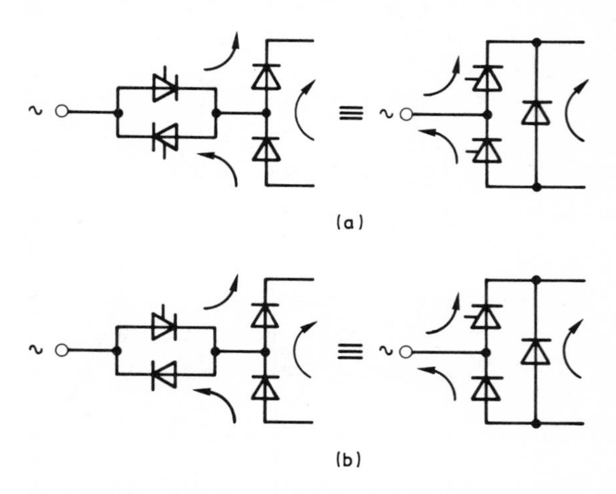

Figure 3.20 The equivalence of a.c. regulators in combination with diode rectifiers to controlled rectifiers: (a) fully controlled; (b) half-controlled.

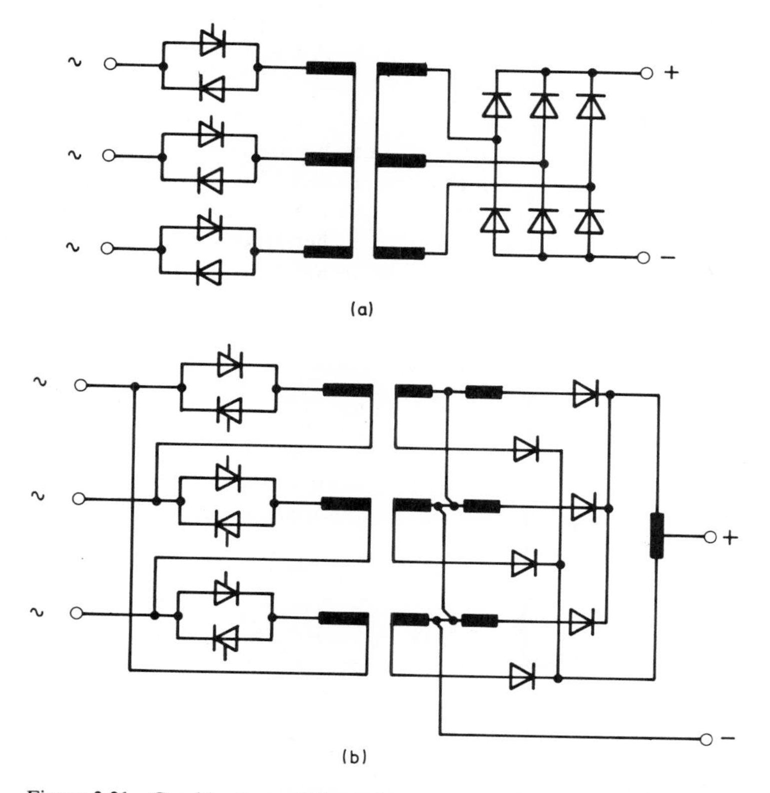

Figure 3.21 Combinations of three-phase a.c. regulators and diode rectifiers equivalent to (a) a half-controlled bridge rectifier and (b) a controlled hexaphase rectifier with free-wheel diode.

INPUT POWER FACTOR AND HARMONICS

The power factor presented to the a.c. supply by a thyristor a.c. regulator is an inherent function of the operation of any lossless series regulator. At full output, the input power factor is simply that of the load on a sinusoidal supply (Figure 3.16):

$$k_{p_{i\,max}} = k_{p_o} = \frac{P_{o\,max}}{V_i I_{o\,max}} \tag{3.4}$$

For any load that can be represented by a series circuit including constant resistance,

$$P_o = I_o^2 R$$

so for any output

$$k_{p_i} = \frac{P_i}{V_i I_i} = \frac{P_o}{V_i I_o} = \frac{R I_o}{V_i} \tag{3.5}$$

Hence $$k_{p_i} = k_{p_o} \frac{I_o}{I_{o\,max}} = k_{p_o} \sqrt{\left(\frac{P_o}{P_{o\,max}}\right)} \tag{3.6}$$

The input power factor with any load is associated with both phase displacement and harmonics. Harmonic currents are generated by a.c. regulators in proportions comparable with those generated by controlled rectifiers and of such orders as are compatible with the circuit configuration. Thus a single-phase full-wave regulator generates all the odd harmonics, a three-wire six-thyristor regulator generates all the odd harmonics other than the triplens, and a three-phase thyristor–diode regulator generates even harmonics in addition. The variations of the low-order harmonics with delay angle, computed by Fourier analysis of the voltage waveforms of Figures 3.4, 3.9 and 3.12, are shown in Figures 3.22, 3.23 and 3.24.

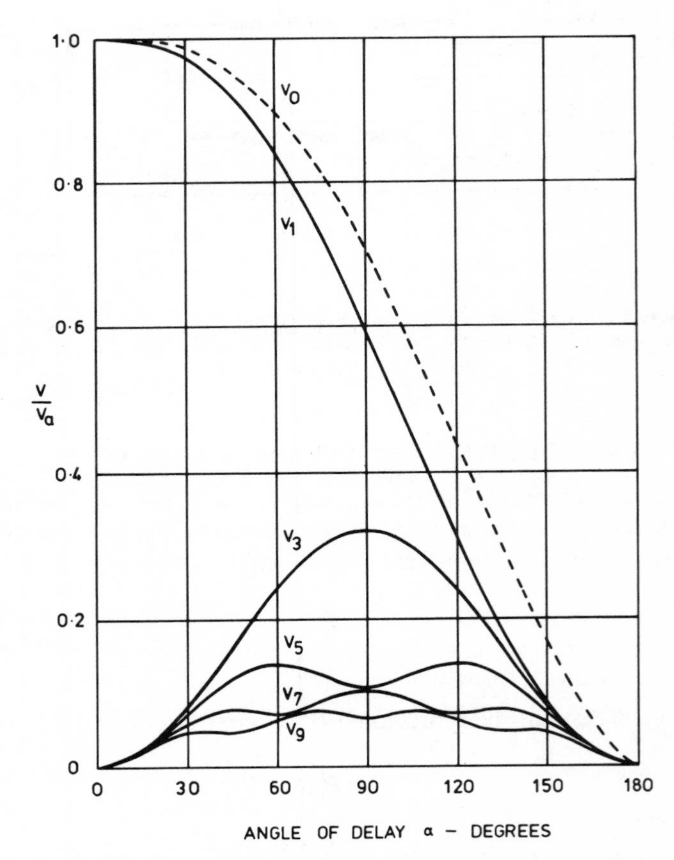

Figure 3.22 Harmonic components of the output voltage of a single-phase a.c. regulator with resistive load.

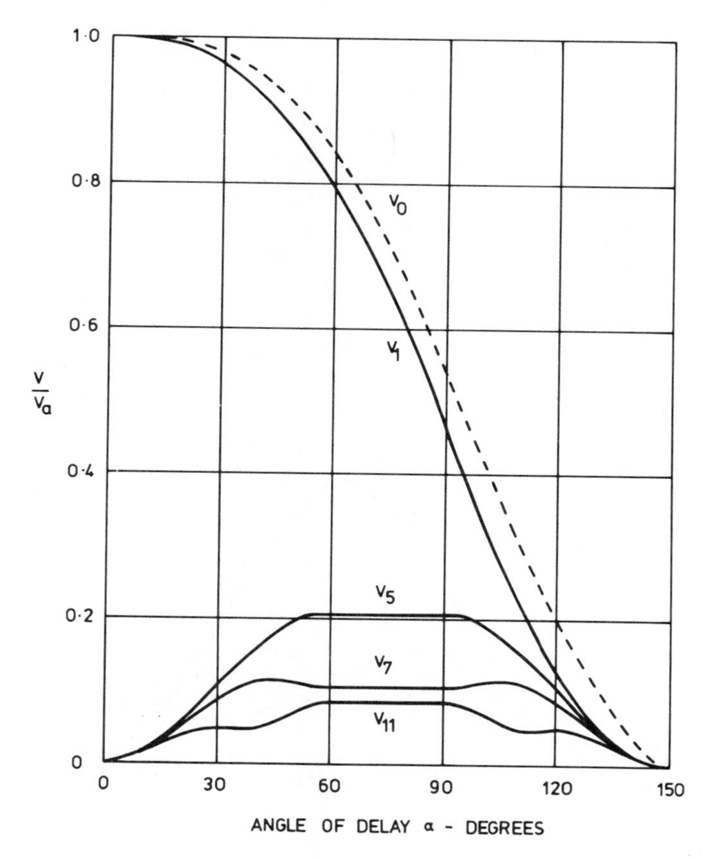

Figure 3.23 Harmonic components of the output voltage of a three-phase, three-wire, six-thyristor regulator with resistive load.

INTEGRAL-HALF-CYCLE CONTROL

Thyristor regulators with resistive heating loads are often controlled by a method alternative to delay angle control, in which the load is connected to the supply for a variable number of complete cycles or half-cycles in each of a succession of relatively long periods, as illustrated in Figure 3.25, the control system being designed to fire the thyristors as early as possible in each conducting half-cycle so that the instantaneous voltage at the instant of switching is minimized. This 'zero-voltage' switching, coupled with the inherent property of the thyristors of switching off at almost zero current, achieves a very large reduction in harmonics of the supply frequency, and so avoids the problems associated with harmonic currents and radio-frequency interference in delay-angle-controlled regulators.

Where the load is substantial in relation to the supply capacity, this mode of control may cause problems by disturbing the supply voltage at subharmonics of

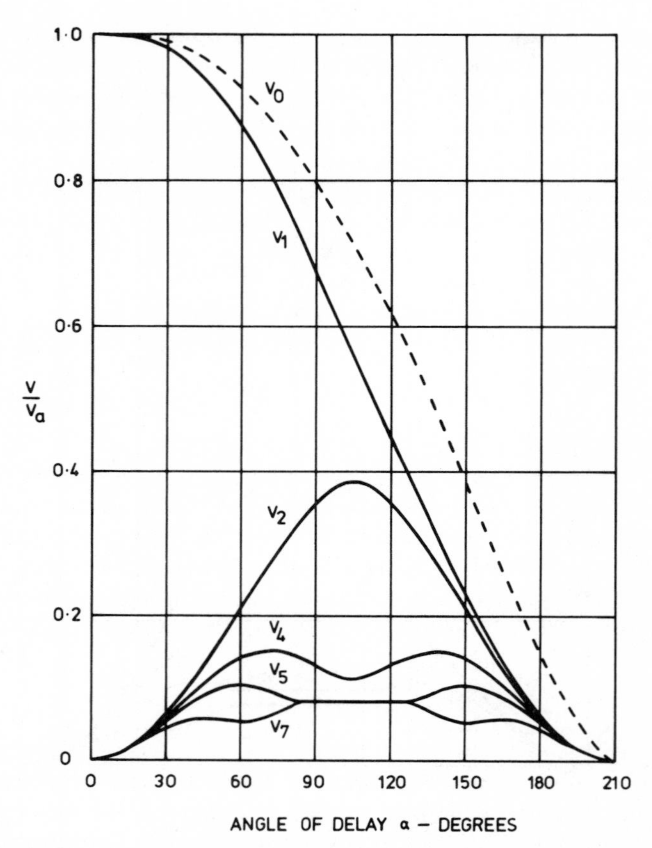

Figure 3.24 Harmonic components of the output voltage of a three-phase thyristor-diode regulator with resistive load.

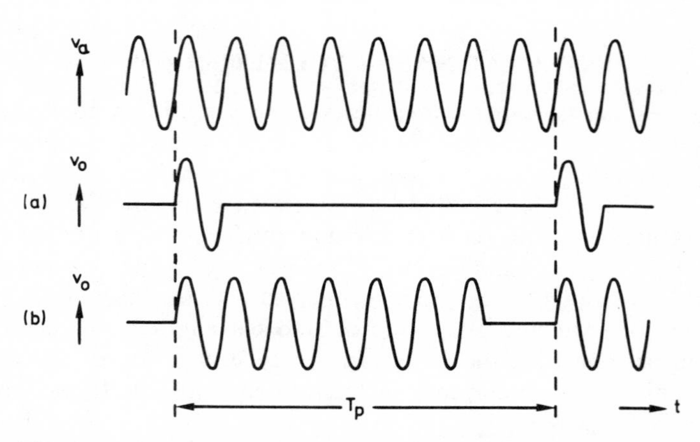

Figure 3.25 Integral-half-cycle operation in a single-phase a.c. regulator: (a) 35 % r.m.s. output voltage; (b) 90 % r.m.s. output voltage.

the supply frequency: lighting systems, for example, can be seriously affected by quite small voltage fluctuations. Technical difficulties arise if the load is coupled to the regulator through a transformer, owing to the repeated transient magnetization conditions, and in general the system is limited to furnaces and other simple heating loads.

Referring to Figure 3.25, if the total operating period T_p contains, in general, N half cycles ($= 2fT_p$) and the number of conducting half-cycles per period is n, the r.m.s. output voltage is

$$V_o = V_i \sqrt{\left(\frac{n}{N}\right)} \tag{3.7}$$

More usefully, the output power is

$$P_o = \frac{V_i^2 n}{RN} = P_{o\,max} \frac{n}{N} \quad \text{if } R \text{ is constant} \tag{3.8}$$

The resolution of the integral-half-cycle system, in terms of power, is thus $1/N$. The choice of N or T_p is a compromise between this resolution and the adverse effects which a large value of T_p has in response time and temperature ripple in a heating load. The rigidity of this compromise is mitigated in some systems by allowing the operating period to vary according to the output required.

THYRISTOR TAP-CHANGERS

An alternative technique for avoiding, or at least reducing, the large proportions of harmonics generated by the phase-controlled thyristor regulator, without the low-frequency load fluctuations entailed in integral-half-cycle control, is to use the thyristors, in conjunction with a transformer, or auto-transformer, in a tap-changing configuration rather than in a series regulator.

The simplest arrangement, shown in Figure 3.26, provides two levels of output voltage, according to which of the two pairs of thyristors is fired, and, assuming that the firing signals are appropriately synchronized to the voltage and current half-cycles, the circuit introduces no harmonics, or electromagnetic interference, and causes no reduction of the input power factor. Provided that one pair of

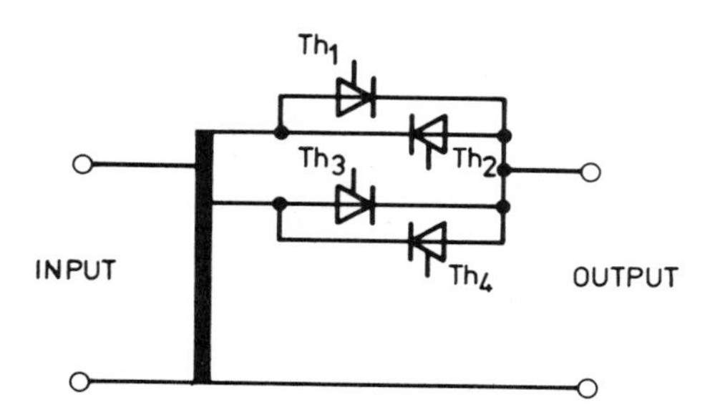

Figure 3.26 Basic thyristor tap-changer.

thyristors or the other is maintained in a conducting state at any one time, the thyristors are subjected only to the voltage between the transformer taps, and so can be rated (in principle at least) commensurately with the range of control provided rather than with the total voltage.

Several expedients may be adopted to achieve continuous, or more nearly continuous, control of the output voltage. At the expense of greater complexity, the number of taps may be increased to the extent necessary to provide the required range with adequate resolution, an extra pair of thyristors being added for each tap. An alternative configuration, which uses more thyristors for a given number of steps but needs voltage ratings related only to the individual winding voltages, is the series arrangement shown in Figure 3.27. In an elegant variant of this latter circuit, the winding voltages are graded in a binary sequence, enabling a good resolution to be obtained with the minimum number of windings and thyristors; for example, winding voltages in the proportions 1:2:4:8:16 will afford a range of 31 % with a resolution of 1 %.

A further approach is to combine the tap-changing principle with that of phase control, switching from the lower-voltage to the higher-voltage tap in every half-cycle at an instant that can be varied over a range of up to 180°; the resulting output voltage waveform is shown in Figure 3.28. This represents a compromise, giving continuous control, but generating harmonics in proportion to the range of voltage variation; thus (in a single-phase circuit) a range of 20 % in voltage control would imply a maximum third-harmonic component of 6.4 % of maximum voltage instead of 30 % for a simple series regulator with a maximum delay angle of 77°. Yet another possibility is to apply the principle of integral-half-cycle

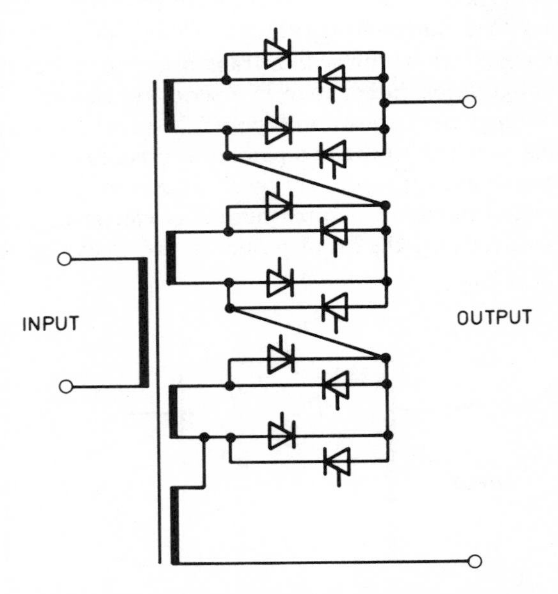

Figure 3.27 Multi-step series tap-changer regulator.

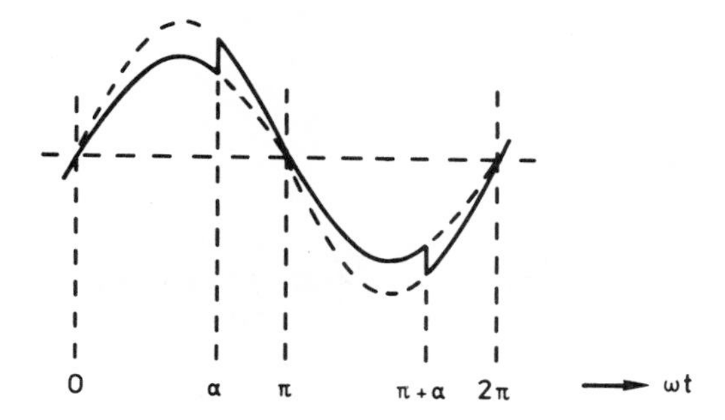

Figure 3.28 Output voltage waveform of a phase-controlled thyristor tap-changer.

control, connecting the load cyclically to two taps with time-ratio control on a time-base which is sufficiently long in comparison with the supply period to afford the resolution required.

In general, all these principles can be applied to switching circuits on the primary side of the transformer, so long as the total range of control is fairly small, and in some cases this may be advantageous, in that it enables the thyristors to operate at higher voltage and lower current.

In all systems of this type, control of the thyristors is essentially a matter of firing one pair of thyristors or another, according to the desired potential of the output line. In practice, control systems have to be designed with some care, and some form of logic has generally to be employed, to ensure that, while the thyristors are always in a condition to conduct when required, there is no possibility of a short circuit across a transformer winding through two thyristors—e.g., through Th_1 and Th_4 in Figure 3.26. This and the relatively large number of thyristors needed have so far tended to discourage the wide use of tap-changing systems.

CHAPTER 4

D. C. Switching Regulators

A switching regulator varies its average output, relative to its input, by varying the proportion of its operating time during which the output is connected to the input. While this broad definition logically embraces controlled rectifiers and a.c. regulators described in other chapters, the regulators considered in this chapter are those which provide a d.c. output from a d.c. supply, and, lacking the facility of natural commutation afforded by an a.c. supply, employ forced commutation where necessary to effect the required switching functions.

An elementary switching regulator is illustrated in Figure 4.1, wherein the unspecified switching device S operates with a regular periodic time T_p, and is closed for a time $T_a = \gamma T_p$ in each period. It will be assumed throughout this discussion that switching devices have no losses—that is, they conduct with zero voltage-drop, pass no leakage current when blocking and switch from either state to the other in zero time. Thus the instantaneous output voltage in Figure 4.1 is either the input voltage V_i or zero, and the mean and r.m.s. output voltages are

$$\left.\begin{aligned} \bar{V}_o &= V_i \frac{T_a}{T_p} = \gamma V_i \\ V_o &= V_i \sqrt{\left(\frac{T_a}{T_p}\right)} = \sqrt{\gamma} V_i \end{aligned}\right\} \tag{4.1}$$

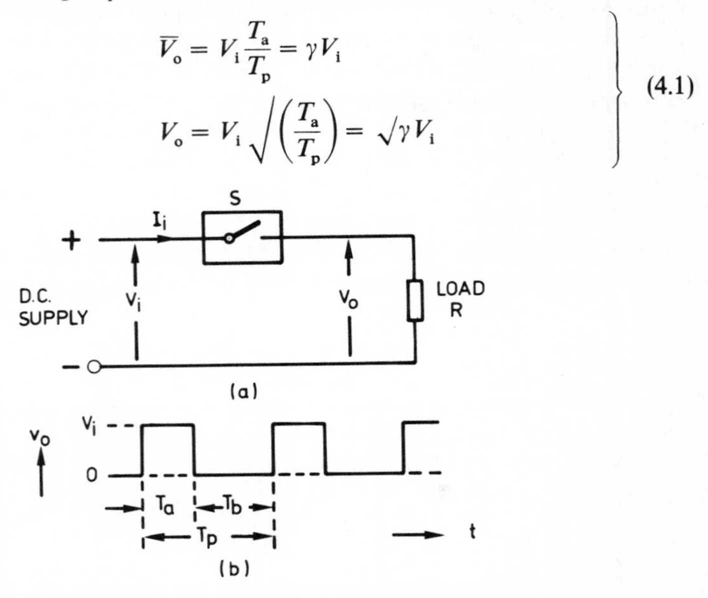

Figure 4.1 (a) Elementary d.c. switching regulator; (b) Waveform of output voltage.

Such a simple circuit is of limited practical use, since it can supply only a resistive load, and the large proportion of ripple in the output is not normally acceptable.

To smooth the output requires the addition of inductance, and possibly a capacitor, as in Figure 4.2. A second switching device is also needed, to carry the current flowing in the inductance when it is not drawn from the supply: S_1 and S_2 are synchronized in operation so that while one is conducting the other is blocking. The waveform of voltage at the output of the series switch S_1, i.e., the voltage v_o across the shunt switch S_2, is now the same as that of the output voltage in Figure 4.1, but the current in the load, instead of rising to V_i/R when S_1 is closed, tends to remain steady, because of the smoothing effect of L_o, at a level $\overline{V}_o/R$. Similarly, when S_1 is blocking and S_2 is conducting, the same current flows in the free-wheeling path through S_2.

If it is assumed for the moment that the output current is smoothed completely by an infinite inductance, then

$$\left.\begin{aligned} \text{Output voltage } \overline{V}_o &= \gamma V_i \\ \text{Output power } P_o &= \overline{V}_o I_o \\ \text{Input current } \overline{I}_i &= \gamma I_o \\ \text{Input power } P_i &= V_i \overline{I}_i = V_i \gamma I_o = P_o \end{aligned}\right\} \quad (4.2)$$

The equality of input and output power is a necessary consequence of assuming zero losses, and the equations demonstrate that the switching regulator, with smoothing, achieves in d.c. terms what a transformer achieves in an a.c. circuit—namely, a transformation of voltage accompanied by an inverse transformation of (mean) current. The truth of this statement is not directly affected by the fact that the input current may contain a high proportion of ripple, since it is assumed to flow in a d.c. source having no resistance, and is therefore not associated with any loss. The discontinuity of the input current means in fact that a low-impedance source is necessary, and in many practical circumstances an input filter (L_i, C_i) has to be provided; the input–output relationships may then be understood entirely in terms of substantially smooth voltages and currents.

In practice, the inductance in the load circuit is frequently not large enough to exclude from the output current an appreciable ripple component, which may be

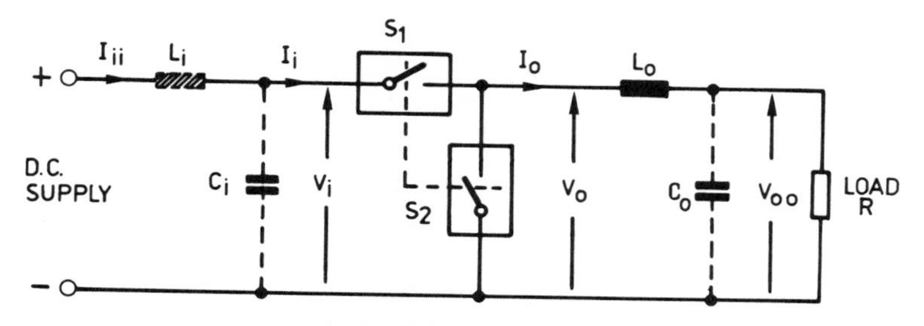

Figure 4.2 Switching regulator with smoothed output.

significant in several ways. Firstly it may significantly affect the total r.m.s. output current, and therefore the I^2R losses in the load and in the regulator; secondly it increases the instantaneous current which the series switch S_1 must be capable of switching off; thirdly, if the output current becomes discontinuous, the control chracteristic will in many cases be altered.

Two cases of practical interest may be identified broadly as that in which the voltage across the dissipative part of the load is substantially smoothed, as by a battery, a smoothing capacitor or a separately excited d.c. motor, and that in which the only smoothing is that provided by the inductance, as in the case of a regulator connected to a d.c. series motor.

Operation with smoothed output voltage is illustrated by the waveforms of Figure 4.3. The voltage applied to the inductance during the 'on' period of S_1 is $(V_i - \overline{V}_o)$, and during the 'off' period it is $-\overline{V}_o(V_{oo} = \overline{V}_o)$. The total excursion of current in L_o is therefore

$$\hat{I}_o - \check{I}_o = \left(\frac{V_i - \overline{V}_o}{L_o}\right) T_a = \frac{\overline{V}_o}{L_o} T_b \tag{4.3}$$

Also

$$\hat{I}_o + \check{I}_o = 2\overline{I}_o \tag{4.4}$$

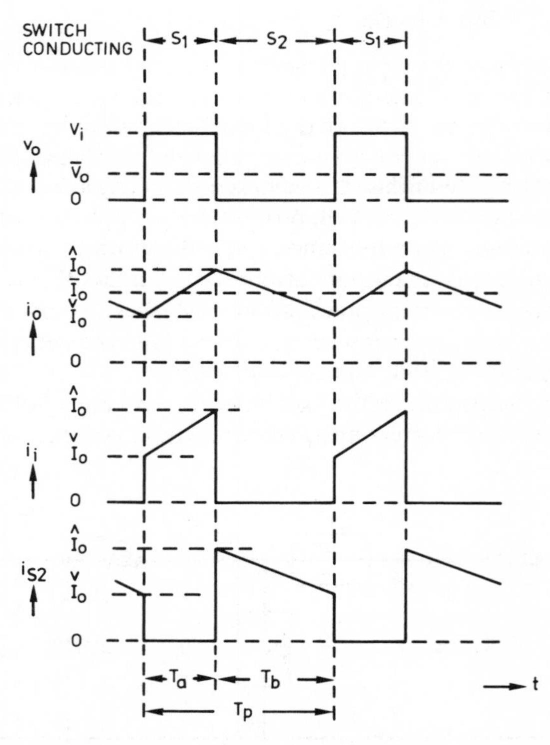

Figure 4.3 Waveforms in the d.c. switching regulator of Figure 4.2.

From these expressions the various current values of interest may be found.
Adding (4.3) and (4.4), the peak current is

$$\hat{I}_o = \bar{I}_o + \frac{\bar{V}_o}{2L_o} T_b = \bar{I}_o + \frac{V_i T_p}{2L_o}\gamma(1-\gamma) \tag{4.5}$$

Similarly,

$$\check{I}_o = \bar{I}_o - \frac{\bar{V}_o}{2L_o} T_b = \bar{I}_o - \frac{V_i T_p}{2L_o}\gamma(1-\gamma) \tag{4.6}$$

The r.m.s. ripple current is

$$I_{or} = \frac{\hat{I}_o - \check{I}_o}{2\sqrt{3}} = \frac{V_i T_p}{2\sqrt{3}L_o}\gamma(1-\gamma) \tag{4.7}$$

and the total r.m.s. current is

$$I_o = \sqrt{(\bar{I}_o^2 + I_{or}^2)} \tag{4.8}$$

The output voltage V_o across S_2 contains a ripple component whose r.m.s. value is given by

$$V_{or} = \sqrt{(V_o^2 - \bar{V}_o^2)} = V_i\sqrt{[\gamma(1-\gamma)]} \tag{4.9}$$

The maximum values of the expressions for $\hat{I}_o$ and I_{or} above, of interest for design purposes, occur when $\gamma = 0.5$, assuming $\bar{I}_o$ to be independent of γ.

Harmonic content of output and input

The distribution of output harmonics is easily found by Fourier analysis of the waveform of the voltage v_o across S_2 (Figure 4.2) (see Appendix 4(i)). A complete spectrum of harmonic voltages is present, except in the extreme cases $\gamma = 0$ and $\gamma = 1$ and when γ has particular values for which individual components fall to zero. The magnitude of the nth harmonic of the switching frequency is

$$V_{on} = V_i \frac{\sqrt{2}\sin n\gamma\pi}{n\pi} \tag{4.10}$$

If the impedance of the output capacitor or the load is negligible, the magnitude of any harmonic component of the output ripple current is simply a function of the harmonic voltage and the impedance of L_o:

$$I_{on} = \frac{V_{on}}{\omega_n L_o} = \frac{V_{on}}{n\omega_1 L_o} \tag{4.11}$$

Graphs showing the functions of (4.9) and (4.10), and the form of those of (4.7) and (4.11) are given in Figures 4.4 and 4.5 respectively.

At the input, if the output current is completely smoothed, the waveform of current is similar to that of v_o, so the relative magnitudes of the harmonics are the same:

$$I_{ir} = I_o\sqrt{[\gamma(1-\gamma)]}$$

$$I_{in} = I_o\frac{\sqrt{2}\sin n\gamma\pi}{n\pi} \tag{4.12}$$

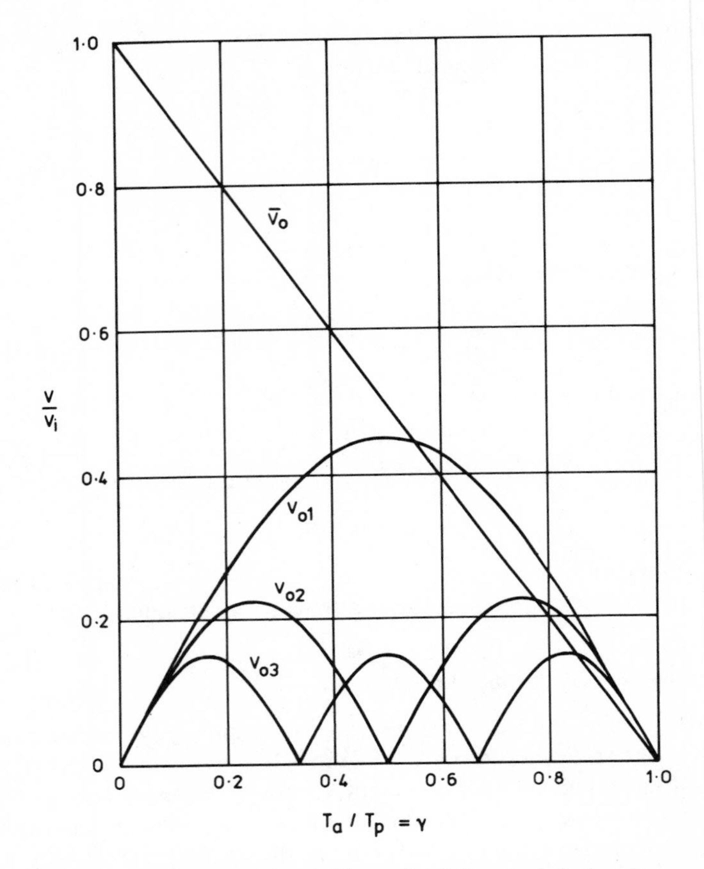

Figure 4.4 Variation with switching ratio of the output ripple voltage from a d.c. (step-down) switching regulator.

In various circumstances, both the ripple voltage in the smoothed output and the ripple current in the smoothed input are likely to be of interest. As with a rectifier, the ripple voltage across the load at any particular harmonic frequency is approximately the product of the harmonic voltage before smoothing and the attenuation due to the filter $L_o - C_o$: i.e.

$$V_{oon} \approx V_{on}\left(\frac{X_{Cn}}{X_{Ln} + X_{Cn}}\right) = \frac{V_{on}}{(\omega_n/\omega_0)^2 - 1} \tag{4.13}$$

where ω_0 is the resonant frequency of the filter, $= 1/\sqrt{(L_o C_o)}$. From the point of view of the total ripple voltage, it will rarely be necessary to consider more than the fundamental frequency, although higher frequencies may be significant in regard to interference, for example with communications or signalling systems.

In the same way the ripple current in the d.c. supply is approximately the product of harmonic components of the unsmoothed input current and the

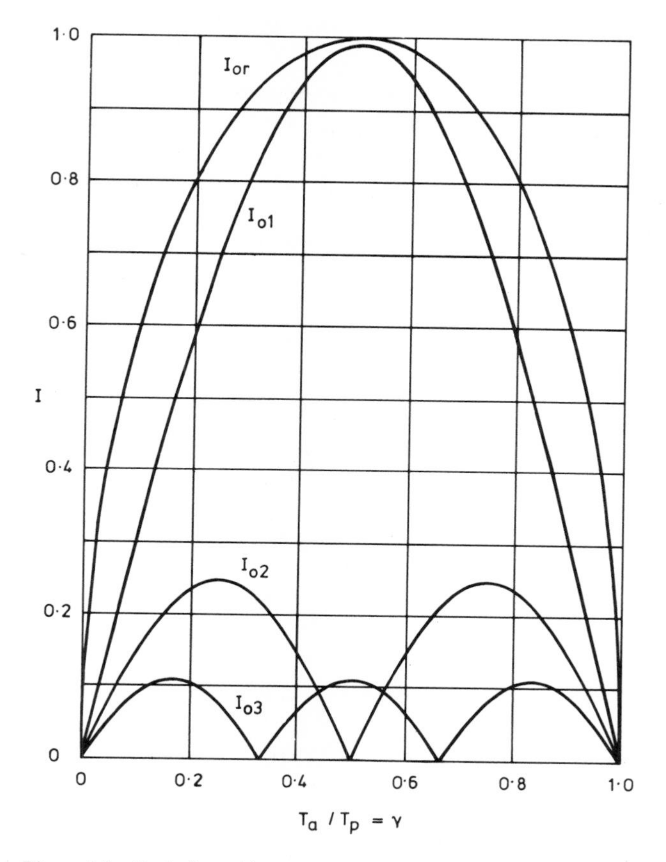

Figure 4.5 Variation with switching ratio of the output ripple current from a d.c. (step-down) switching regulator with smoothed output. (Arbitrary current scale.)

attenuation in the input filter $L_i - C_i$:

$$I_{iin} \approx I_{in}\left(\frac{X_{Cn}}{X_{Ln}+X_{Cn}}\right) = \frac{I_{in}}{(\omega_n/\omega_0)^2 - 1} \tag{4.14}$$

It should be noted that even when no particular attenuation is required a check on the resonant frequencies of the supply and load circuits is nevertheless desirable, since a considerable magnification is possible in some circumstances.

Practical step-down regulator

The waveforms shown in Figure 4.3 indicate that under the conditions represented—i.e., the normal conditions in a voltage-reducing, or step-down, regulator—the function required of S_2 can be performed by a free-wheel diode, and only S_1 needs to be capable of interrupting forward current, or to be provided

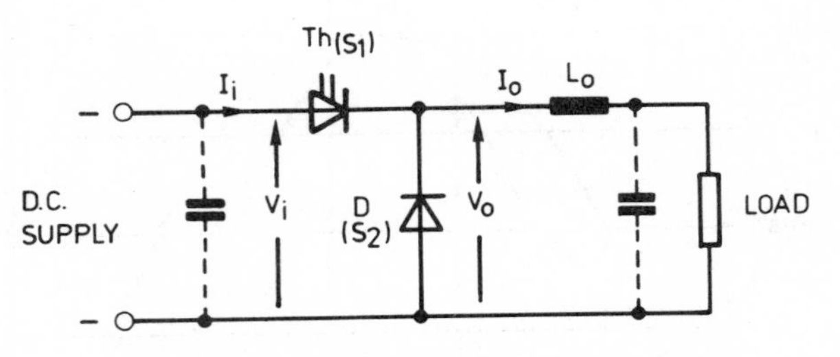

Figure 4.6 Practical step-down regulator.

with a means of forced commutation. Without particularizing as to the means of commutation, a practical step-down regulator is thus of the form shown in Figure 4.6; the symbol used for Th (S_1) in this diagram represents, here and elsewhere, a thyristor provided with any suitable means of forced commutation.

Discontinuous output current

The statement that a diode in the place of S_2 will function as described above rests upon the tacit assumption that the current that flows in it during the free-wheeling periods is never expected to reverse. However, if the instantaneous current excursion $\hat{I}_o - \check{I}_o$ is independent of $\overline{I}_o$ (4.5), there will be some value of $\overline{I}_o$, a function of L_o, below which $\check{I}_o$ as given by (4.6) is negative, and in this region, if S_2 is a diode, the output current in fact becomes discontinuous. Putting $\check{I}_o = 0$ in (4.6) gives a critical value for the product $\overline{I}_o L_o$:

$$\overline{I}_o L_o = \frac{V_i T_p}{2}\gamma(1-\gamma) \tag{4.15}$$

In the region of discontinuous output current, when i_o has fallen to zero and neither switch is conducting, the smoothed output voltage V_{oo} appears across S_2, and the waveforms are as illustrated in Figure 4.7. The mean output voltage is not now, as in (4.1), a simple function of γ, but depends on T_c as well as T_a. In terms of γ and the circuit parameters, the output voltage may be determined as follows:

$$\hat{I}_o = \frac{(V_i - \overline{V}_o)\gamma T_p}{L_o} \quad \text{and} \quad \overline{I}_i = \frac{\gamma \hat{I}_o}{2} = \overline{I}_o \frac{\overline{V}_o}{V_i}$$

$$\therefore \quad \frac{V_i(1 - \overline{V}_o/V_i)\gamma T_p}{L_o} = \frac{2\overline{I}_o \overline{V}_o}{\gamma V_i} \tag{4.16}$$

whence

$$\frac{\overline{V}_o}{V_i} = \frac{1}{1 + \dfrac{2\overline{I}_o L_o}{\gamma^2 T_p V_i}} \tag{4.17}$$

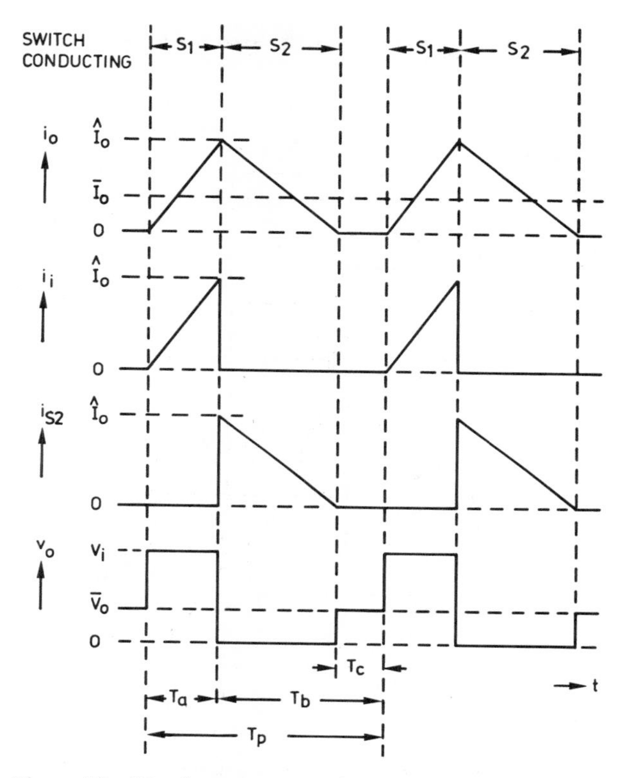

Figure 4.7 Waveforms in a step-down regulator with discontinuous output current.

Alternatively, replacing $\overline{I}_o$ by $\overline{I}_i V_i / \overline{V}_o$,

$$\frac{\overline{V}_o}{V_i} = 1 - \frac{2\overline{I}_i L_o}{\gamma^2 T_p V_i} \tag{4.18}$$

(a slight approximation if the output is not completely smoothed).

Regulator without output voltage smoothing

If the output circuit is represented simply as an inductance and a resistance in series, the sections of the output current waveform are exponential instead of linear, as illustrated in Figure 4.8.

During the 'off' period of S_1,

$$i_o = \hat{I}_o e^{-Rt/L_o}$$

$$\therefore \check{I}_o = \hat{I}_o e^{-RT_b/L_o} \tag{4.19}$$

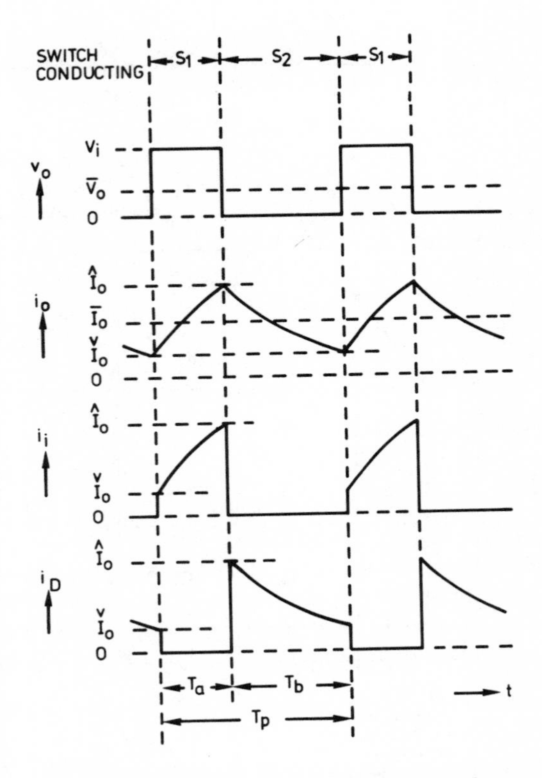

Figure 4.8 Waveforms in a step-down regulator with LR output circuit (unsmoothed output voltage).

During the 'on' period,

$$i_o = \check{I}_o + \left(\frac{V_i}{R} - \check{I}_o\right)(1 - e^{-Rt/L_o})$$

$$\therefore \; \hat{I}_o = \check{I}_o + \left(\frac{V_i}{R} - \check{I}_o\right)(1 - e^{-RT_a/L_o}) \tag{4.20}$$

Then from (4.19) and (4.20),

$$\hat{I}_o = \hat{I}_o e^{-RT_b/L_o} + \left(\frac{V_i}{R} - \hat{I}_o e^{-RT_b/L_o}\right)(1 - e^{-RT_a/L_o})$$

$$= \frac{V_i}{R}(1 - e^{-RT_a/L_o}) + \hat{I}_o e^{-RT_p/L_o}$$

whence $$\hat{I}_o = \frac{V_i}{R}\left(\frac{1 - e^{-RT_a/L_o}}{1 - e^{-RT_p/L_o}}\right)$$

and
$$\check{I}_o = \frac{V_i}{R}\left(\frac{1-e^{-RT_a/L_o}}{1-e^{-RT_p/L_o}}\right)e^{-RT_b/L_o} \tag{4.21}$$
$$= \frac{V_i}{R}\left(\frac{e^{RT_a/L_o}-1}{e^{RT_p/L_o}-1}\right)$$

Without a smoothed output voltage, the minimum output current $\check{I}_o$ as given by (4.21) never drops to zero. The r.m.s. ripple current may if necessary be derived from the expressions of (4.19) and (4.20); such an effort is, however, rarely justified, since the harmonic content of V_o is the same as in the case of a smoothed output voltage and the impedance of the load circuit to harmonics differs only by the inclusion of the load resistance. Thus the error in estimating the total r.m.s. current as if the output voltage were smoothed will be small so long as the load time-constant is not much shorter than T_p.

Modes of control

There is no fundamental reason why any of the three periods T_a, T_b or T_p should be regarded as fixed, since only the ratio T_a/T_p affects the output voltage directly, and a practical system may well operate with a fixed repetition frequency, or with a fixed 'on' time, or with some variable regime. In general the choice will be affected by some or all of the following:

(i) external constraints upon frequency, such as the need to avoid interference with communications systems;
(ii) required performance—i.e. speed of response;
(iii) limitations imposed by preferred circuit techniques;
(iv) efficiency, as affected by switching and associated losses;
(v) component cost;
(vi) size and weight.

Apart from the first two, these factors are to a considerable degree interrelated. Increasing the frequency, for example, enables the smoothing inductor to be made smaller and cheaper, but tends to reduce efficiency by increasing losses in the forced-commutation circuits, and may necessitate the use of more expensive thyristors and other components. An optimum design represents a compromise between such conflicting considerations.

STEP-UP REGULATORS

So long as the switches S_1 and S_2 are conceived in appropriate terms, nothing in the basic operation of the switching regulator as discussed above requires the direction of current and power flow to be from the higher-voltage circuit to the lower, and the fundamental relationships are readily adaptable if the load is replaced by a d.c. supply and the d.c. supply by a load circuit including a capacitor or other means of energy storage, as in Figure 4.9. To avoid unnecessary

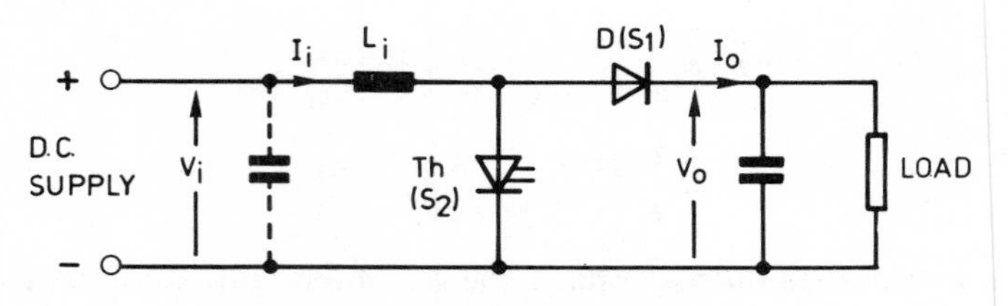

Figure 4.9 Step-up switching regulator.

complication it will be assumed that both the input and output voltages are smooth, and the waveforms of current are therefore linear, as in Figure 4.10, which is essentially a rearrangement of Figure 4.3 with the currents reversed. In this mode of operation the functions of S_1 are performed by a simple diode, and it is S_2 which has to be capable of interrupting forward current.

The excursions of current in the input inductor may be evaluated in a manner similar to that employed for the step-down regulator:

$$\hat{I}_i - \check{I}_i = \frac{(V_o - V_i)T_a}{L_i} = \frac{V_i T_b}{L_i} \tag{4.22}$$

$$\hat{I}_i + \check{I}_i = 2\bar{I}_i \tag{4.23}$$

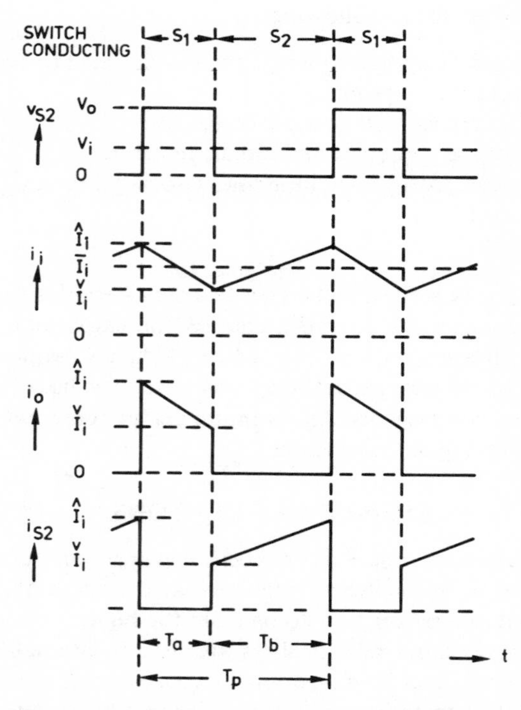

Figure 4.10 Waveforms in the step-up regulator of Figure 4.9.

If $T_a/T_p = \gamma$, as before (note that in this case T_a is the time for which the diode conducts),

$$V_i = \gamma V_o, \quad \text{or} \quad V_o = \frac{V_i}{\gamma} \tag{4.24}$$

Then

$$\hat{I}_i = \overline{I}_i + \frac{V_i T_p}{2L_i}(1-\gamma) = \overline{I}_i + \frac{V_o T_p}{2L_i}\gamma(1-\gamma) \tag{4.25}$$

and

$$\check{I}_i = \overline{I}_i - \frac{V_i T_p}{2L_i}(1-\gamma) = \overline{I}_i - \frac{V_o T_p}{2L_i}\gamma(1-\gamma) \tag{4.26}$$

For continuous conduction in L_i,

$$L_i \overline{I}_i \not< \frac{V_o T_p}{2}\gamma(1-\gamma) = \frac{V_i T_p}{2}(1-\gamma) \tag{4.27}$$

Given continuous conduction, the r.m.s. ripple current in the input inductor is

$$I_{ir} = \frac{\hat{I}_i - \check{I}_i}{2\sqrt{3}} = \frac{V_i T_p}{2\sqrt{3}\,L_i}(1-\gamma) \tag{4.28}$$

and the harmonic components of the input current are

$$I_{in} = \frac{\sqrt{2}\,V_o \sin n\gamma\pi}{n^2 \pi \omega_1 L_i} \tag{4.29}$$

(cf. equations (4.10) and (4.11)).

The output voltage with discontinuous input current is obtained as follows (waveforms are illustrated in Figure 4.11):

$$\hat{I}_i = \frac{V_i T_b}{L_i} \quad \text{and} \quad \overline{I}_i - \overline{I}_o = \frac{\hat{I}_i(1-\gamma)}{2} \tag{4.30}$$

$$\therefore \quad \frac{V_i(1-\gamma)T_p}{L_i} = \frac{2(\overline{I}_i - \overline{I}_o)}{1-\gamma} = \frac{2\overline{I}_o(V_o/V_i - 1)}{1-\gamma} \tag{4.31}$$

whence

$$\frac{V_o}{V_i} = 1 + \frac{(1-\gamma)^2 T_p V_i}{2\overline{I}_o L_i} \tag{4.32}$$

Alternatively,

$$\frac{V_o}{V_i} = \frac{1}{1 - \dfrac{(1-\gamma)^2 T_p V_i}{2\overline{I}_i L_i}} \tag{4.33}$$

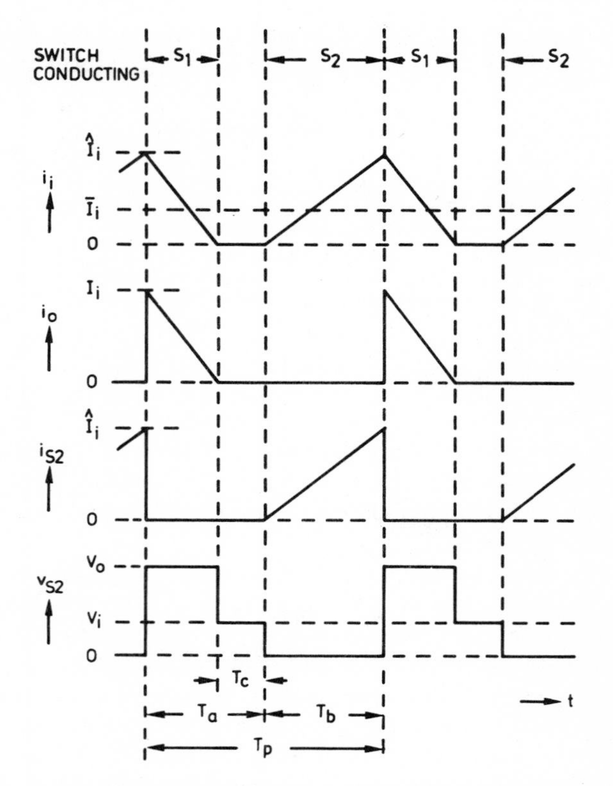

Figure 4.11 Waveforms in a step-down regulator with discontinuous input current.

BI-DIRECTIONAL AND FOUR-QUADRANT REGULATORS

The elements of the step-down regulator (Figure 4.6) and the step-up regulator (Figure 4.9) can be combined, as in Figure 4.12, to form a bi-directional converter—i.e. one in which power and current can flow from either d.c. circuit to the other, depending on the balance of voltages (V_1 must exceed V_2). Thus the armature voltage and speed of a d.c. motor might be controlled up to the limit of a

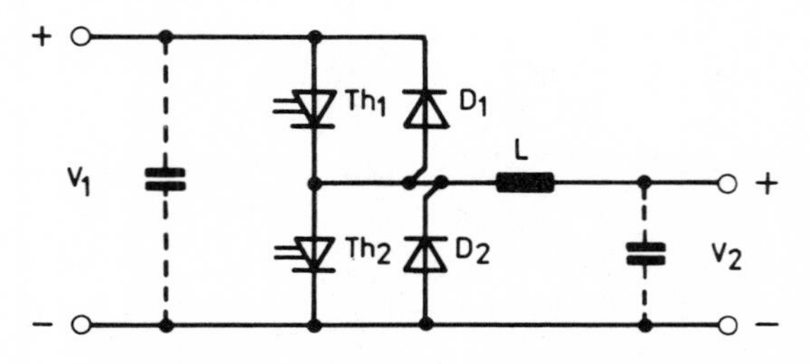

Figure 4.12 Bi-directional switching regulator.

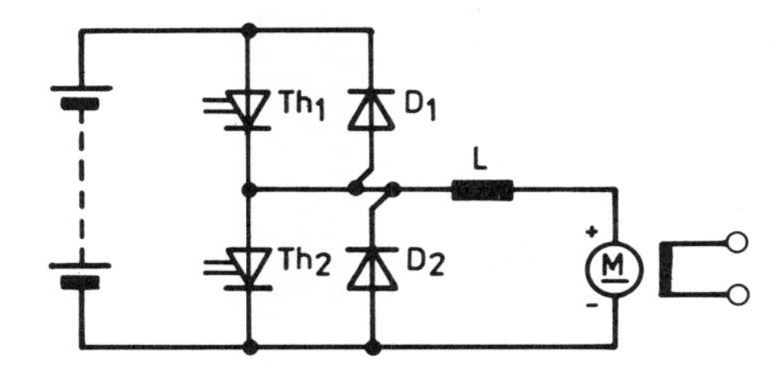

Figure 4.13 Bi-directional switching regulator used to control a d.c. motor with either direction of current.

given supply voltage regardless of the direction of armature current (and therefore of torque) by using the circuit of Figure 4.13. (Alternatively, one thyristor only, with its associated forced-commutation circuit, could be used in conjunction with a contactor to switch the circuit into the configuration appropriate to the direction of power flow: the combined circuit has the advantage of faster response.)

In the bi-directional converter, assuming that the thyristors are supplied with gate current for the whole of their appropriate periods, the question of operation with discontinuous current does not arise. When Th_1 receives gate current, for example, its cathode is effectively connected to the positive side of the higher-voltage supply (Figure 4.12) whether current is flowing in it or in the opposite direction through D_1; similarly when Th_2 is fed with gate current its anode is connected to the negative side with either direction of current. Thus the combination of a thyristor and a diode in parallel functions in this arrangement in effect as a bi-directional switch, and it is possible for the current in the inductor to alternate without its affecting the basic relationship $\overline{V}_2 = \gamma \overline{V}_1$.

In the terminology used in connection with naturally commutating converters, the circuit of Figure 4.12 is capable of operating in two quadrants—two directions of current for a single polarity of voltage. To permit operation in four quadrants without recourse to reversing contactors, additional switching elements are necessary, leading to the bridge circuit of Figure 4.14. In this arrangement, if Th_1 and Th_2 operate with a switching ratio γ_1 and Th_3 and Th_4 operate with a switching ratio γ_2, the mean potential of point X, relative to the

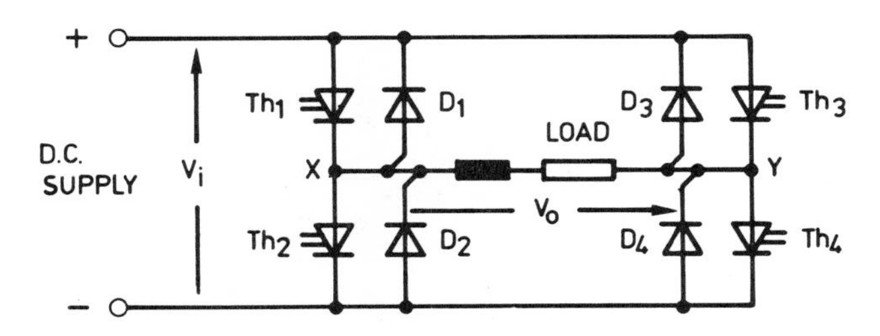

Figure 4.14 Four-quadrant switching regulator.

negative supply terminal, is $\gamma_1 V_1$, and that of point Y $\gamma_2 V_i$, regardless of the output current; the mean output voltage is thus

$$\overline{V}_o = V_i(\gamma_2 - \gamma_1) \tag{4.34}$$

and can be controlled continuously from $-V_i$ to $+V_i$. A typical application for such a circuit is the control of a separately excited d.c. motor with either direction of rotation or torque.

MULTI-PHASE SWITCHING REGULATORS

As a means of reducing the ripple generated at input and output, generally at high power levels, a switching regulator may comprise two or more channels in parallel, operating out of phase so that harmonic cancellation occurs in a manner analogous to the operation of a multiplex rectifier.

A two-phase step-down regulator of this type is shown in Figure 4.15: at (a) the outputs of the two channels are combined through an interphase reactor, which is followed by a single smoothing inductor, while at (b) a similar effect is obtained with two smoothing inductors. The waveforms of Figure 4.16 illustrate the operation of the circuit with an interphase reactor, assuming continuous and smooth current in the smoothing inductor. If the interphase reactor is so designed that its magnetizing current is small compared with the output current, the currents in the two halves of the winding are nearly equal, and each channel therefore carries approximately half the output current continuously, and functions in the manner described above with reference to a single channel.

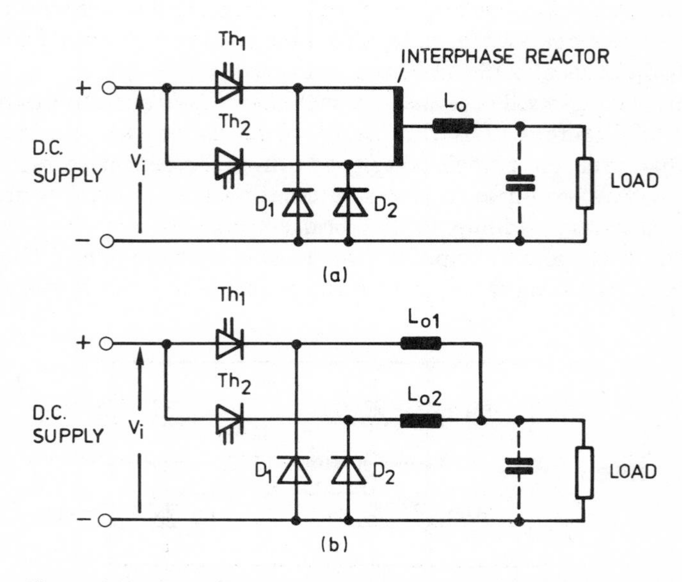

Figure 4.15 Two-phase step-down switching regulators: (a) with interphase reactor; (b) with separate smoothing inductors.

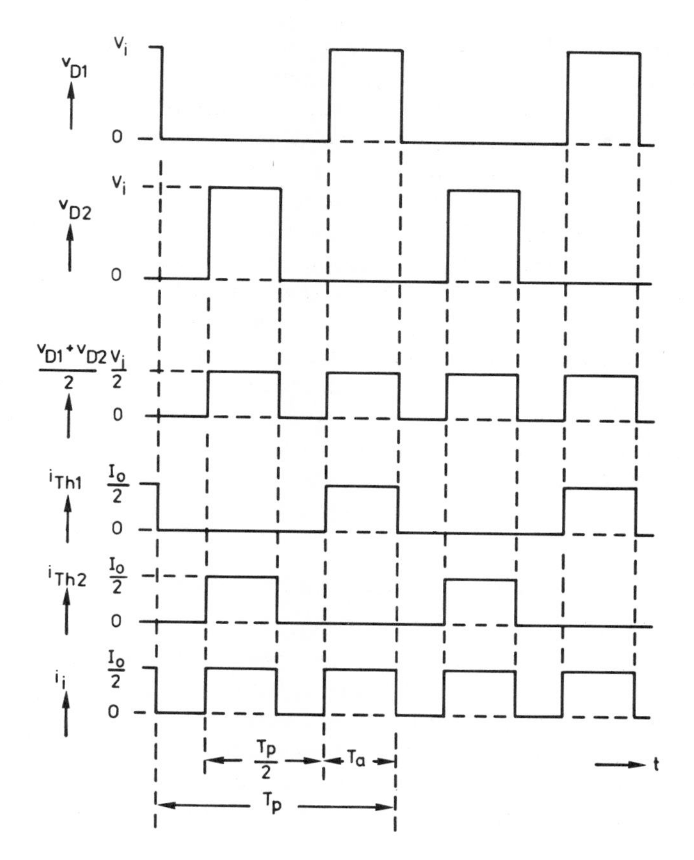

Figure 4.16 Waveforms in a two-phase switching regulator (Figure 4.15).

Figure 4.16 shows the voltages v_{o1} and v_{o2} produced across the two free-wheel diodes with a particular value of γ (about 0.3): as a result of transformer action in the interphase reactor, which ensures that the voltages across the two halves of the winding are substantially equal, the voltage at the input to the smoothing inductor is the average of the diode voltages, i.e. $(v_{o1} + v_{o2})/2$ and the magnitude of the resultant ripple voltage is half that of the ripple voltage across either diode. Similarly, the addition of the two thyristor currents at the input results in a ripple current whose magnitude is only half that which would be produced by a single-channel regulator with the same output current. A similar reduction occurs for other values of γ, unless $\gamma = 0.5$, in which case the fall of voltage across one diode coincides with the rise of voltage across the other, and both the output voltage and the input current are ideally completely smooth. The average output voltage of the complete regulator is the same as that of either channel operating individually—i.e. γV_i.

The principle may be extended to any number of channels. With m channels, operating with a time displacement between one channel and the next of T_p/m, i.e. with a phase displacement of $2\pi/m$ radians at the fundamental switching frequency f_1, the output and input waveforms of the complete regulator repeat at a frequency mf_1, and the harmonic components of the input current and the output voltage are therefore restricted to frequencies of mf_1 and multiples thereof. These components are the result of the co-phasal addition of the components of those frequencies in the inputs and outputs of the individual channels, since the phase displacement at the nth harmonic is $2\pi n/m$, which is effectively zero for values of n which are multiples of m: for all other values of n the resultant is zero. The relative amplitudes of the harmonics that remain are thus the same as for any one of the individual channels, and the overall result is to multiply the fundamental ripple frequency by m and to reduce the maximum magnitude of the ripple by the same factor. By comparison, a single-channel regulator operating at m times the switching frequency would produce the same fundamental ripple frequency, but would not give any reduction in ripple magnitude.

COMMUTATION CIRCUITS

So far in this chapter the fundamental principles of d.c. switching regulators have been discussed without reference to the details of the methods employed to effect the necessary forced commutation of thyristor current. The operation of typical circuits may be illustrated by some selected examples.

As explained in Chapter 1, forced commutation circuits are essentially of two kinds—those in which a substantial reverse voltage is applied to the thyristor to be turned off and those in which the thyristor is shunted by a reverse-connected diode and the discharge current from the commutating capacitor is limited by an inductor. The circuits most commonly used in simple switching regulators are of the first kind.

The turning-off of a conducting thyristor by triggering an auxiliary thyristor to connect a suitably charged capacitor across it is basically straightforward, and variety in practical circuits is principally associated with the means whereby the capacitor is charged to the appropriate voltage and with the appropriate polarity when required.

A simple circuit which, if not particularly useful, serves to illustrate some of the problems to be considered, is shown in Figure 4.17, with representative voltage waveforms in Figure 4.18. The mode of operation is as follows: when the main thyristor Th_1 is fired at $t = 0$, current flows from the positive supply terminal to the load, and the commutating capacitor C is charged positively (i.e. with the upper plate in the diagram positive) through R, nearly to the supply voltage.

Subsequently, at t_2, the auxiliary (turn-off) thyristor Th_2 is fired and the capacitor is connected across Th_1, applying a reverse voltage to it for a period T_q, $t_2 - t_3$, which depends on the initial charge on the capacitor and the load current i_o: taking the initial voltage on C as equal to V_i, and neglecting the effect of

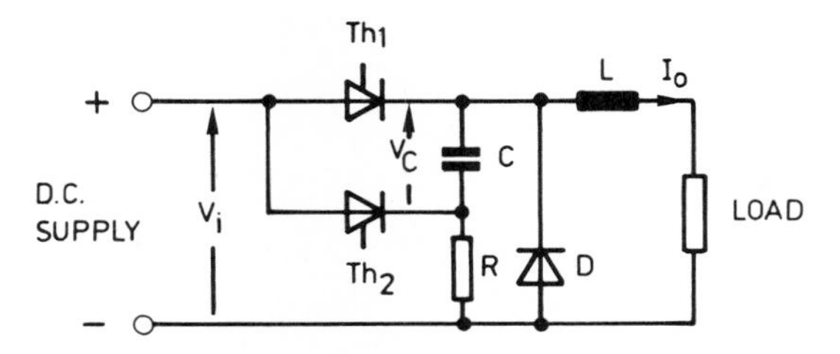

Figure 4.17 Step-down switching regulator with resistive capacitor-charging circuit.

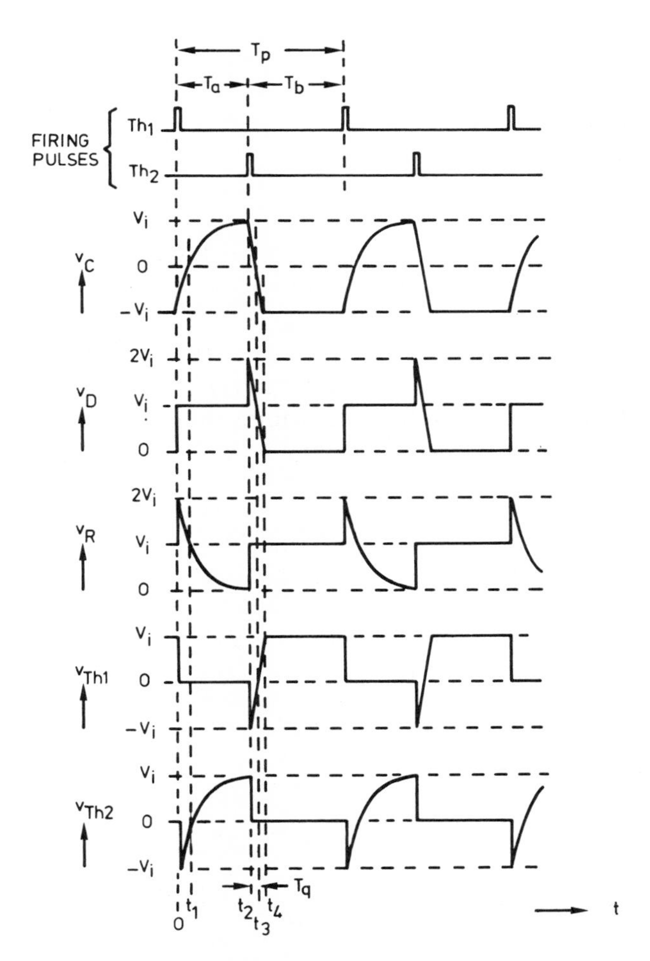

Figure 4.18 Waveforms in a d.c. switching regulator with resistive capacitor-charging circuit (Figure 4.17).

stored charge in Th_1,

$$T_q = \frac{CV_i}{i_o} \tag{4.35}$$

Assuming the circuit to be so designed and operated that this turn-off interval is adequate, Th_1 then assumes a blocking state and the load current, now flowing in the capacitor, charges it with reversed polarity until at t_4 the voltage across it reaches $-V_i$ and further charging is prevented by the diversion of the output current to the free-wheel diode D.

Although the capacitor current ceases at t_4, Th_2 continues to conduct current through R until Th_1 is again fired to repeat the cycle as from $t = 0$. The capacitor is then connected across Th_2 to turn it off by the application of reverse voltage during the interval 0–t_1 as the charge is reversed by current through R. The turn-off interval in this case is a function of the time constant RC:

$$T_{q(Th2)} = -RC \log_e 0.5 \approx 0.7\, RC \tag{4.36}$$

This is not usually a critical design factor, since the current in R is less than the maximum value of i_o.

The main practical objections to this type of circuit are the power loss entailed in the charging of the capacitor in preparation for turning off the main thyristor and the limitations imposed on the operating conditions by the effect of this loss on the efficiency of the circuit. The capacitor is charged exponentially through R during the conducting period T_a of Th_1, and if it is to be charged nearly to the input voltage T_a must be at least several times the time-constant RC; C being determined in accordance with V_i, i_o and the turn-off time of Th_1, the maximum permissible value of R is proportional to the minimum required value of T_a, which for a given range of γ is inversely proportional to operating frequency. During the 'off' period, R is connected across the supply through Th_2, and this represents a power loss dependent on γ and approximately proportional to $1/R$. There is thus a conflict between operating frequency and range of control on the one hand and efficiency on the other.

Suppose $T_{a\,min} = gRC$ (g is likely to be at least 3)

$$\text{then (from (4.28))}\ R = \frac{T_{a\,min}}{gC} = \frac{T_{a\,min}}{g i_o T_q} \tag{4.37}$$

$$\text{Dissipation in } R \approx \frac{(1-\gamma)V_i^2}{R}$$

$$= g(1-\gamma)V_i i_o \frac{T_q}{T_{a\,min}} \tag{4.38}$$

Figure 4.19 shows a commutating circuit in which losses associated with charging the capacitor are avoided completely (apart from inevitable imperfections in components). Th_1 is again the main thyristor; Th_2–Th_5 are auxiliary thyristors, fired in diagonal pairs—i.e. Th_2 simultaneously with Th_5 and Th_3 with

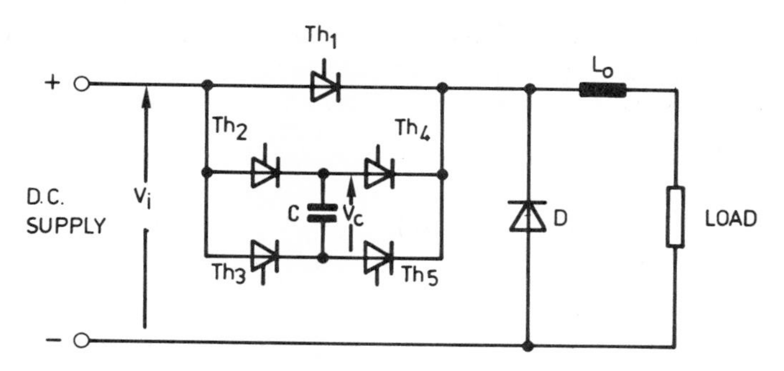

Figure 4.19 Switching regulator with 'lossless' commutation circuit.

Th_4. If Th_2 and Th_5, say, have been fired, the capcitor will be charged positively to V_i (the upper plate in the diagram positive) and all the auxiliary thyristors will have regained their blocking state. If now Th_1 is conducting and is to be turned off, Th_3 and Th_4 are fired, and the capacitor is thereby connected with the appropriate polarity across Th_1. At the end of the commutation process, corresponding to t_4 in Figure 4.18, the capacitor is again charged to V_i, but negatively, and the turn-off process may then be repeated by firing Th_2 and Th_5, which brings the circuit back to its original condition without the need for any distinct capacitor-charging operation. In comparison with the circuit of Figure 4.17, this arrangement not only functions with greater efficiency and with less restriction on frequency, but imposes no constraint upon the minimum conducting period of the main thyristor.

The more commonly used circuit of Figure 4.20 operates by virtue of a distinct capacitor-recharging process, but uses a resonant circuit to achieve a short recharging time with relatively high efficiency. The cycle may be considered to start with C charged positively to V_i as a result of previous conduction of the auxiliary thyristor Th_2. The firing of the main thyristor Th_1, as well as connecting the output to the positive input terminal, at the same time completes the resonant circuit L_1–C through diode D_1. A (very nearly) sinusoidal pulse of current with a duration $\pi\sqrt{(L_1C)}$ thereupon flows round this loop; as a result C is charged to

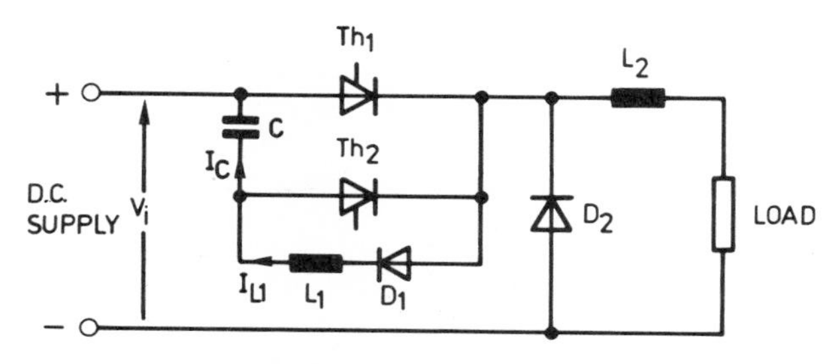

Figure 4.20 Switching regulator with resonant charge-reversal circuit.

nearly the original voltage but with reversed polarity—i.e. with its upper plate negative; waveforms representing this process are shown in Figure 4.21. Assuming the loss in the resonant circuit to be low—i.e. assuming its Q factor to be reasonably high, but finite—

$$i_{L1} \approx \frac{V_i}{\sqrt{(L_1/C)}} e^{-\omega_0 t/2Q} \sin \omega_0 t$$

and

$$v_C \approx V_i e^{-\omega_0 t/2Q} \cos \omega_0 t \tag{4.39}$$

where $\omega_0 = 1/\sqrt{(L_1 C)}$.

Thus when the current in the loop falls to zero after a time interval π/ω_0, the capacitor is charged to a voltage $-e^{-\pi/2Q} V_i$. Diode D_1 prevents a reversal of the current, and the capacitor retains its charge, now of the polarity required to turn off Th_1, until Th_2 is fired again and the cycle repeated. The waveforms of Figure 4.22 illustrate the operation of the complete circuit. The turn-off interval is slightly reduced by the loss of capacitor voltage entailed in the charge-reversal process, and becomes

$$T_q = \frac{C V_i e^{-\pi/2Q}}{i_o} \tag{4.40}$$

As with other circuits in which the commutating capacitor is charged during the conducting period of the main thyristor, this mode of operation sets a lower limit to the 'on' time, which must not be less than $\pi\sqrt{(L_1 C)}$. The choice of L_1, for

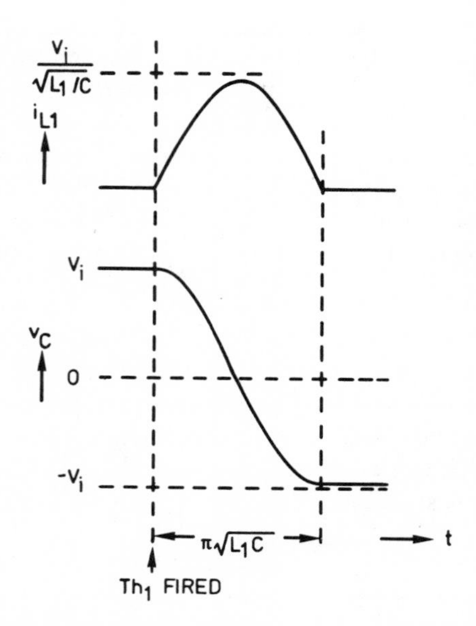

Figure 4.21 Waveforms in the charge-reversal circuit of Figure 4.20.

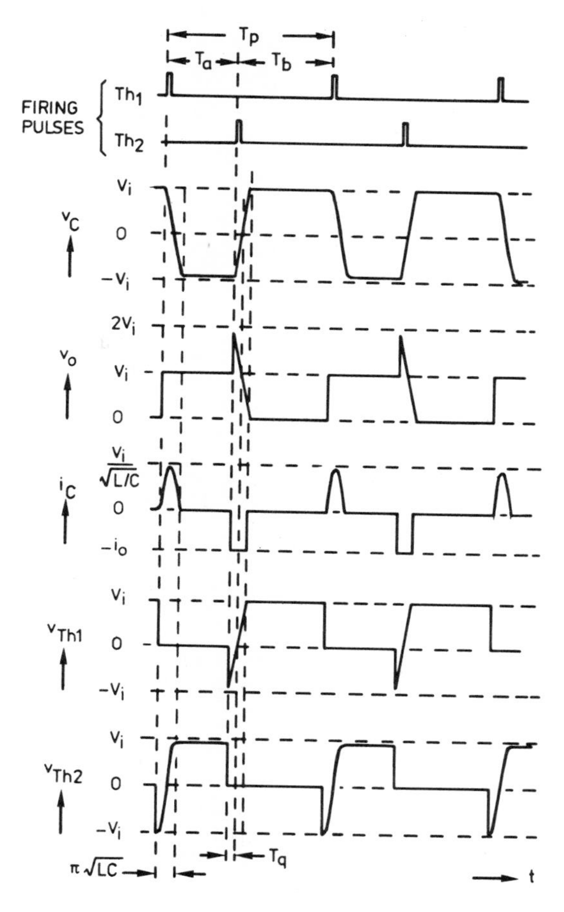

Figure 4.22 Waveforms in a d.c. switching regulator with resonant capacitor charging (Figure 4.20).

a given value of C, rests on a compromise between this limitation and the current ratings of C and Th_1, in that the peak charging current (which adds to the output current in Th_1) is inversely proportional to the charging time.

A feature of all circuits in which the main thyristor is turned off by the application of a substantial reverse voltage is the contribution of that voltage to the output of the regulator. Apart from its considerable effect on the peak output voltage, reference to the waveform of v_o in Figure 4.18 or 4.22 will show that the mean output voltage is strictly not γV_i but

$$\bar{V}_o = V_i \frac{T_a + 2T_q}{T_p} \tag{4.41}$$

(approximately, if the capacitor is not charged quite to V_i). Thus in addition to the effect of any limitation on T_a, the charge on the capacitor gives rise to a lower limit on the mean output voltage:

$$\bar{V}_{\text{o min}} = V_\text{i} \frac{2T_\text{q}}{T_\text{p}} = \frac{2V_\text{i} V_C C}{i_\text{o} T_\text{p}} \tag{4.42}$$

where V_C is the capacitor voltage—i.e. V_i (ideally) in the circuits of Figure 4.17 and 4.19, or $V_\text{i}\text{e}^{-\pi/2Q}$ in that of Figure 4.20.

Since $\bar{V}_{\text{o min}}$, from (4.42), is inversely proportional to the output current, there is implied a minimum output power, which can be related directly to the operating frequency $f\,(= 1/T_\text{p})$ and the apparent stored energy represented by the charged capacitor in series with the d.c. supply:

$$\bar{P}_{\text{o min}} = \tfrac{1}{2}C(V_\text{i} + V_C)^2 f \tag{4.43}$$

Commutation with inverse parallel diode

Compared with the commutation circuits described above, those in which the main thyristor is shunted by a reversely connected diode have several advantages. They do not subject the load to peak voltages in excess of the supply voltage, the turn-off interval does not increase in inverse proportion to the load current and the effect of the commutation circuit on the minimum output obtainable is much less significant.

A commutating circuit of this type, in effect a modification of the circuit of Figure 4.20, is shown in Figure 4.23, with representative operating waveforms in Figure 4.24. Operation is started by firing Th_1; in addition to connecting the load to the positive terminal of the supply, this completes a circuit through D_2, L_1 and C, and as a result a substantially sinusoidal pulse of current flows to charge the capacitor to a voltage approaching twice the supply voltage. Actually,

$$\hat{V}_C = V_\text{i}(1 + \text{e}^{-\pi/2Q}) \tag{4.44}$$

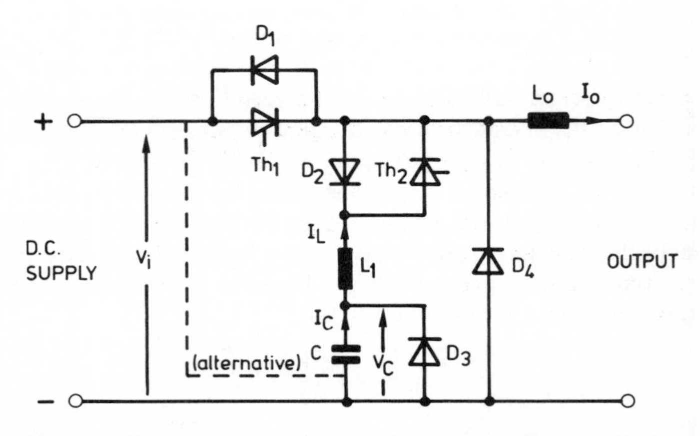

Figure 4.23 Switching regulator with inverse-diode commutation circuit.

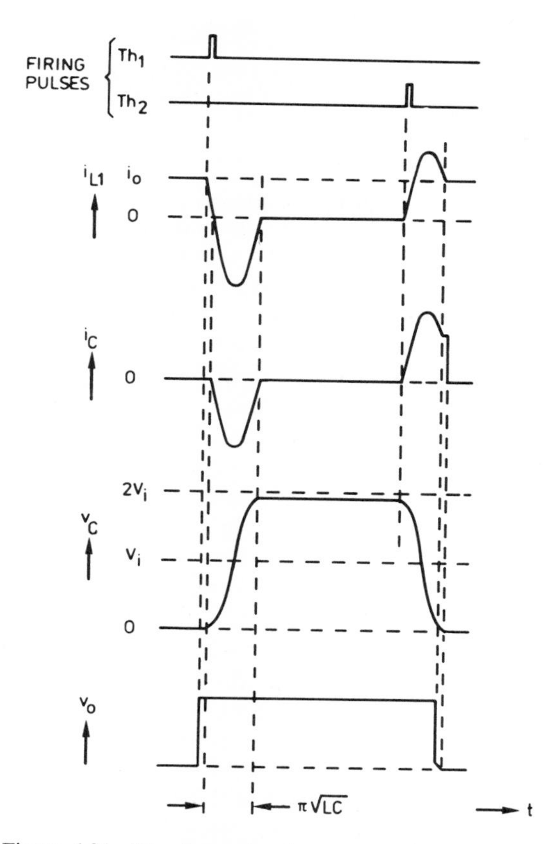

Figure 4.24 Waveforms in the commutation circuit of Figure 4.23.

To turn off Th_1, the auxiliary thyristor Th_2 is fired, completing a circuit to discharge the capacitor through D_1, the peak current being

$$\hat{I}_C = \frac{\hat{V}_C}{\sqrt{(L_1/C)}} e^{-\pi/(4Q)} = \frac{V_i}{\sqrt{(L_1/C)}} (e^{-\pi/(4Q)} + e^{-3\pi/(4Q)}) \qquad (4.45)$$

(It is particularly important in this circuit that the d.c. source should be of low impedance, since it forms part of the commutation circuit; it is possible, however, to avoid this requirement by substituting the alternative connection shown in the broken line in Figure 4.23.)

The nominal 'on' time of Th_1, T_a, must not be less than $\pi\sqrt{(L_1 C)}$ to allow the full pulse of capacitor charging current to flow, and the output voltage pulse is not terminated until the capacitor discharge current falls to the level of i_o. The minimum effective 'on' time is thus something less than $2\pi\sqrt{(L_1 C)}$, typically of the order of four times the desired value of T_q, at any output current.

Towards the end of the discharge period of the capacitor, when i_C has fallen to i_o, i_C remains equal to i_o until v_C reaches zero, when i_{L1} commutates to D_3 and thereafter decays to zero as a result of losses or continues to flow (as indicated in Figure 4.24) until Th_1 is fired again.

Switching regulators without auxiliary thyristors

A class of thyristor switching regulators exists, and is of some significance, in which commutation is effected, not through the agency of an auxiliary thyristor, but automatically by virtue of resonance in a suitably arranged commutation circuit. There is a greater diversity of such circuits than can reasonably be described here, but a typical principle, based on a linear resonant circuit, may be exemplified by considering the two-thyristor circuit previously described with reference to Figures 4.23 and 4.24 with the auxiliary thyristor Th_2 and its parallel diode D_2 removed—i.e., with L_1 connected directly to the cathode of Th_1. The effect is then as if Th_2 were always fired exactly at the end of the first half-sine-wave of current in the capacitor following the firing of Th_1, so that the 'on' time is almost invariable at a little less than the period of oscillation of L_1 and C.

Circuits of this kind suffer from the fundamental limitation that the only mode of control available is that of frequency variation, and from the disadvantage of high volt-ampere ratings in the reactive components, since at maximum mean output voltage L_1 and C are carrying a continuous alternating current of an r.m.s. value not very different from the maximum d.c. load current. There is some compensation, however, in the simplicity of the system.

A more elegant arrangement embodies a saturable reactor, preferably with a sharply saturating square-loop core, in place of the linear inductor L_1. The rapid transition from the high-inductance unsaturated state to the low-inductance saturated condition amounts in effect to a switching operation, and permits much more independent determination of the 'on' time and the turn-off interval of the thyristor. Figure 4.25 shows a simple circuit of this ferro-resonant type, omitting, in this case, the diode in parallel with the thyristor; waveforms illustrating its operation are shown in Figure 4.26.

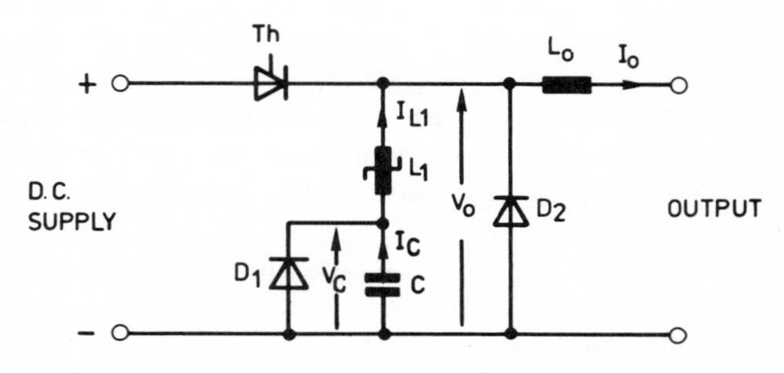

Figure 4.25 Switching regulator with ferro-resonant commutation circuit.

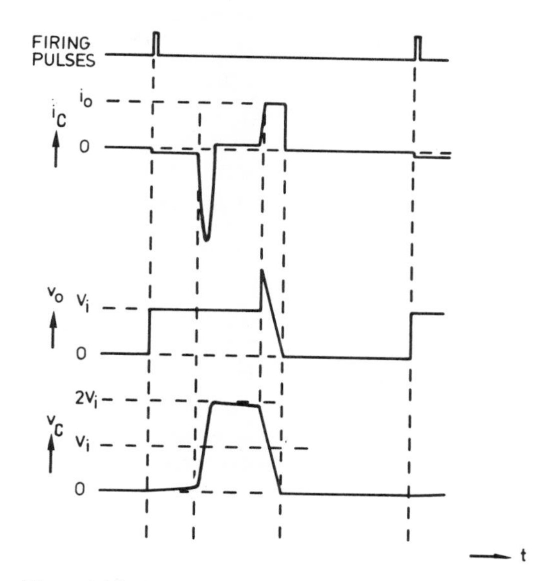

Figure 4.26 Waveforms in the regulator of Figure 4.25.

Referring to Figure 4.25, assuming C to be initially discharged and the instantaneous output voltage zero, firing the thyristor applies the input voltage to L_1, which from the previous cycle has been left in what will be termed a state of positive saturation, associated with a positive (upward) direction of i_{L1}. The resulting current in L_1 is a small negative (downward) magnetizing current as the flux in the core moves towards saturation in the opposite direction, and v_C rises only minimally. After a time interval such that the voltage–time integral across the winding corresponds to twice the saturation flux of the core, L_1 saturates in the negative direction, and a half-sine-wave of current, determined in magnitude and duration by the saturated inductance of the reactor and the value of C, charges C to approximately $2V_i$; this reverses the voltage across L_1, which resumes its unsaturated condition as the core flux moves back towards its original positive saturation value. When the core again saturates, the capacitor is connected more-or-less directly to the cathode of the thyristor, and assuming the capacitance to be adequate the normal turn-off process ensues, and the circuit is left in its original condition.

By suitable dimensioning of the saturable reactor, the saturating time, and hence the 'on' time of the thyristor, can be determined at any desired value, within reason, while the turn-off interval is governed only by the capacitance. There are a number of variants of this system, notably one attributable to R. E. Morgan, in which the saturable reactor carries an additional winding in series with the load, and the charging of the capacitor is thereby enhanced by transformer action.

Two- and four-quadrant regulators

Functionally, two- and four-quadrant regulators such as those illustrated in Figures 4.12 and 4.14 are hardly distinguishable from voltage-fed forced-commutation inverters, and similar considerations apply to their commutation circuits. For an example of a suitable commutating system, reference may be made to the 'McMurray' inverter discussed in Chapter 5 (Figure 5.14).

APPENDIX 4(i)

FOURIER ANALYSIS OF SWITCHING REGULATOR WAVEFORMS

The waveform representing ideally the output voltage of a step-down regulator, or the thyristor voltage in a step-up regulator, is shown in Figure 4.27, arranged for convenience so that the harmonic analysis yields only cosine terms. If the waveform, of unit amplitude, is represented as

$$f(\omega t) = a_0 + a_1 \cos \omega t + a_2 \cos 2\omega t + \ldots + a_n \cos n\omega t + \ldots$$

then

$$a_0 = \gamma \text{ and } a_n = \frac{2}{\pi}\int_0^{\pi} f(\omega t) \cos n\omega t \, d\omega t \tag{4.46}$$

i.e.,

$$a_n = \frac{2}{\pi}\int_0^{\gamma\pi} \cos n\omega t \, d\omega t$$

$$= \frac{2}{\pi}\left[\frac{\sin n\omega t}{n}\right]_0^{\gamma\pi} = \frac{2}{n\pi} \sin n\gamma\pi \tag{4.47}$$

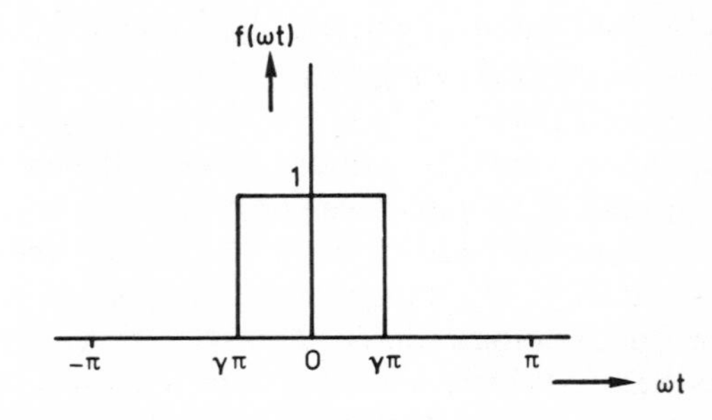

Figure 4.27 Generalized switching regulator waveform.

CHAPTER 5

Static Power Inverters

Inversion, i.e. the conversion of d.c. to a.c. power, is closely related to rectification, in that in general it entails the connection of the a.c. to the d.c. circuit through switching devices which are closed for appropriate periods relative to the a.c. voltage waveform. In rectification with diodes or thyristors, the opening and closing of the switching devices is effected, wholly or partly (depending on whether the rectifier is uncontrolled or controlled), by the a.c. supply voltage through the process known as natural commutation. The same process can be applied to inversion, as described in Chapter 2, when it is a question of feeding power from a d.c. source into a relatively large a.c. power system. When, however, it is required to provide a.c. power for a load from a d.c. supply, as the only power source, the existence of the a.c. system voltages necessary for natural commutation cannot be assumed *a priori*, and a different approach is required to contrive that the switching devices are turned off at the appropriate times.

There are, in fact, two basic approaches to this problem which between them cover the majority of inverters in use—disregarding, that is, various unorthodox or specialized designs which it is not the intention to consider here. In the first, the principle of natural commutation is set aside and replaced by that of forced commutation, whereby a means is assumed, or provided, of enabling the currents flowing in the switching devices to be commutated at any instants dictated by the control system, without regard to the voltages existing in the circuit: where the switching devices are thyristors, with no inherent ability to interrupt an established current, this implies the addition of forced-commutation circuits including capacitors, and, in most cases, auxiliary thyristors. In the second approach, the inverter is designed to operate by virtue of natural commutation on the assumption that the necessary a.c. output voltage exists, and the range of loading conditions and the operating strategy are constrained in such a way that the assumption is satisfied on a stable and repetitive basis. It will be seen that these two approaches lead to two distinct classes of inverter, known respectively as voltage-fed and current-fed inverters, each with its particular attributes, and a hybrid class which may be described as current-fed with voltage-fed characteristics.

VOLTAGE-FED INVERTERS

As in the case of d.c. switching regulators, it is convenient first to consider the principles of forced-commutation inverters without specific reference to the

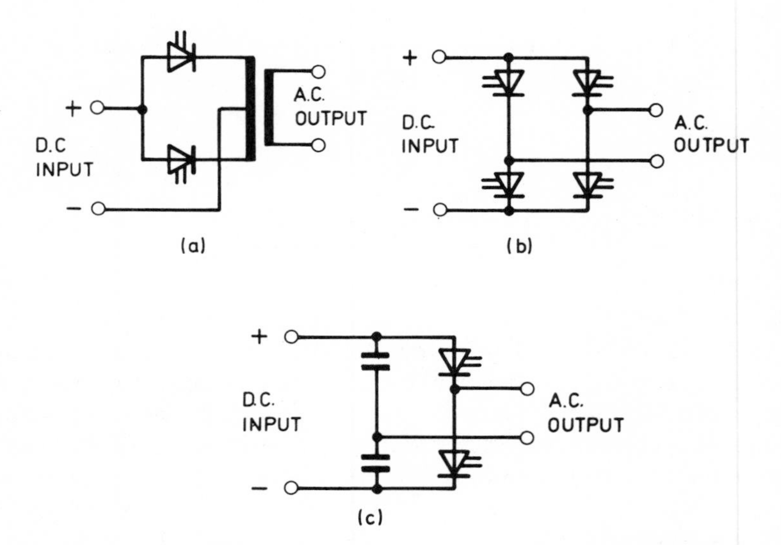

Figure 5.1 Basic forms of single-phase thyristor inverter: (a) bi-phase; (b) bridge; (c) half-bridge.

details of the forced-commutation circuits themselves, and the symbol introduced in Figure 4.6 will again be used to represent a thyristor provided with any suitable means of achieving commutation, when required, in the manner discussed in Chapter 1.

The configurations of inverters are essentially the same as those employed in rectifiers. Figure 5.1 shows the basic single-phase arrangements—the bi-phase, bridge and half-bridge circuits. Discussion will initially be directed to the bridge circuit, which is not complicated by the necessity for a transformer and can operate in some modes which are not possible in the bi-phase circuit, or in the half-bridge, on its own. The basic principles are, however, the same for all three.

Basic operation

The simplest mode of inversion is the generation of an alternating voltage of square waveform across a resistive load, as illustrated in Figure 5.2. The thyristors are turned on and off in diagonal pairs—i.e. Th_1 and Th_4 alternately with Th_2 and Th_3—each pair for one half-period of the desired output, so that the d.c. supply is connected across the load alternately in opposite directions. In practice, however, an inverter is rarely, if ever, required to supply a purely resistive load, and it is essential that it should be capable of operating in a satisfactory and predictable manner with loads which include greater or lesser amounts of reactance.

If the resistive load in Figure 5.2 is replaced by one which includes inductance in a series circuit, operation in the terms described above becomes impossible, since the turn-off of one pair of thyristors and the turn-on of the other pair at the end of a half-cycle of the voltage waveform implies an instantaneous reversal of the load

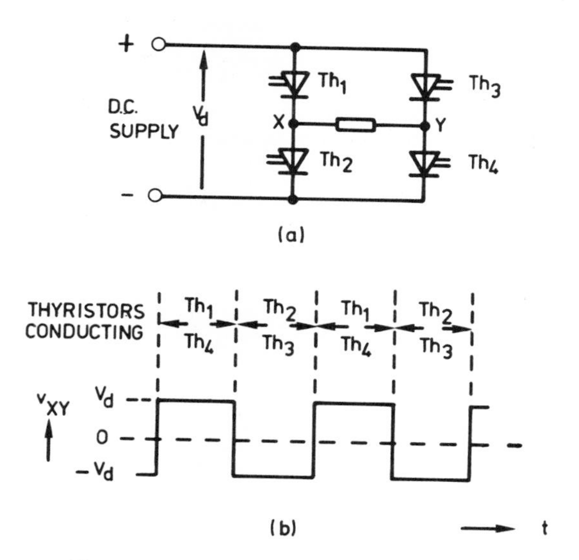

Figure 5.2 Bridge inverter with resistive load.

current which cannot be achieved without the generation of an infinite voltage across the inductance.

To put it another way, commutation of the output current is not possible, because there are not alternative paths for current in a given direction; operation as described for a resistive load is possible only with a load that permits the half-cycles of output current to coincide exactly with the half-cycles of voltage. Strictly speaking this does not exclude all loads that include reactance, but it does exclude virtually all practical loads, and, if it could not be circumvented, would represent an intolerable limitation on the usefulness of the circuit.

Reactive feedback diodes

The Normal solution to the problem of accommodating loads in which the current is not in step with the voltage is to add a diode in parallel with each thyristor, as shown in Figure 5.3. This prevents the instantaneous output voltage

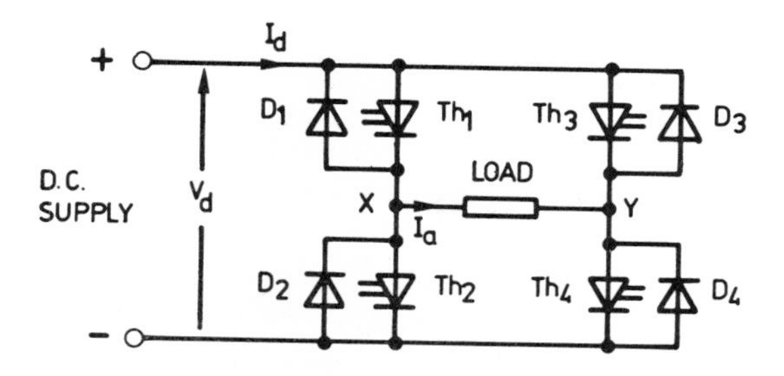

Figure 5.3 Bridge inverter with reactive feedback diodes.

from appreciably exceeding the supply voltage, and provides paths for load currents in opposition to the instantaneous output voltage.

In this arrangement (as in the case of the two-quadrant d.c. switching regulator described in Chapter 4) the combination of a thyristor and a parallel-connected diode has the effect of a fully controllable bi-directional switch: if, in Figure 5.3, Th_1 is in a conducting state and Th_2 is blocking, and load current is flowing from X to Y, point X is connected to the positive supply terminal through Th_1, while if the current is reversed point X is again connected to the positive terminal through D_1. Thus each output terminal is effectively connected to one side of the d.c. supply or the other, depending only upon the states of the thyristors, and the incorporation of what are commonly referred to as 'reactive feedback' diodes results in full control over the instantaneous output voltage of the inverter regardless of the behaviour of the load. As far as performance is concerned, this direct determination of output voltage is the distinguishing attribute of the voltage-fed inverter.

Bridge inverter with inductive load

Figure 5.4 illustrates the behaviour of the single-phase bridge inverter, provided with reactive feedback diodes, with a load that can be represented by inductance

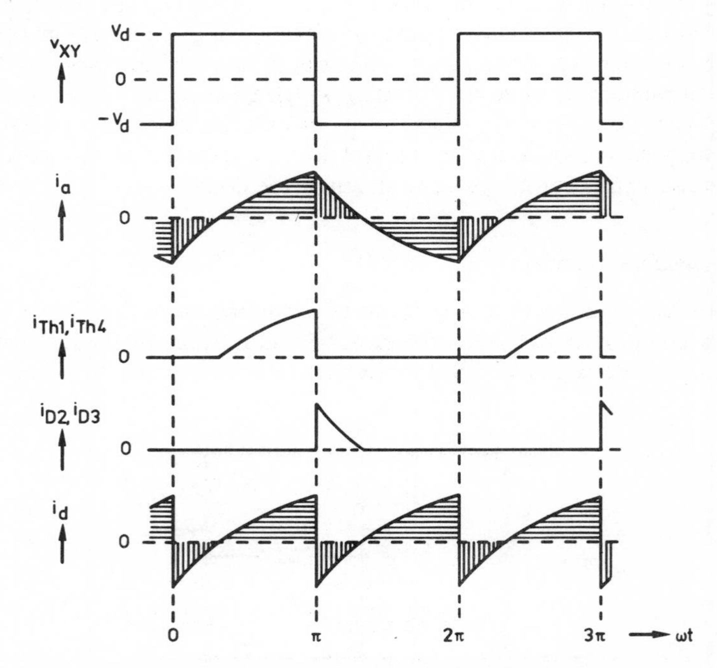

Figure 5.4 Voltage and current waveforms in a bridge inverter (Figure 5.3) with series *LR* load: ≡ thyristor current; ||| diode current.

and resistance in series. In accordance with the explanation in the previous paragraph, the out put voltage waveform is determined entirely by the pattern in which the thyristors are turned on and off; the voltage waveform, in conjunction with the impedance of the load, determines the current waveform, and the phase relationship between the current and voltage waveforms determines the apportionment of current between the thyristors and the diodes. The voltage waveform being square, the current waveform in this case comprises a sequence of exponentials.

For the purpose of considering the current loading of the thyristors and diodes in quantitative terms, it is convenient to postulate a load that draws a sinusoidal current, which, while it does not accurately represent many real loads supplied by a square-wave voltage source, is as representative as any, and is not too unlike some loads of practical importance. To make the discussion more general, one pair of arms of the bridge will be considered initially—a so-called half-bridge, which may constitute part of a complete bridge or may be a complete inverter in the half-bridge configuration already referred to (Figure 5.1(c)).

With an in-phase load (Figure 5.5), since the voltage and current zeros coincide, complete half-sine-waves of current flow alternately in the two thyristors, while

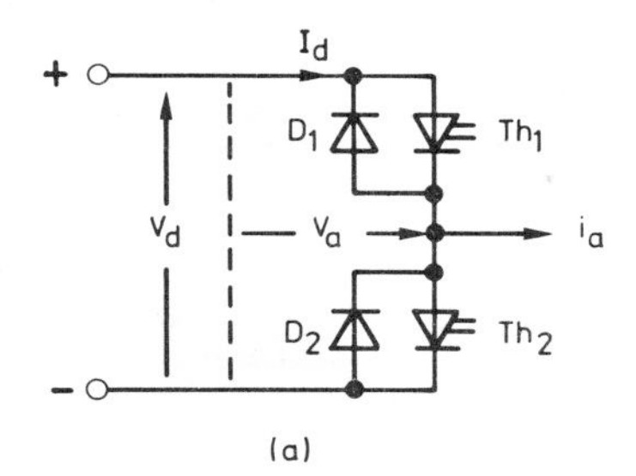

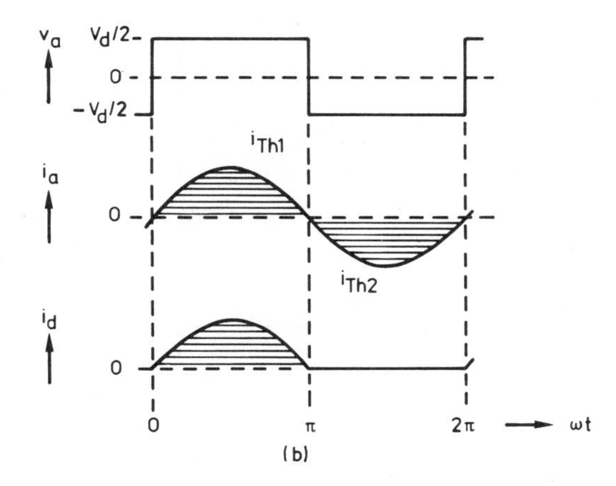

Figure 5.5 Inverter half-bridge with sinusoidal in-phase output current: ≡ thyristor current.

the diodes carry no current at all. If I_a is the output current, the mean thyristor current is

$$\overline{I}_{Th} = \overline{I}_d = \frac{\overline{I}_a}{2} = \frac{\sqrt{2}}{\pi} I_a \tag{5.1}$$

The output voltage of the half-bridge has an amplitude $V_d/2$ and hence an r.m.s. fundamental component $V_{a1} = \sqrt{2}\, V_d/\pi$. The output power is therefore

$$P_o = V_{a1} I_a = \frac{\sqrt{2}\, V_d}{\pi} I_a = \frac{V_d \overline{I}_a}{2} \tag{5.2}$$

This equates with the input power—necessarily, since it has been assumed that there are no losses.

With a purely reactive load—i.e. with the load current in quadrature with the output voltage (Figure 5.6)—quarter-cycles of load current flow consecutively in the thyristors and the diodes. Then,

$$\overline{I}_{Th} = \overline{I}_D = \frac{\overline{I}_a}{4} = \frac{I_a}{\sqrt{2}\pi} \tag{5.3}$$

and

$$\overline{I}_d = \overline{I}_{Th} - \overline{I}_D = 0 \tag{5.4}$$

In the case of a load drawing a current at a lagging phase-angle ϕ (Figure 5.7) the mean thyristor current is

$$\overline{I}_{Th} = \frac{1}{2\pi} \int_{\phi}^{\pi} \hat{I}_a \sin(\omega t - \phi)\, \mathrm{d}\omega t = \overline{I}_a \left(\frac{1 + \cos\phi}{4} \right) \tag{5.5}$$

while the mean diode current is

$$\overline{I}_D = \frac{1}{2\pi} \int_{\pi}^{\pi+\phi} \hat{I}_a \sin(\omega t - \phi)\, \mathrm{d}\omega t = \overline{I}_a \left(\frac{1 - \cos\phi}{4} \right) \tag{5.6}$$

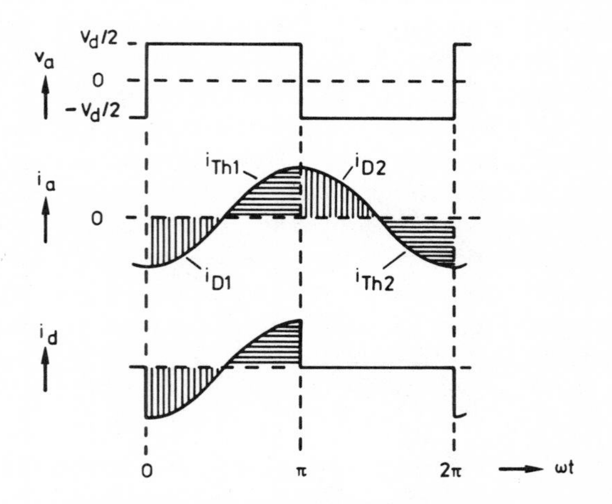

Figure 5.6 Waveforms in an inverter half-bridge (Figure 5.5(a)) with sinusoidal quadrature (lagging) output current: ≡ thyristor current; ||| diode current.

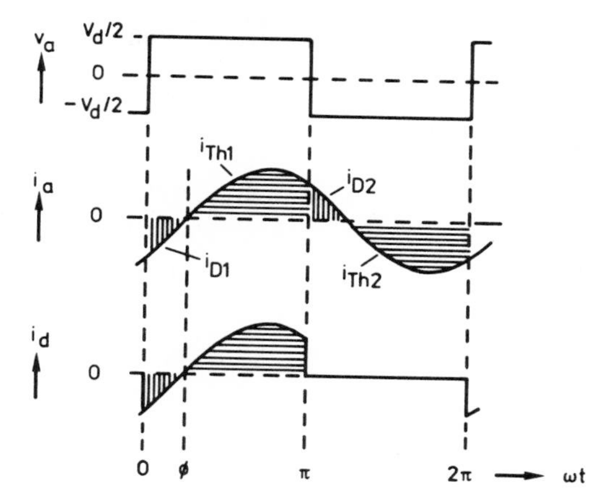

Figure 5.7 Waveforms in an inverter half-bridge (Figure 5.5(a)) with sinusoidal lagging output current: ≡ thyristor current; ||| diode current.

then the mean input current is

$$\overline{I}_{\mathrm{d}} = \overline{I}_{\mathrm{Th}} - \overline{I}_{\mathrm{D}} = \overline{I}_{\mathrm{a}} \frac{\cos\phi}{2} \tag{5.7}$$

Comparing the latter expressions with those for purely in-phase and reactive loads, $\overline{I}_{\mathrm{d}}$ in (5.7) corresponds to the in-phase component of load current $I_{\mathrm{a}} \cos\phi$ (as it must, to satisfy the power equation) while a mean current $\overline{I}_{\mathrm{a}}(1-\cos\phi)/4$ can be regarded as circulating between the thyristors and the diodes as a 'reactive' component corresponding to that expressed in (5.3); since, however, the 'active' and 'reactive' components of mean current on the d.c. side can only add arithmetically, the circulating current in the diodes is proportional to $(1-\cos\phi)$ and not to $\sin\phi$; the loading of the reactive feedback diodes is thus not commensurate with the reactive current drawn by the load.

In the same way that rectification of a sinusoidal input voltage in a rectifier gives rise to a ripple voltage at its output, rectification of a sinusoidal output current in the inverter produces a ripple component in the input current, which is increased (relative to the mean direct current) in the case of a reactive load, by the alternation of current between the thyristors and the diodes. The input current waveform in a complete single-phase bridge inverter with sinusoidal output current at phase-angle ϕ is illustrated in Figure 5.8; the mean input current (the sum of the mean input currents to the two half-bridges) is

$$\overline{I}_{\mathrm{d}} = \overline{I}_{\mathrm{a}} \cos\phi = \frac{2\sqrt{2}}{\pi} I_{\mathrm{a}} \cos\phi \tag{5.8}$$

while the total r.m.s. input current, independently of ϕ, is equal to I_{a}. The r.m.s. input ripple current is therefore

$$I_{\mathrm{dr}} = \sqrt{(I_{\mathrm{d}}^2 - \overline{I}_{\mathrm{d}}^2)} = I_{\mathrm{a}} \sqrt{\left(1 - \frac{8\cos^2\phi}{\pi^2}\right)} \tag{5.9}$$

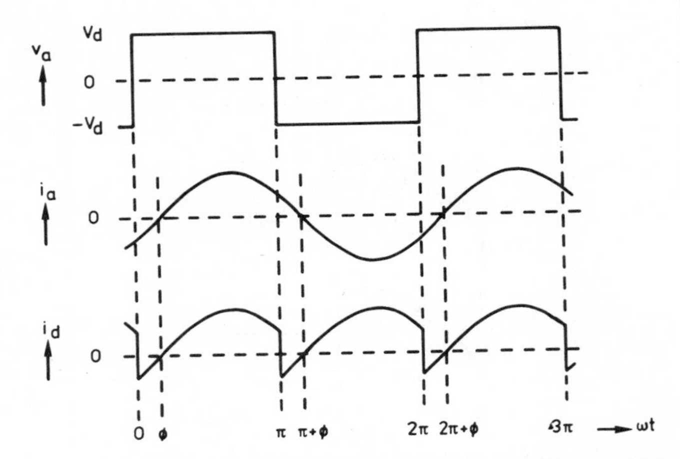

Figure 5.8 Waveforms in a single-phase bridge inverter with sinusoidal lagging output current.

In most cases it is necessary to make specific provision for the ripple current drawn by inverter, by connecting a capacitor across the input terminals. The present discussion assumes that the impedance of the d.c. supply is sufficiently low to obviate any significant ripple voltage; otherwise the output voltage waveforms would be modified.

Bridge inverter with leading load

With a leading sinusoidal output current, the behaviour of the inverter is generally similar to that described above with a lagging load, with the slight modifications illustrated in Figure 5.9; for a phase angle ϕ expressions (5.5)–(5.9) are equally applicable. An important difference, however, is that with leading output current the current in each thyristor falls naturally to zero after a conduction period $(\pi - \phi)/\omega$ and the thyristor experiences the voltage-drop of its associated diode as a reverse voltage for the remainder of the output voltage half-cycle. Provided, therefore, that the period ϕ/ω constitutes an adequate turn-off interval for the thyristor (with limited reverse voltage) the inverter will operate in this mode without forced commutation.

This ability of the voltage-fed inverter to operate in an uncomplicated and efficient manner with leading output current is of limited value in practice, since there are not many applications in which a leading load phase angle can be guaranteed under all forseeable steady-state and transient operating conditions. An important exception, however, is the generation of medium-frequency power for series-tuned induction heating loads, in which it is legitimate to adjust the operating frequency over a narrow range below the resonant frequency in order to maintain the necessary phase advance. This particular application is considered in a little more detail in Appendix 5(i).

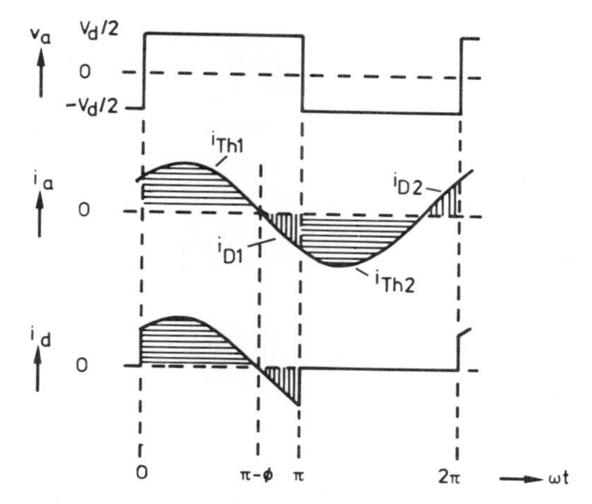

Figure 5.9 Waveforms in an inverter half-bridge (Figure 5.5(a)) with sinusoidal leading output current: ≡ thyristor current; ||| diode current.

Three-phase bridge inverter

By further analogy with conventional rectifier circuits, an inverter to generate a three-phase output may comprise a bridge of six thyristors, together with associated reactive feedback diodes. Such a bridge may conveniently be considered as an assembly of three half-bridges, each driven in the same manner as in the single-phase bridge, but with outputs mutually displaced in phase by $2\pi/3$ radians (Figure 5.10).

The output voltage waveforms in this circuit are illustrated in Figure 5.11, which shows the phase voltages v_X, v_Y and v_Z relative to the mid point of the d.c. supply and the corresponding interphase voltages. The phase voltages, thus defined, are the same as those in the single-phase inverter, and include a third-

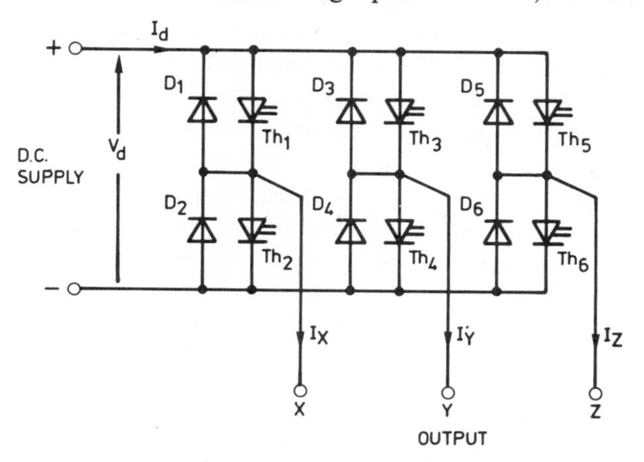

Figure 5.10 Three-phase bridge inverter.

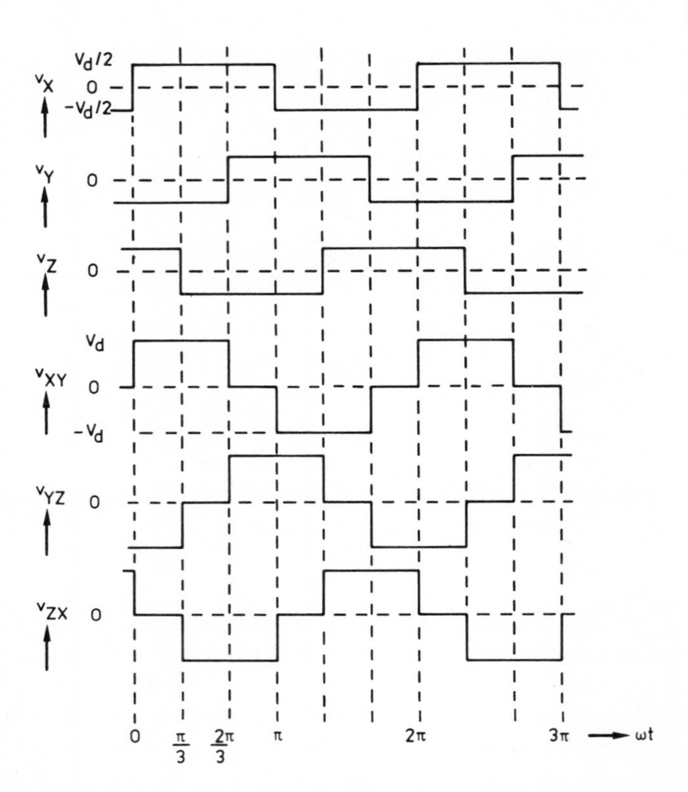

Figure 5.11 Output voltage waveforms in a three-phase bridge inverter (Figure 5.10).

harmonic component of one third of the amplitude of the fundamental component; the third harmonic does not, however, appear in the interphase output voltages, since these voltages are composed of pairs of phase voltages added with a phase displacement of $\pi/3$ at the fundamental frequency, and therefore of π at the third-harmonic frequency. The whole series of triplen harmonics is similarly eliminated, and the harmonic content that remains is that represented by the series $n = 6r \pm 1$, where r is any positive integer, the nth harmonic having an amplitude $1/n$ relative to the fundamental component. From inspection of the waveforms it can be seen that the interphase output voltage has a mean value of $2V_d/3$ and an r.m.s. value of $\sqrt{(2/3)}\ V_d$, while harmonic analysis shows that the fundamental r.m.s. component is $3/\pi$ times the total r.m.s. voltage—i.e. $\sqrt{6}\ V_d/\pi$.

Alternatively, the fundamental-frequency interphase voltage may be calculated as the resultant of the outputs of two half-bridges—i.e.

$$2\cos(\pi/6) \times \sqrt{2}\ V_d/\pi$$

If a balanced load is connected to the three-phase inverter, the neutral point of the load assumes a potential which at any instant is the mean of the three phase potentials, as shown in Figure 5.12, and an alternating voltage, of square

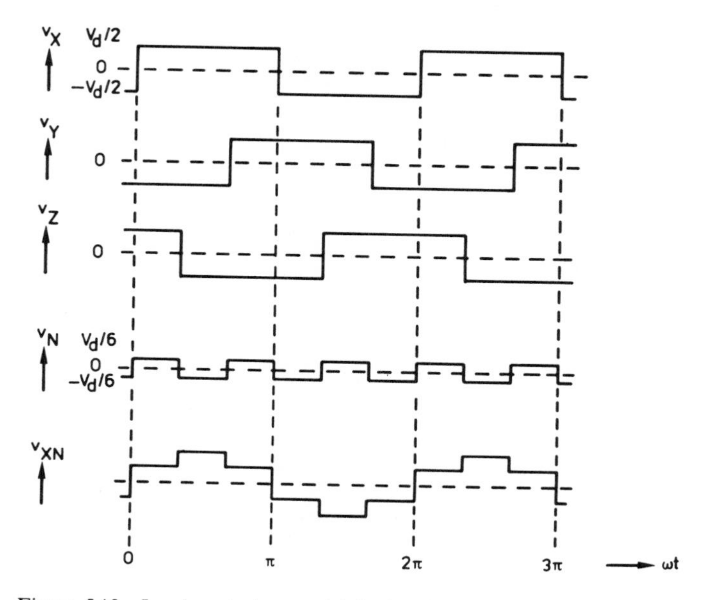

Figure 5.12 Load neutral potential (v_N) and phase voltage (v_{XN}) waveforms in a three-phase bridge inverter (Figure 5.10).

waveform, at three times the inverter output frequency, appears between the load neutral point and the midpoint of the d.c. supply; this voltage contains all the triplen-frequency components eliminated from the inverter output voltage. The load phase voltages contain the same harmonics as the interphase voltages, in the same proportions, but in a different pattern of phase relationships to the fundamental components.

With sinusoidal output currents and a load phase angle ϕ (lagging), the input current is of the form shown in Figure 5.13, representing the sum of the currents drawn by the three half-bridges. The mean input current (cf. (5.7)) is

$$\bar{I}_d = \bar{I}_a \frac{3\cos\phi}{2} = I_a \frac{3\sqrt{2}\cos\phi}{\pi} \tag{5.10}$$

The r.m.s. input current may be found by integration over a period $\pi/(3\omega)$ e.g. from 0 to $\pi/3$ in Figure 5.13:

$$i_d = \hat{I}_a \sin\left(\omega t + \frac{\pi}{3} - \phi\right) \tag{5.11}$$

$$I_d = \sqrt{\left(\frac{3}{\pi}\int_0^{\pi/3} i_d^2 \, d\omega t\right)}$$

$$= \hat{I}_a \sqrt{\left(\frac{3}{\pi}\left[\frac{\omega t}{2} - \tfrac{1}{4}\sin 2\left(\omega t + \frac{\pi}{3} - \phi\right)\right]_0^{\pi/3}\right)}$$

$$= I_a \sqrt{\left(1 + \frac{3\sqrt{3}}{2\pi}\cos 2\phi\right)} \tag{5.12}$$

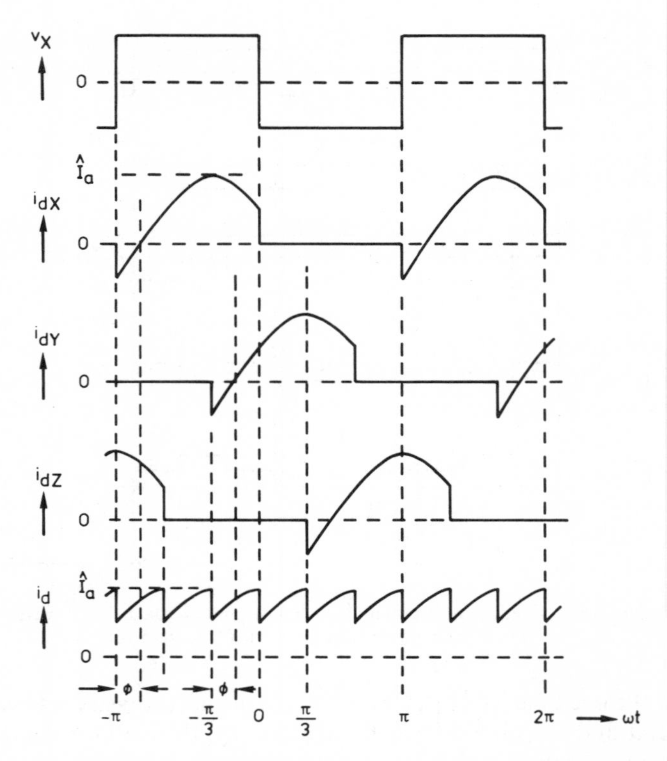

Figure 5.13 Current waveforms in a three-phase bridge inverter (Figure 5.10) with sinusoidal lagging output current.

The r.m.s. input ripple current is then

$$I_{dr} = \sqrt{(I_d^2 - \overline{I}_d^2)}$$

$$= I_a\sqrt{\left[1 - \frac{3\sqrt{3}}{2\pi} - \left(\frac{18}{\pi^2} - \frac{3\sqrt{3}}{\pi}\right)\cos^2\phi\right]} \tag{5.13}$$

Methods of forced commutation

Because of the essential configuration of the main circuit, forced-commutation systems in voltage-fed inverters have to be of the type in which the thyristor is shunted by a diode and the capacitor discharge current is controlled by an inductor forming, with the capacitor, a resonant circuit which determines the turn-off interval. It is possible (and may possibly be advantageous) to treat each arm of the inverter separately and apply the kind of technique used for the single main thyristor of a single-quadrant d.c. switching regulator; in general, however, more economical and efficient arrangements can be contrived by taking advantage of the convenient disposition of the main circuit components and the cyclic mode of operation, so that a single forced-commutation circuit is shared by two (or possibly more) inverter arms.* It will suffice for present purposes to describe in detail one notable example, attributed to W. McMurray.

The McMurray inverter

In the McMurray inverter, a single commutating capacitor is used for each half-bridge, with a separate commutating thyristor for each arm, as shown in Figure 5.14, where Th_3 and Th_4 are the main thyristors, and Th_1 and Th_2 are auxiliary commutating thyristors. The operation of this circuit may be illustrated by reference to Figure 5.15. Suppose that Th_3 is conducting current from the positive supply terminal to the load, which is presumed to include an appreciable series inductance, so that the load current does not change significantly in the course of the commutation process: as a result of previous operation the commutating capacitor C is charged with a voltage, in excess of the supply voltage, of the polarity shown. When it is required to turn off Th_3, Th_1 is fired, causing C to discharge through L with a sinusoidal pulse of current of an amplitude, determined by the charge on C and the values of C and L, substantially in excess of the instantaneous load current. During the period for which the instantaneous capacitor current exceeds the load current, the excess flows in D_1,

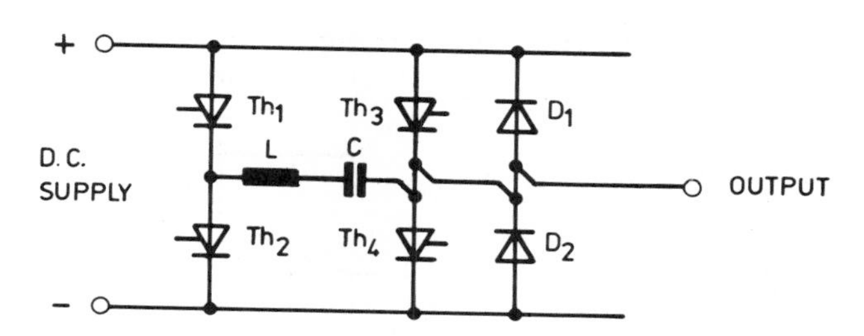

Figure 5.14 Commutation system of McMurray inverter.

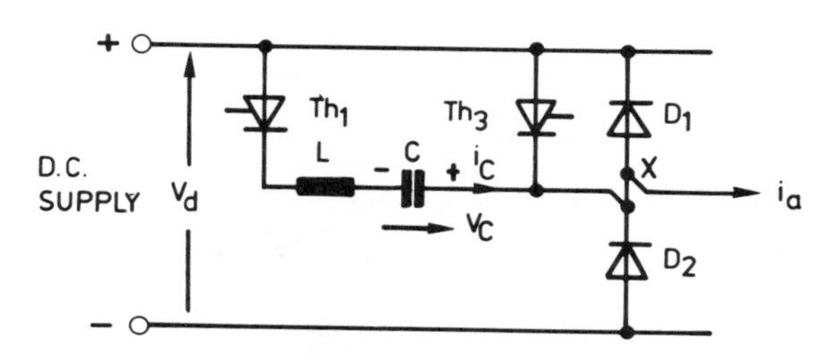

Figure 5.15 Illustrating the operation of the McMurray inverter: Th_3 about to be turned off.

* As noted in Chapter 4, an identical approach serves for the two-quadrant d.c. switching regulator.

while Th_3 is non-conducting and experiences the forward voltage drop of D_1 as a small reverse voltage: the circuit being so designed that this period constitutes an adequate turn-off interval for the thyristor, the latter remains non-conducting thereafter, and the load current is eventually transferred to D_2.

Waveforms illustrating the above process are drawn in Figure 5.16. As C discharges, the voltage across it falls, and at t_2 reverses, aiming at a final voltage opposite and, because of losses, somewhat less than equal to its original voltage. At t_3, however, when i_C falls to the level of i_a, and the difference current $i_C - i_a$ switches from D_1 to D_2, an additional step of voltage, equal to V_d, is introduced into the oscillatory circuit represented by L and C; as a result of this, the reduction in the charge on C due to losses is offset, so that (assuming a stable repetitive condition to have been established) the final charge on the capacitor is equal to its initial charge, and the circuit is prepared for a similar operation to turn off Th_4 when Th_2 is fired.

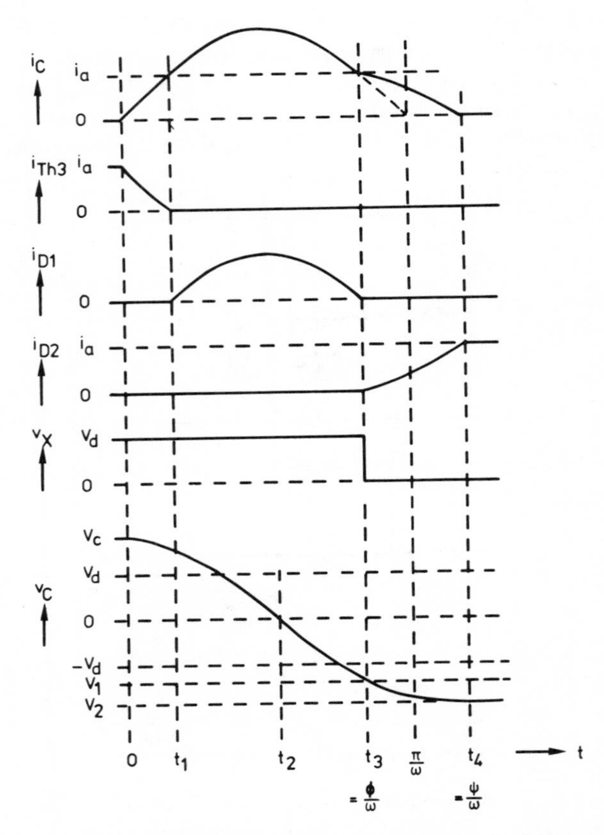

Figure 5.16 Waveforms in the commutating circuit of Figure 5.14.

Somewhat unusually, this circuit exhibits a degree of automatic compensation for varying load current. If i_a increases, t_3 (Figure 5.16) is reached earlier in the discharge period of the capacitor, and the effect of the voltage step V_d at that point is enhanced, with the result that the capacitor voltage is increased step-by-step until equilibrium is re-established at a higher level (allowance has to be made for the limited rate of adjustment). The operation of the circuit does not, however, have to be dependent upon the automatic switching of v_X from the positive to the negative supply potential at t_3: the same effect may be obtained, even though i_c is still in excess of i_a, by firing Th_4 (Th_3 in the alternate half-cycle) at an appropriate time in advance of t_3.

A more detailed analysis of the complete commutation process is given in Appendix 5(ii).

The bi-phase inverter

The operation of the bi-phase voltage-fed inverter (Figure 5.17) is essentially similar to that of the half-bridge. In the same way, each parallel combination of

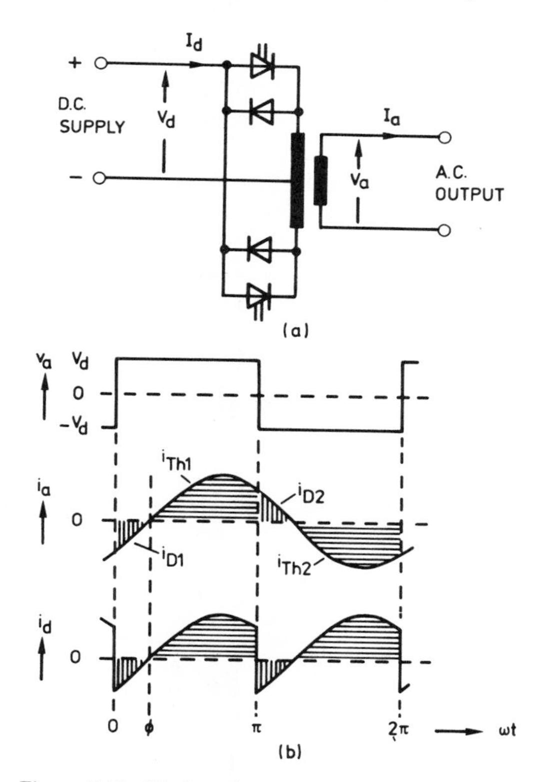

Figure 5.17 Bi-phase inverter with sinusoidal lagging output current: ≡ thyristor current; ||| diode current.

thyristor and diode functions effectively as a bi-directional switch, and the periodic firing and turning-off of the two thyristors connects the supply alternately to the two ends of the transformer primary winding, producing at the secondary terminals a square wave of voltage unaffected by the flow of load current. It remains only to observe that the turning-off of a thyristor while load current is flowing in an unchanging direction results in a transference of current not merely from the thyristor to the opposite diode but from one half of the primary winding to the other. Straightforward operation as illustrated by the waveforms of Figure 5.17 is possible, therefore, only if the two half-windings are closely coupled.

Because current flows for only half the total period in each section of the primary winding, the bi-phase connection entails an appreciable penalty in terms of transformer utilization, and it is normally used only for low input voltages, in order to minimize the conduction losses in the thyristors and diodes.

Output voltage control

It is a common requirement to vary the output voltage of an inverter relative to its input voltage, for the purpose of maintaining constant output voltage despite variations in input voltage and voltage drops, or for other reasons. Apart from the obvious possibility of employing separate voltage regulators, this can conveniently be accomplished by modifying the timing of the thyristor conduction periods in the inverter itself.

If the outputs of two similar inverters, each provided with reactive feedback diodes and producing an output voltage of square waveform unaffected by its load, are connected in series in the manner illustrated in Figure 5.18, and the firing

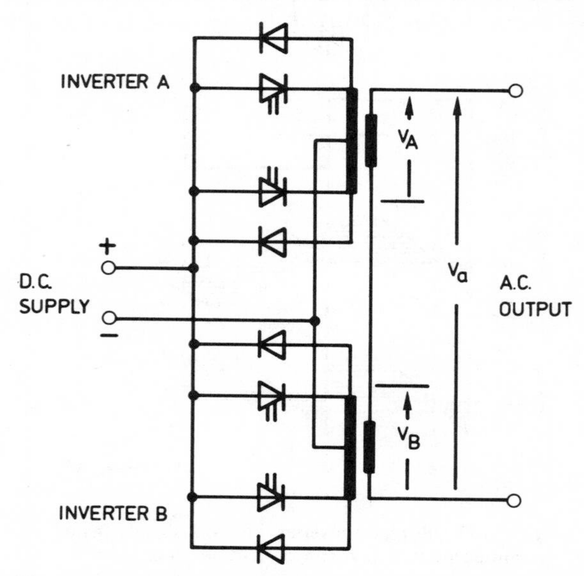

Figure 5.18 Two bi-phase inverters with outputs connected in series to provide output voltage control.

pulses of one inverter—say inverter B—are delayed by an interval α/ω with respect to those of the other inverter, A, the fundamental component of the combined output voltage is reduced by virtue of the phase displacement between the fundamental components of the two inverter outputs:

$$\frac{V_{o1}}{V_{o1\,max}} = \cos\frac{\alpha}{2} \tag{5.14}$$

The two inverter outputs and the combined output are shown in the waveforms of Figure 5.19, from which it can be seen (albeit of less general interest) that the mean and r.m.s. values are reduced in the ratios

$$\frac{\overline{V}_o}{\overline{V}_{o\,max}} = 1 - \frac{\alpha}{\pi}, \qquad \frac{V_o}{V_{o\,max}} = \sqrt{\left(1 - \frac{\alpha}{\pi}\right)} \tag{5.15}$$

In a single-phase bridge inverter, the outputs of the two half-bridges are effectively in series, and a result similar to that described above is obtained, without the need for an output transformer, simply by displacing the firing pulses of the two half-bridges. More specifically,

$$V_{a1} = \frac{2\sqrt{2}}{\pi} V_d \cos\frac{\alpha}{2} = V_{a1\,max} \cos\frac{\alpha}{2}$$

Alternatively,

$$V_{a1} = \frac{2\sqrt{2}}{\pi} V_d \sin\frac{\lambda}{2} = V_{a1\,max} \sin\frac{\lambda}{2} \qquad (5.16)^*$$

where λ is the pulse width, $(\pi - \alpha)$.

Figure 5.20 shows waveforms illustrating the operation of the bridge inverter of Figure 5.3 with what is commonly termed pulse-width control; a load which draws a sinusoidal current with a lagging phase-angle ϕ is assumed.

It can be observed in Figure 5.20 that during periods of zero instantaneous output voltage the load current circulates through one diode and one thyristor, and also that the operation of the bridge is asymmetrical; the latter effect is to be expected (and occurs equally in the arrangement of Figure 5.18) since the

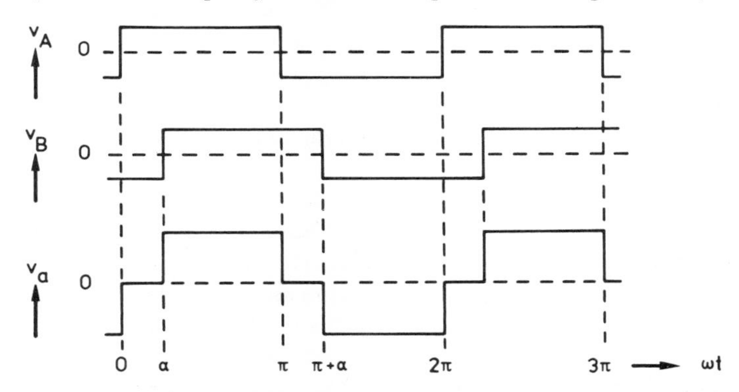

Figure 5.19 Output voltage waveforms in the inverter arrangement of Figure 5.18

* For the harmonic analysis of a square wave, see Appendix 5(iii).

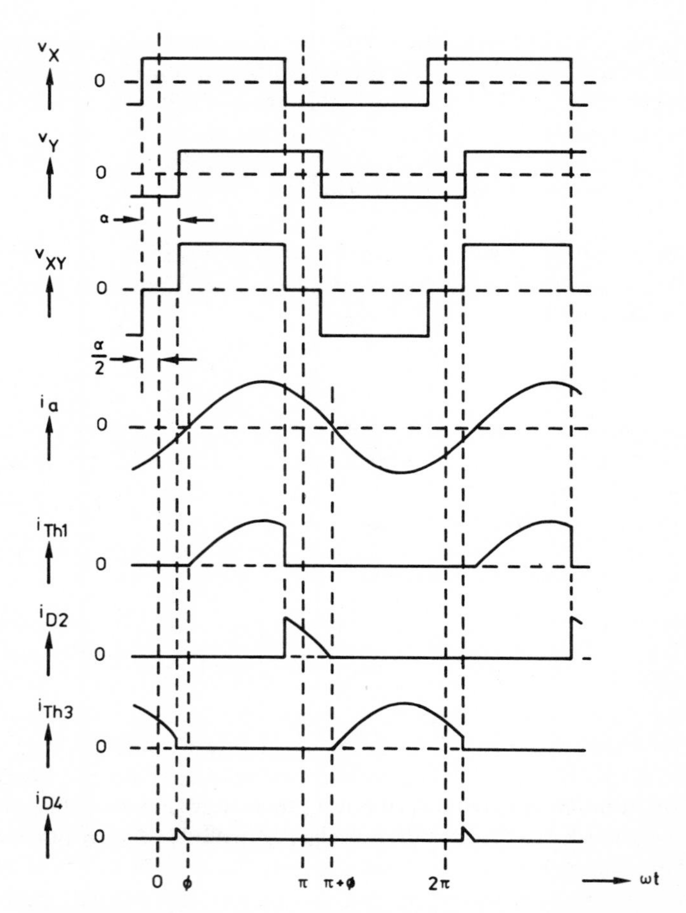

Figure 5.20 Waveforms in a single-phase bridge inverter (Figure 5.3) with phase displacement (pulse width) output voltage control.

fundamental component of the combined output voltage lags the voltage at X (Figure 5.3), and leads that at Y, by an angle $\alpha/2$, so that the output current phase-angles seen by the inverter at X and Y are modified by the pulse-width control to $(\phi + \alpha/2)$ and $(\phi - \alpha/2)$ respectively.

While the concept of phase displacement has been used for convenience in introducing the mode of voltage control described above, it is the variation of the output voltage waveform that is ultimately significant, however produced. An alternative method of contriving an identical result in the single-phase bridge is illustrated in Figure 5.21; the difference in this mode of operation as compared with the previous one lies in the different pattern of circulating-current paths during the periods of zero output voltage, which results in symmetrical loading as between the two half-bridges but an unbalance between the positive and negative arms of each half-bridge.

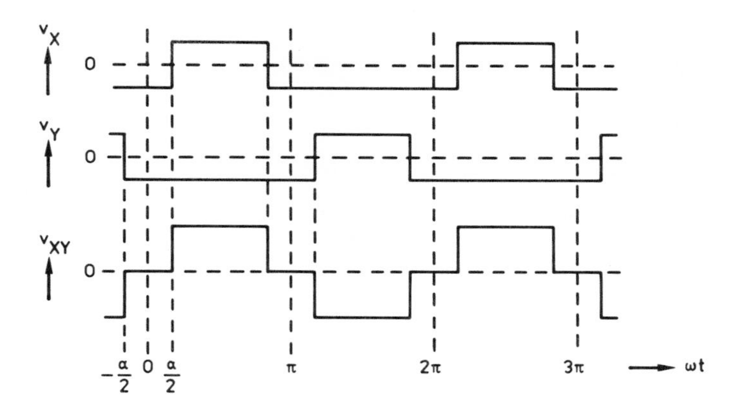

Figure 5.21 Output voltage waveforms in a single-phase bridge inverter (Figure 5.3) with alternative pulse-width voltage control.

Voltage control in three-phase inverters

In the three-phase bridge inverter, a variable phase displacement between the outputs of any two half-bridges is not a possible approach to the problem of output voltage control, since a phase separation of $2\pi/3$ has to be preserved in order to obtain a balanced output. A suitable control strategy may, however, be devised in the following manner.

Considering, with reference to Figure 5.11, the way in which v_{YZ}, as an example, is obtained by subtracting v_Z from v_Y, it is not practicable to reduce the pulse width in the waveform of v_{YZ} by modifying the transitions of the square waves representing v_Y and v_Z, since they must occur at the regular $\pi/3$ intervals if the balanced three-phase output is to be preserved; hence the necessary modification must be effected by introducing additional, intermediate transitions into the phase voltage waveforms. Thus, the leading edge of the v_{YZ} waveform must be retarded not by altering the transition of v_Y at the angle indicated as $2\pi/3$, but by introducing a 'notch' into the waveform of v_Z at that point. To preserve balance and symmetry, similar notches must be introduced, in a suitable polarity, one half-cycle later in v_Z and at corresponding points in v_X and v_Y; the overall result is then as shown in Figure 5.22.

It can be seen from Figure 5.22 that the interphase output will disappear completely when the angle of delay reaches $\pi/3$, at which point the output voltage of each half-bridge (relative to the mid-point of the d.c. supply) becomes a square wave of three times the nominal inverter output frequency. For the purpose of analysis, the interphase output voltage waveform may conveniently be regarded as a combination of two waveforms at a fixed phase displacement of $\pi/3$, as illustrated in Figure 5.23, each corresponding to the single-phase output waveform of Figure 5.19 with an angle of delay in the range $2\pi/3$ to π. In terms of α in Figure 5.22, therefore, the fundamental component of the output voltage is

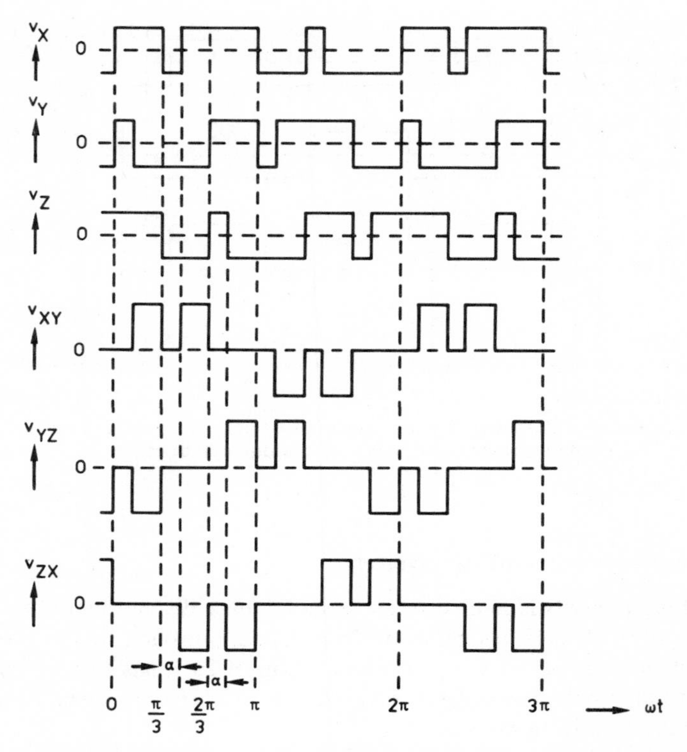

Figure 5.22 Output voltage waveforms in a three-phase bridge inverter (Figure 5.10) with pulse-width voltage control.

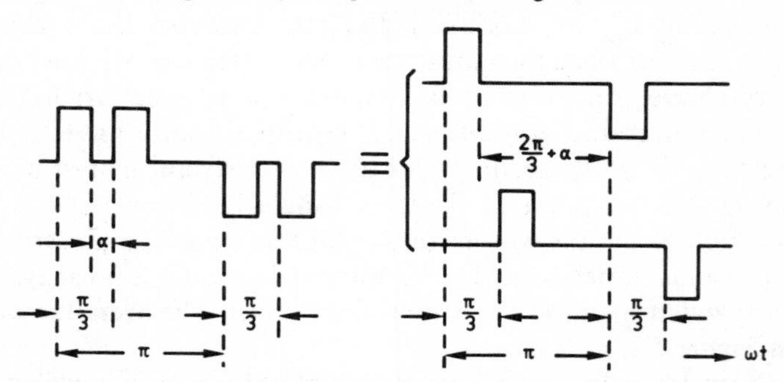

Figure 5.23 Dissection of the controlled output voltage waveform of a three-phase bridge inverter.

given by

$$\left.\begin{aligned} V_{a1} &= \frac{2\sqrt{2}}{\pi} V_d \cos\left(\frac{2\pi/3+\alpha}{2}\right) \times 2\cos\frac{\pi}{6} \\ &= \frac{2\sqrt{6}}{\pi} V_d \cos\left(\frac{\pi}{3}+\frac{\alpha}{2}\right) \\ \text{or} \quad \frac{V_{a1}}{V_{a1\,max}} &= 2\cos\left(\frac{\pi}{3}+\frac{\alpha}{2}\right) \end{aligned}\right\} \quad (5.17)$$

Since pulse-width voltage control entails periods during which the load currents circulate within the inverter bridge, and do not, therefore, flow in the d.c. supply, it results in some cases in an increased input ripple current, albeit not generally very significant in practice.

Output voltage harmonics

The harmonic content of the output voltage of an inverter may be of interest from the point of view of its effect on the load (known or surmised), or for other reasons such as the possibility of interference with communications systems.

The square-wave output voltage of the single-phase inverter without voltage control contains an infinite range of odd harmonics, wherein the nth harmonic is of a magnitude

$$V_{an} = \frac{V_{a1}}{n} = \frac{2\sqrt{2}}{n\pi} V_a \tag{5.18}$$

where V_{a1} is the magnitude of the fundamental component and V_a is the total r.m.s. value. (See Appendix 5(iii).)

Pulse-width control of the output voltage modifies the proportions of harmonics. Referring to the single-phase waveforms of Figure 5.20, when the two constituent square waves v_X and v_Y are combined with a phase-displacement α at the fundamental frequency, the nth-harmonic components are combined with a phase-displacement $n\alpha$, so that they are reduced in the ratio $\cos(n\alpha/2)$. The relative magnitudes of the harmonics therefore become

$$\frac{V_{an}}{V_{a1}} = \frac{\cos(n\alpha/2)}{n\cos(\alpha/2)} \tag{5.19}$$

or, in absolute terms,

$$V_{an} = \frac{2\sqrt{2}}{n\pi} V_d \cos\frac{n\alpha}{2}$$

Alternatively,

$$V_{an} = \frac{2\sqrt{2}}{n\pi} V_d \sin\frac{n\lambda}{2} \tag{5.20}$$

The output voltage waveform of the three-phase bridge inverter, at full output, is similar to that of the single-phase inverter with an angle of delay $\alpha = \pi/3$, and its harmonic magnitudes are therefore obtained by putting $\alpha = \pi/3$ in expression (5.20). The result is zero if n is three or a multiple of three: otherwise, for any odd value of n, $V_{an}/V_{a1} = 1/n$, which is to say that the proportions of all the harmonics present are the same as in the single-phase case.

With voltage control in the manner of Figure 5.22, the approach employed above to the evaluation of the fundamental component may equally be applied to the harmonic components, since the phase displacement between the harmonics contained in the two constituent pulses of the output voltage waveform, mutually displaced by $\pi/3$ at the fundamental frequency (Figure 5.23), is $n\pi/3$, which is effectively $\pi/3$ at all the frequencies present in the interphase output.

Thus

$$\left.\begin{aligned} V_{an} &= \frac{2\sqrt{2}}{n\pi} V_d \cos n\left(\frac{2\pi/3+\alpha}{2}\right) \times 2\cos\frac{\pi}{6} \\ &= \frac{2\sqrt{6}}{n\pi} V_d \cos n\left(\frac{\pi}{3}+\frac{\alpha}{2}\right) \\ \text{and} \qquad & \\ \frac{V_{an}}{V_{a1}} &= \frac{\cos n(\pi/3+\alpha/2)}{n\cos(\pi/3+\alpha/2)} \end{aligned}\right\} \tag{5.21}$$

The above results may equally well be obtained, if preferred, by direct Fourier analysis.

In Figure 5.24 and 5.25 the functions $(1/n)\cos n\alpha/2$ and $(2/n)\cos n(\pi/3+\alpha/2)$ are plotted against α to show the variations in the fundamental and low-order harmonic components, in the single- and three-phase cases, relative to the maximum value of the fundamental component. It will be observed that pulse-

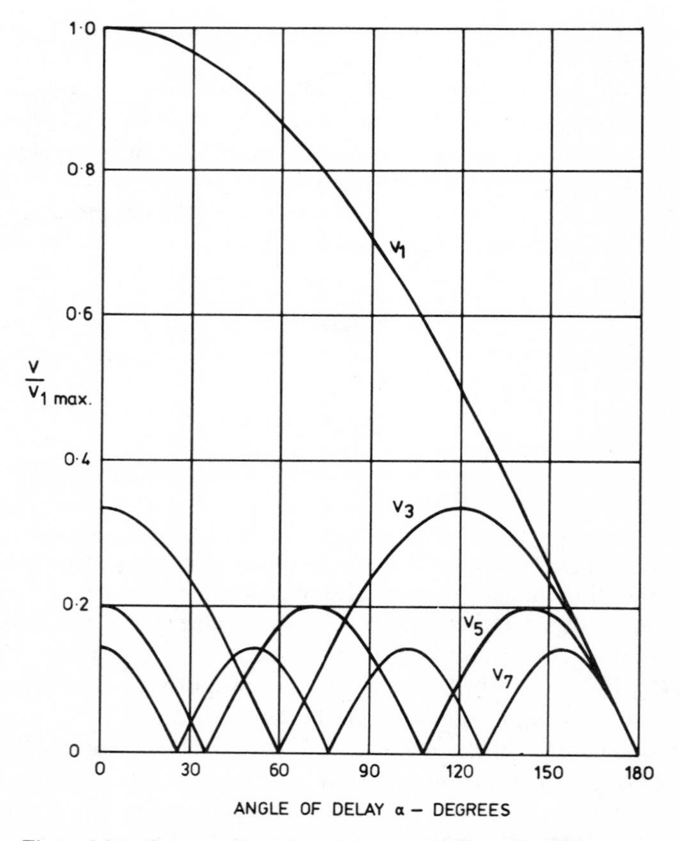

Figure 5.24 Output voltage harmonics in a voltage-controlled single-phase bridge inverter.

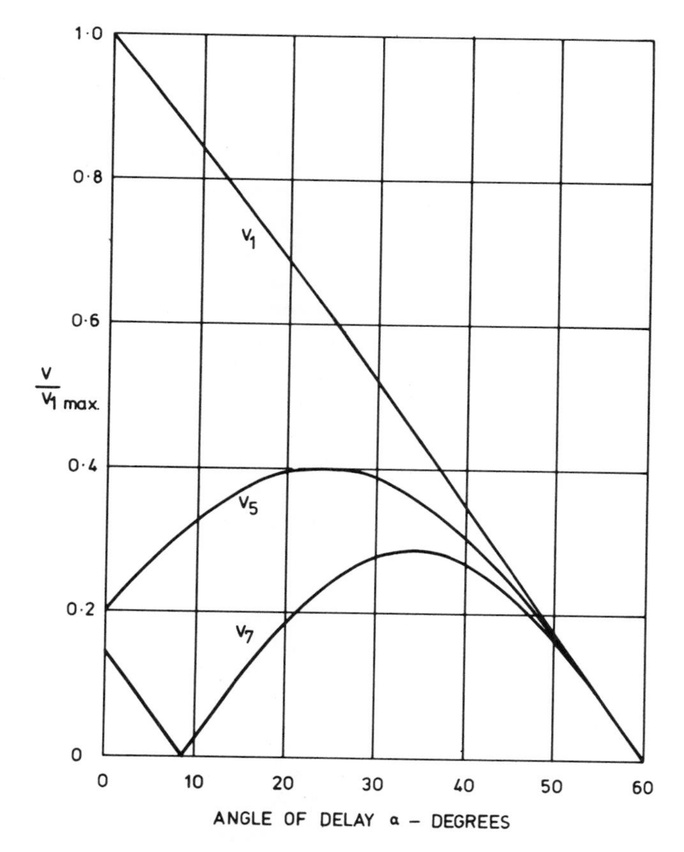

Figure 5.25 Output voltage harmonics in a voltage-controlled three-phase bridge inverter.

width voltage control can cause an appreciable increase in particular harmonics relative to the fundamental, in the single-phase case, or, in the three-phase case, a considerable increase in absolute terms; it is, for this reason, of limited value when a wide range of control is required and the removal of harmonics by filtering is not practicable, as for example in an a.c. motor drive.

Reduction of output harmonics

The harmonic content of the output voltage waveform of the kind of voltage-fed inverter considered above is frequently unacceptable. For general-purpose power supplies, for example, a near-sinusoidal waveform with a total r.m.s. harmonic content of not more than 5% of the fundamental value, and with no individual harmonic component greater than 3%, is often specified. In other cases the requirements may be less stringent, but none the less important. The means whereby the harmonic content can be contained within the necessary limits represents a very important aspect of practical inverter design.

To translate the rectilinear waveform generated by the switching processes in an inverter into an output voltage waveform more closely approaching a sine wave requires, essentially, a harmonic filter—that is, a low-pass filter which provides a relatively large attenuation at the significant harmonic frequencies while affecting the fundamental component only minimally. In practice, it is difficult to design a satisfactory filter when the wanted and unwanted frequencies are so little separated as the fundamental and the third or fifth harmonic components, particularly when allowance is made for possible harmonic currents drawn by the load, which make their own contribution to the harmonic voltage spectrum through the output impedance of the filter. The result is commonly a complicated filter circuit which contains a great deal of material, in the form of inductive and capacitive components, requires the careful adjustment of several tuned circuits and has a poor transient response to changes in load current as well as introducing appreciable regulation and losses. If the output frequency is variable, a filter capable of discriminating between the fundamental component and low-order harmonics becomes virtually an impossibility.

The difficulties associated with harmonic filtering can be considerably alleviated, or even avoided altogether, if the more objectionable harmonics are reduced or removed by other means. The lower-order harmonics can in fact be reduced by a number of methods, most of which entail increasing the number of inverter commutations in each cycle of output.

If 'reduction' of harmonics is taken to include the avoidance of unnecessary increases as a result of voltage control, useful improvements can be achieved with little or no extension of the techniques already described. For example, a three-phase inverter, instead of incorporating a three-phase bridge controlled in the manner described above, with its enhanced fifth and seventh harmonics, may be made up of three single-phase bridges with their outputs connected in star (Figure 5.26). The relative harmonic levels in the three-phase output are then the same as

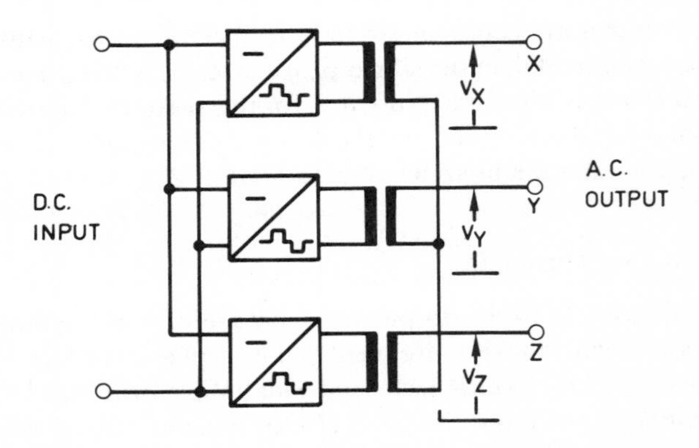

Figure 5.26 Combination of three single-phase inverters used to produce a three-phase output.

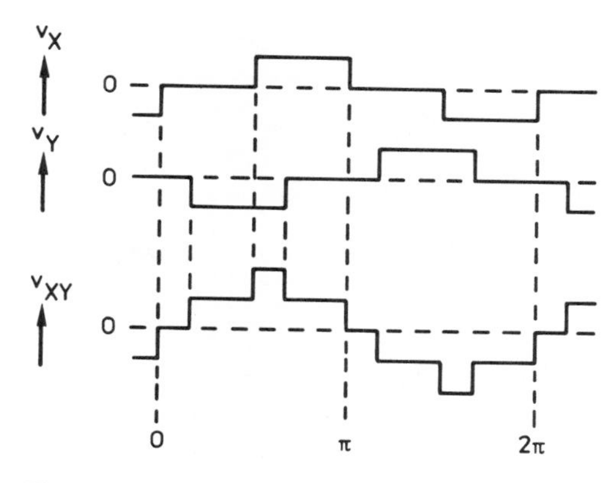

Figure 5.27 Synthesis of the output voltage waveform in the three-phase inverter assembly of Figure 5.26.

those in the single-phase case (Figure 5.24) apart from the elimination of the triplen harmonics. Figure 5.27 illustrates the synthesis of the interphase output voltage waveform, which can have a lower harmonic content than that shown in Figure 5.22 by virtue of its five available voltage levels as compared with three.

Most methods of reducing harmonics entail an increase in the frequency of commutation. The principal techniques are described in the following sections.

Phase multiplication

Phase multiplication, in a three-phase inverter, eliminates, ideally, a series of harmonics in a manner directly analogous to that in which certain input harmonic currents in a rectifier are eliminated by increasing the pulse number. A basic phase-multiplication technique is illustrated in Figure 5.28. Two three-phase bridge (six-pulse) inverters are operated with a fixed time displacement of $\pi/(6\omega)$ between their driving signals, so that there is a phase displacement of $\pi/6$ between their outputs at the fundamental frequency; the output of one inverter is shifted in phase by means of a suitably connected transformer so that the two fundamental-frequency outputs are brought into phase, at equal voltages, and the two outputs are added.

In the arrangement of Figure 5.28, let it be assumed that the output of inverter A lags that of inverter B by $\pi/6$; the delta–star transformer is connected to produce a phase advance of $\pi/6$, and the fundamental components are thus added in phase. The relative phase lag of a harmonic from inverter A, however, is proportional to its order n, that is, $n\pi/6$, while the phase shift of a harmonic in the transformer is either an advance or a lag of $\pi/6$ depending on its phase sequence. The harmonic voltages produced by the six-pulse inverter, as previously shown, are in the series $n = 6r \pm 1$, where r is a positive integer; it can further be shown that those of order $(6r - 1)$ have negative phase sequence, while those of order $(6r + 1)$ have positive phase sequence. The total phase displacement of the nth harmonic from inverter

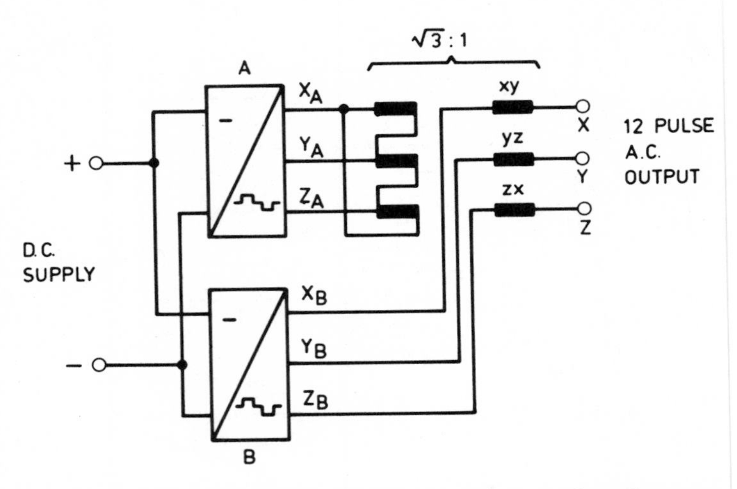

Figure 5.28 Phase-multiplication technique providing a twelve-pulse output from two six-pulse inverters.

A, relative to that from B, is thus

$$\left.\begin{aligned} \phi_n &= \frac{n\pi}{6} + \frac{\pi}{6} \qquad (n = 6r - 1) \\ \phi_n &= \frac{n\pi}{6} - \frac{\pi}{6} \qquad (n = 6r + 1) \end{aligned}\right\} \quad (5.22)$$

or for any value of n, $\phi_n = r\pi$

This means that those harmonics for which r is even are added in phase, like the fundamental component, but that those for which r is odd are added with a phase displacement effectively of π, and are therefore eliminated. That is to say, the 5th, 7th, 17th, 19th etc. are eliminated, while the 11th, 13th, 23rd, 25th etc. are unaffected. This process of eliminating certain harmonic voltages is exactly analogous to the elimination of certain harmonic currents in the twelve-pulse rectifier, and the resulting voltage waveform, shown in Figure 5.29, is similar to the current waveform of Figure 2.29.

Particular harmonics may be eliminated by combining the phase-displaced outputs of several inverters without using phase-shifting transformers, but this technique generally reduces the utilization of the inverter components, and is not widely used.

Multiple pulse-width control

Once the principle of breaking up the interphase output voltage waveform of a three-phase bridge inverter into two pulses in each half-cycle (Figure 5.22) is appreciated, it will be evident that it can readily be extended to produce a greater number of pulses per half-cycle. The possible advantage of doing this, in a three-

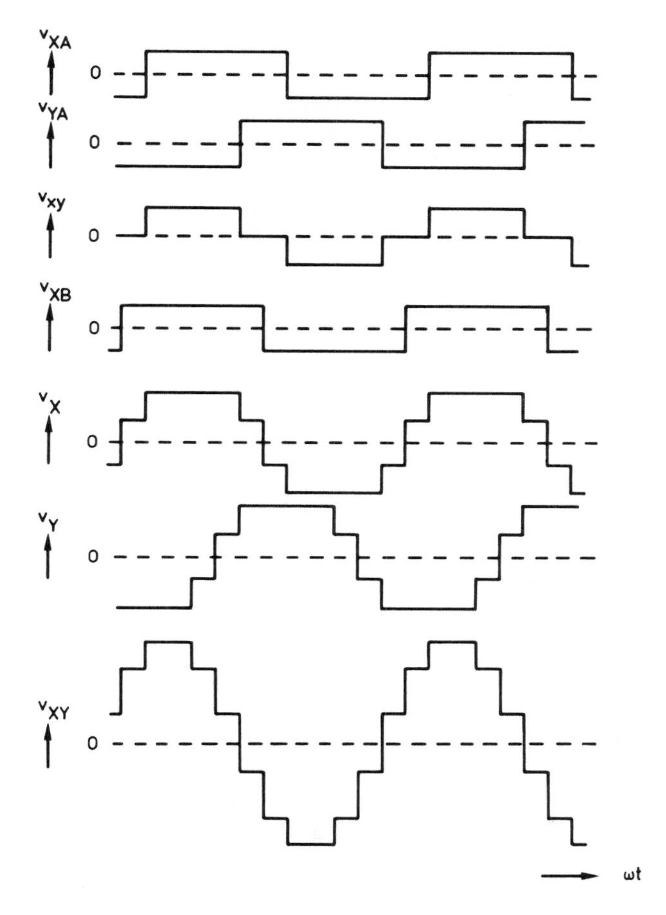

Figure 5.29 Output voltage waveforms in the twelve-pulse inverter of Figure 5.28.

phase inverter particularly, can be illustrated in a general way by means of a graphical analysis of the waveform.

If the full-output voltage waveform of the three-phase bridge is represented for convenience as in Figure 5.30, the nth-harmonic component is given by Fourier analysis as

$$\hat{V}_n = \frac{2}{\pi} \int_{\pi/6}^{5\pi/6} \sin n\omega t \, d\omega t \tag{5.23}$$

For the fifth harmonic, as an example, this is represented in Figure 5.31 by the nett hatched area under that part of the fifth-harmonic wave which is bounded by the complete rectilinear waveform—i.e. between $\pi/6$ and $5\pi/6$ (areas below the zero axis are reckoned as negative). The same process applied to the reduced-voltage waveform of Figure 5.22 is illustrated in Figure 5.32, which shows a doubled nett

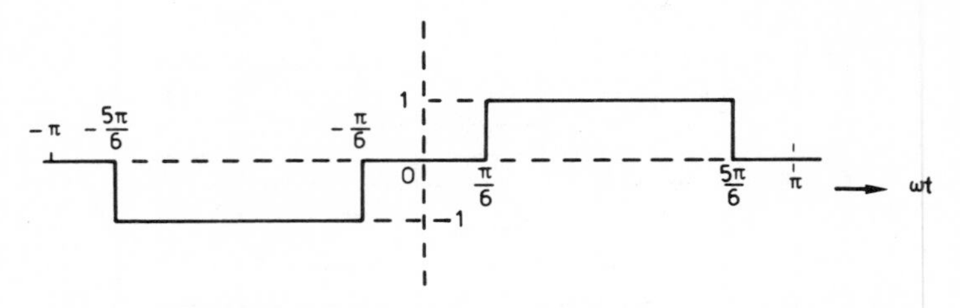

Figure 5.30 Full-output voltage waveform of a three-phase bridge inverter.

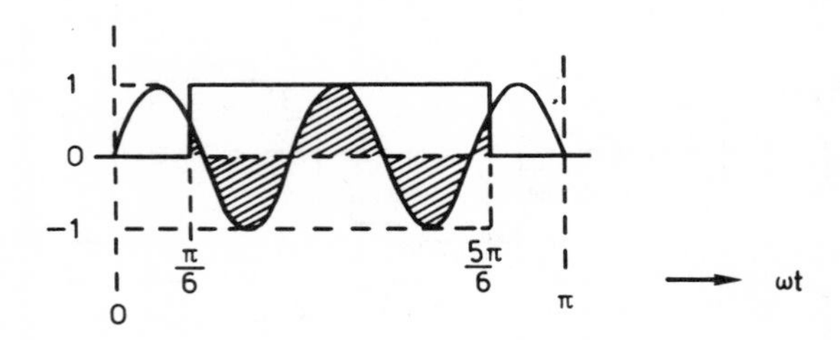

Figure 5.31 Graphical evaluation of the fifth harmonic in the waveform of Figure 5.30.

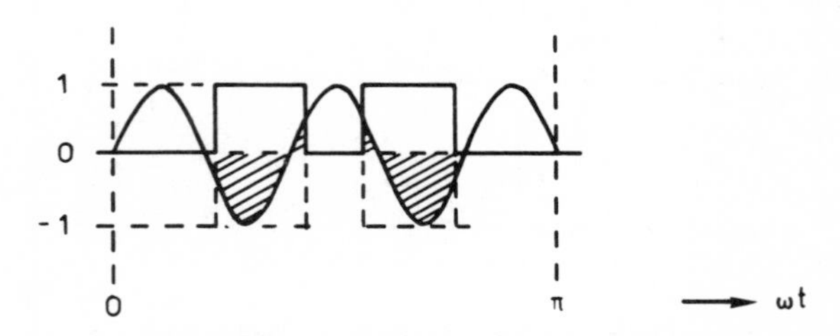

Figure 5.32 Graphical evaluation of the fifth harmonic in the reduced-voltage waveform of Figure 5.22.

area under the fifth-harmonic wave, and thus a doubled fifth-harmonic amplitude, with an angle of delay of 24°.

Comparison of Figure 5.32 with Figure 5.31 suggests that any increase in the low-order harmonic content as a result of voltage control will be minimized if the character of the full-output waveshape is preserved as far as possible. A considerable improvement in fact results from the modification illustrated by the waveforms of Figure 5.33, producing only four pulses in each half-cycle of the output voltage waveform; the effect on the fifth harmonic is illustrated graphically in figure 5.34, and a mathematical evaluation based on the phasor addition of the harmonics residing in the four pulses (Figure 5.35) gives the

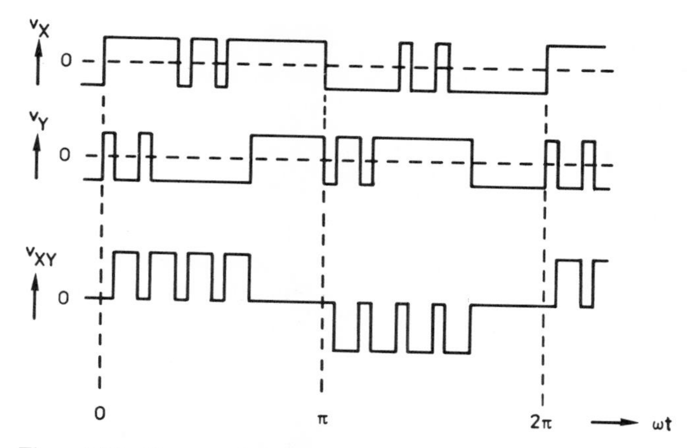

Figure 5.33 Output voltage waveform of a three-phase bridge inverter with multiple pulse-width voltage control.

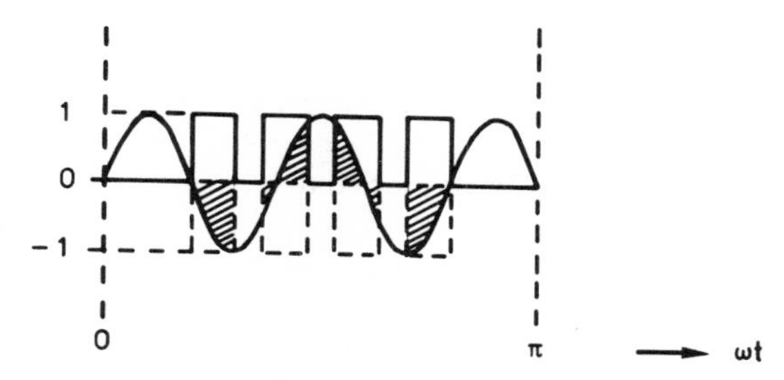

Figure 5.34 Graphical evaluation of the fifth harmonic in the reduced-voltage waveform of Figure 5.33.

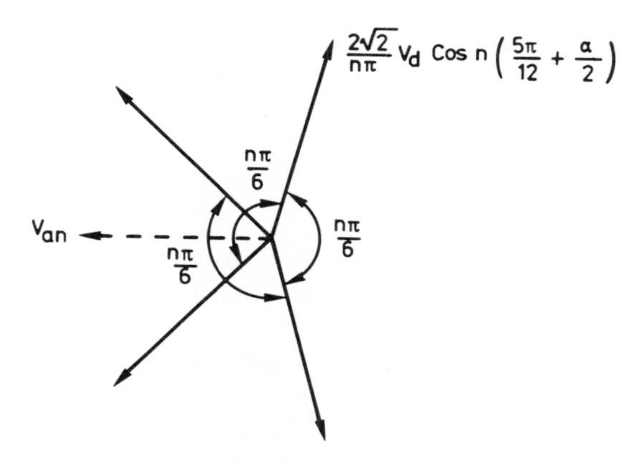

Figure 5.35 Phasor addition of four harmonic components in the inverter output voltage waveform of Figure 5.33.

expression (cf. (5.21))

$$V_{\mathrm{an}} = \frac{2\sqrt{2}}{n\pi} V_{\mathrm{d}} \cos n\left(\frac{5\pi/6+\alpha}{2}\right)\left(2\cos\frac{n\pi}{12} + 2\cos\frac{n\pi}{4}\right)$$

$$\equiv \frac{4\sqrt{6}}{n\pi} V_{\mathrm{d}} \cos n\left(\frac{5\pi}{12}+\frac{\alpha}{2}\right)\cos\frac{n\pi}{12} \tag{5.24}$$

For the fifth harmonic, this amounts to approximately half the maximum value given by (5.21) for the simpler waveform. A larger number of pulses per half-cycle tends to extend the improvement to the higher-order harmonics.

Selected harmonic reduction

In the technique known as selected harmonic reduction, additional steps are introduced into the output waveform of an otherwise unmodified inverter in such a way as to eliminate a limited number of particular low-order harmonics. In a single-phase inverter, for example, the third and fifth harmonics might be eliminated, leaving the seventh as the lowest order to be filtered out, or in a three-phase inverter, with no third harmonic, the fifth and seventh might be removed.

The method is best illustrated by reference to the half-bridge inverter element, with two instantaneous output voltage levels; voltage control by the normal phase-shifting technique can then be considered additionally without any possibility of re-introducing the harmonics that have been eliminated by the basic waveform modification.

Since there are only two levels of instantaneous output voltage, a waveform modification introduced by means of additional commutations must consist effectively in the addition of pulses, or notches, of twice the amplitude of the basic square wave, as shown in Figure 5.36. From the previous discussion of the variation of harmonic levels with pulse width, and Figure 5.24, it can be inferred that a pulse width and a position on the ωt axis relative to the basic square wave—i.e. β and ψ in Figure 5.36—can be chosen such that a particular harmonic component of the notch is equal in magnitude but opposite in phase to the corresponding component of the square wave, and the harmonic is excluded from the resultant waveform.

In the square wave, the amplitude of the nth harmonic, relative to the fundamental component, is $1/n$. In the double-amplitude notch, the nth harmonic has a relative amplitude of $(2/n)\sin(n\beta/2)$. Cancellation will therefore occur, if the phase relationship is correct, when

$$\sin\frac{n\beta}{2} = 0.5 \tag{5.25}$$

It can be seen from inspection of the waveforms, and confirmed by Fourier analysis, that the 'centre line' of the notch needs to be displaced from the zero-crossing of the square wave by one quarter-period of the harmonic, so

$$\psi = \frac{\pi}{2n} \tag{5.26}$$

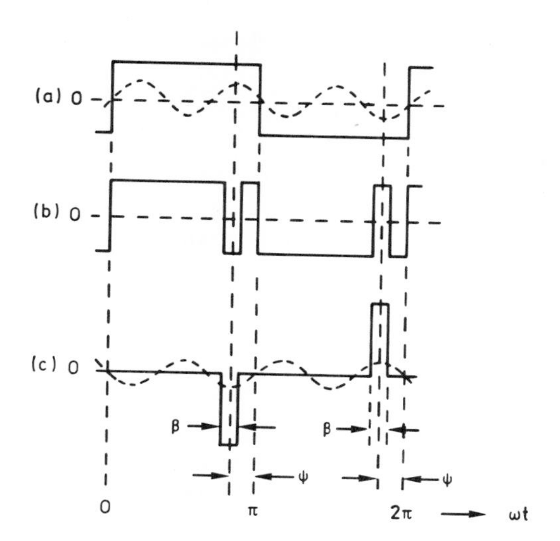

Figure 5.36 Half-bridge output voltage waveform modified by four additional commutations per cycle: (a) basic square-wave; (b) modified waveform; (c) double-amplitude pulse added to (a) to produce (b) (third-harmonic components shown in broken lines).

Figure 5.36 is drawn to illustrate the removal of the third harmonic component, for which, from the above, $\beta = 20^\circ$ and $\psi = 30^\circ$.

The same technique may be applied to remove a different harmonic—the fifth, say—but the values of β and ψ required will obviously not be the same, and one notch in each half-cycle can remove only one harmonic. In the more elaborate waveform of Figure 5.37, the harmonic component required to cancel that of the basic square wave is provided by two notches per half-cycle, and it can be seen that so long as the two notches of each pair are symmetrically spaced from the zero-crossings of the square wave, their resultant harmonic components are in the correct phase regardless of the actual value of ψ (within limits). The phasor addition of the components is illustrated in Figure 5.38; from this, for cancellation of the nth harmonic,

$$\frac{1}{n} = \frac{2}{n} \sin \frac{n\beta}{2} \times 2 \sin n\psi$$

or

$$\sin \frac{n\beta}{2} \sin n\psi = 0.25 \tag{5.27}$$

Equation (5.26) can be solved simultaneously for two values of n (albeit transcendental), and hence two chosen harmonics can be absent from the waveform of Figure 5.37(b). For the third and fifth harmonics, $\beta \approx 9.7^\circ$ and $\psi \approx 28.5^\circ$: for the fifth and seventh, $\beta \approx 5.8^\circ$ and $\psi \approx 19.2^\circ$. By introducing more notches it is possible to extend the principle to eliminate further harmonics, and it has been inferred that a given number of notches per half-cycle can eliminate the same number of harmonic components.

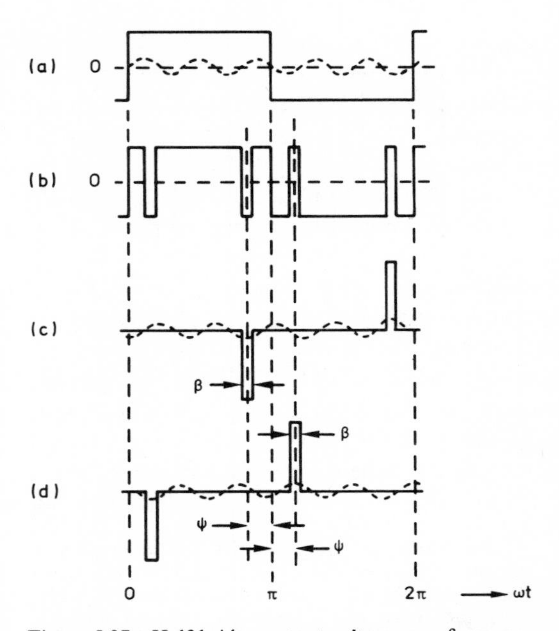

Figure 5.37 Half-bridge output voltage waveform modified by eight additional commutations per cycle: (a) basic square-wave; (b) modified waveform; (c), (d) double-amplitude pulses added to (a) to produce (b) (fifth-harmonic components shown in broken lines).

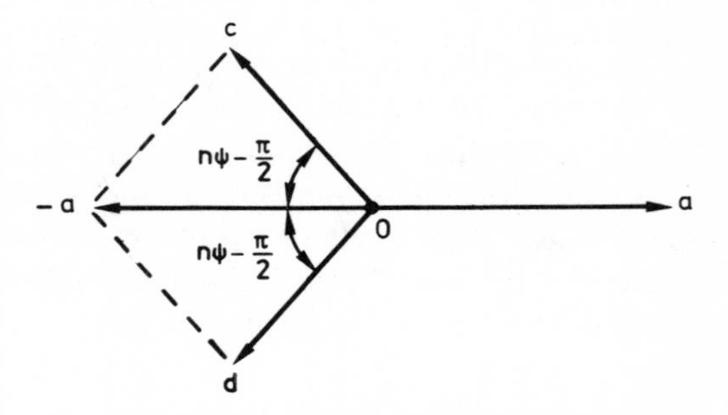

Figure 5.38 Cancellation of harmonic voltage components in the waveforms of Figure 5.37: *a*—*n*th harmonic component of square-wave; *c*, *d*—*n*th harmonic components of notches.

Pulse-width modulation

Generating the commutation patterns required to eliminate an extended series of harmonics by the method of selected harmonic reduction described above, with the necessary accuracy, entails considerable complexity in the control system. Largely for this reason, the basically simpler technique generally known as pulse-width modulation is more often considered as a means of producing an improved waveform prior to filtering.

Pulse-width modulation for this purpose entails generating rectilinear output voltage pulses at a repetition frequency considerably higher than the fundamental frequency and modulating their duration so that the integrated value of each pulse is proportional to the instantaneous value of the required fundamental component at the time of its occurrence: that is, the pulse duration is modulated sinusoidally.

This is illustrated, for an inverter capable of producing three instantaneous levels of output voltage (a bridge inverter, for example), in Figure 5.39. This depicts what might be regarded as an ideal scheme in which the pulse-repetition frequency is an integral multiple of the modulating frequency and the output pulses are symmetrically disposed, on the time axis, about regularly spaced ordinates. The durations of the pulses are proportional to the corresponding ordinates of the modulating sine wave: thus,

$$\frac{\beta_1}{y_1} = \frac{\beta_2}{y_2} = \frac{\beta_r}{y_r} = \text{constant} \tag{5.28}$$

This might be regarded as an extension of the principle of multiple pulse-width control described above, and it can easily be imagined that, if the repetition frequency of the pulses is high enough in comparison with the modulating frequency, the resulting pulse train will have an average effect, for most practical purposes, closely similar to that of the fundamental sine wave, containing little or nothing in the way of low-order harmonics, and that only a small degree of filtering will suffice to produce a virtually sinusoidal waveform. In fact, Fourier analysis confirms that unwanted frequency components below the pulse repetition frequency in the waveform of Figure 5.39 are confined effectively to about three sidebands, and the lowest frequency likely to be noticeable is $(p-5)f_1$, where f_1 is the modulating frequency and the pulse repetition frequency is pf_1. Thus if p is, say, eighteen, the lowest frequency to be removed by the filter is the thirteenth harmonic. If the inverter system is capable of generating pulses at a very

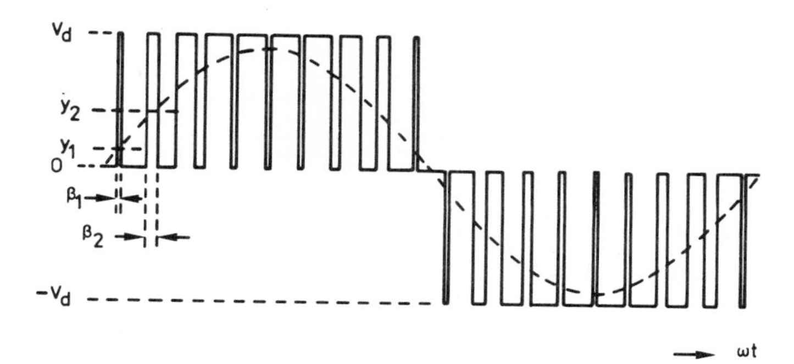

Figure 5.39 Sinusoidal pulse-width-modulation in a single-phase bridge inverter ($m = 0.9$).

high frequency, the requirements placed upon the filter can be so reduced that in material terms it becomes relatively insignificant.

A commonly adopted method of defining the pulse durations, which gives an approximation to the ideal modulation pattern referred to above, is illustrated in Figure 5.40, in this case for a half-bridge. At (a), a modulating sine wave is compared with a triangular wave, the 'carrier', at the required pulse repetition frequency, the relative amplitudes of the two signals being in proportion to those of the required fundamental-frequency output voltage and the d.c. supply voltage of the inverter. The switching instants are determined by the control system at the points of coincidence of the two waveforms: hence, as a matter of geometry, the width of each output pulse (b) nearly corresponds to the average value of the required output potential during the period of the pulse. The approximation in this process lies in the slight irregularity in the timing of the switching instants due to the variation of the modulating signal within the pulse period; this becomes less significant as p is increased. The combination of two half-bridge outputs, generated by co-phasal triangular waves and antiphase modulating signals, produces a three-level waveform with a doubled carrier frequency, as shown in Figure 5.41.

Unlike phase multiplication, methods of reducing harmonics, such as selected harmonic reduction and pulse-width modulation, which depend simply on modifying the output voltage waveform of an inverter by an increased frequency of commutation, have the disadvantage that the commutation losses are correspondingly increased, while the fundamental-frequency output obtainable is somewhat reduced. They are therefore generally less attractive for thyristor

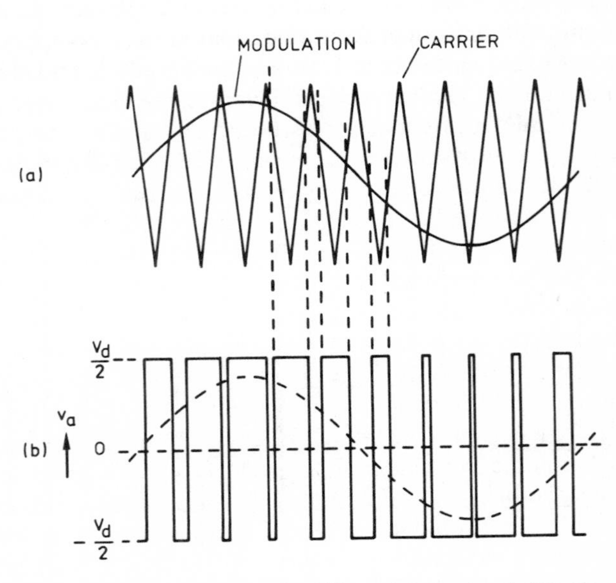

Figure 5.40 Sinusoidal pulse-width modulation in a half-bridge inverter: (a) modulation process; (b) output voltage.

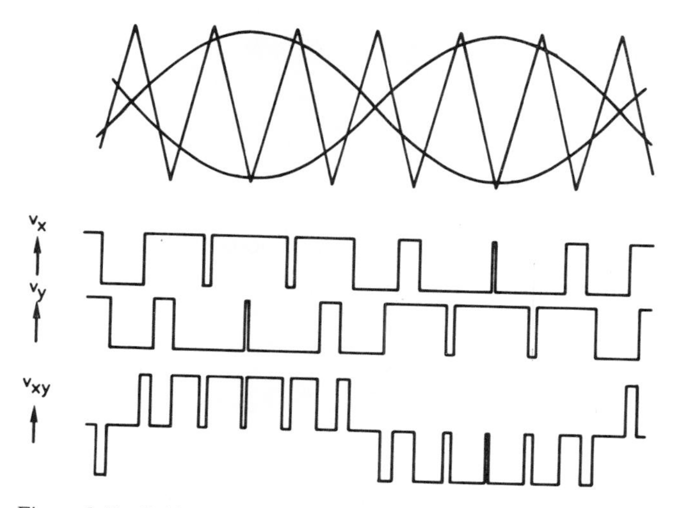

Figure 5.41 Bridge inverter output generated by pulse-width modulation in two half-bridges with co-phasal carriers.

inverters than for transistor inverters, in which commutation losses are less significant, and most useful for low-frequency outputs.

CURRENT-FED INVERTERS

In the so-called voltage-fed inverter so far discussed, the instantaneous output voltage is at all times directly dependent upon that of the d.c. supply, which is of low impedance at all frequencies of interest, and the output current is a function of the load admittance. In the current-fed inverter, the d.c. supply is of high impedance, by virtue of an inductor, and to the extent that the input current is held constant by the inductance the output current waveform is determined by the operation of the inverter, while the output voltage waveform depends upon the nature of the load impedance.

Determination of the output primarily in terms of the current waveform means that the current-fed inverter tends to be unsuitable for some types of load with which the voltage-fed inverter operates quite satisfactorily—loads, for example, which present a high impedance to harmonic currents or a low power factor. On the other hand, it is suitable for certain loads which present problems with voltage-fed inverters, and in such cases it offers notable advantages.

An ideal load for a current-fed inverter, in general, is one which presents a low impedance to harmonic currents and has a power factor close to unity. An example, of considerable practical importance, is an induction heating load; in itself, this is normally highly inductive, but for reasons of economy and efficiency it is usual to connect a capacitor in parallel with the coil to resonate with it, as nearly as possible, at the supply frequency, so that the power factor is corrected nearly to unity, while the capacitor is of low reactance at harmonic frequencies.

The generation of medium-frequency power for induction heating probably represents, in fact, the principal use for the true current-fed inverter. Figure 5.42 shows a single-phase bridge inverter of this type—the so-called tuned parallel inverter—with the power-consuming part of the load represented by a shunt resistance.

Since the function of the d.c. inductor L_d is to maintain a continuous flow of current from the d.c. supply, there is no question here of providing for reactive current flow by connecting feedback diodes in parallel with the thyristors, as in the voltage-fed inverter, and Figure 5.42 represents essentially the complete power circuit. L_c represents a small inductance which may be the stray inductance of the connections, or the leakage inductance of an output transformer if one is included, or may be deliberatedly introduced to limit di/dt in the thyristors.

If it is postulated that a substantially sinusoidal alternating voltage exists across the resonant-circuit load in Figure 5.42, the thyristor bridge will invert power into it from the d.c. supply in exactly the same way that it would into any other a.c. system, so long as the operating conditions are favourable. If the frequency of the alternating voltage is sufficiently low that the turn-off time of the thyristors can be neglected, the necessary conditions are basically those discussed in Chapter 2—namely, that the thyristors should be fired with an angle of delay broadly in the range 90–180° relative to the output voltage waveform, but sufficiently in advance of 180° to ensure complete commutation of the input current, having regard to the commutating inductance represented by L_c. This implies the requirement that, following the firing of one pair of thyristors, a sufficient voltage–time integral must remain in the alternating voltage half-cycle to reverse the current in L_c fully before the output voltage changes polarity, leading to the limiting condition previously expressed in (2.102):

$$\int_{\alpha/\omega}^{\pi/\omega} \hat{V}_a \sin \omega t \, dt \not< 2L_c I_d$$

or

$$\cos \alpha \not< \frac{2\omega L_c I_d}{\hat{V}_a} - 1 \tag{5.29}$$

Operation close to the limit of commutation is illustrated by the waveforms of Figure 5.43; it is assumed in this section that the d.c. inductor is sufficiently large

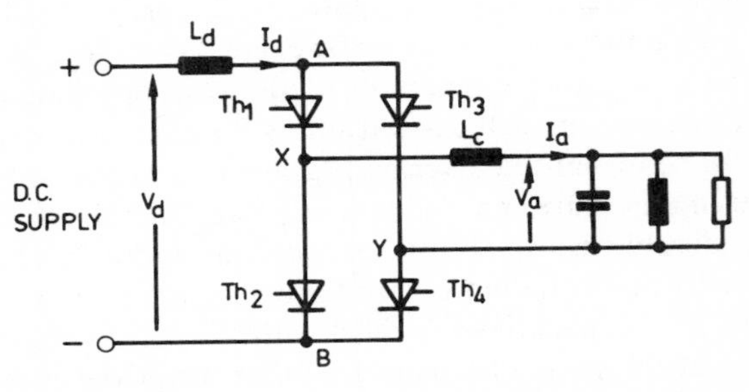

Figure 5.42 Current-fed inverter with parallel-tuned load.

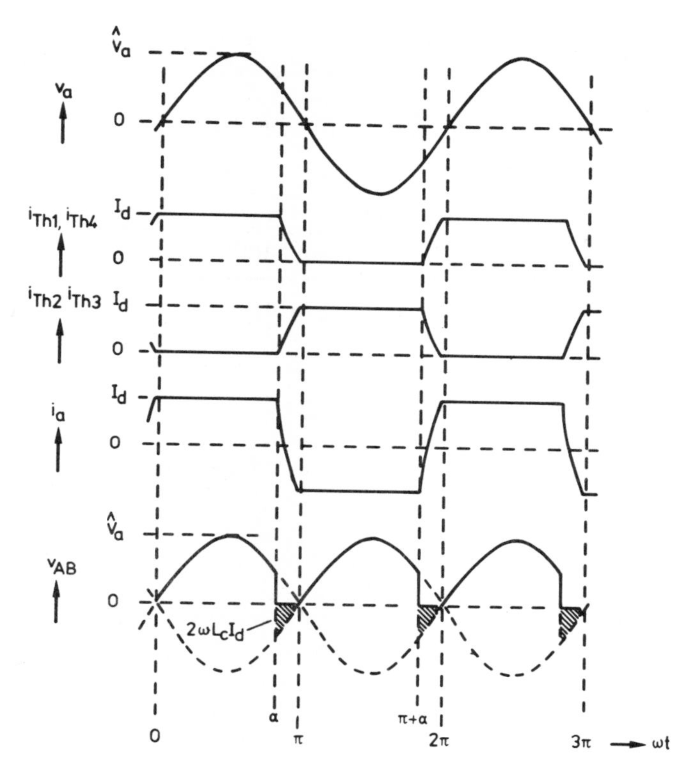

Figure 5.43 Waveforms in a tuned parallel inverter (Figure 5.42) close to the limit of commutation.

that the input current is substantially smooth. The output current from the thyristor bridge is of near-trapezoidal waveform, containing a series of odd harmonics whose amplitudes, relative to that of the fundamental component, at least in the lower orders, are approximately the reciprocals of their orders; from this it can be deduced that, on the assumption that the resonant load circuit is not too heavily damped ($Q > 2$, say) and presents a relatively low impedance to harmonics, the harmonic voltages developed are small enough to justify the original assumption that the alternating voltage was substantially sinusoidal. The system therefore constitutes an inverter capable of supplying a.c. power to a passive load. The d.c. and a.c. voltages are related by the equation

$$V_d = -\frac{2}{\pi}\hat{V}_a \cos\alpha + 4fL_c I_d \tag{5.30}$$

Operation with significant turn-off time

The waveforms of Figure 5.44 illustrate the operation of the tuned parallel inverter when, as is more usual, the thyristor turn-off time is not to be neglected.

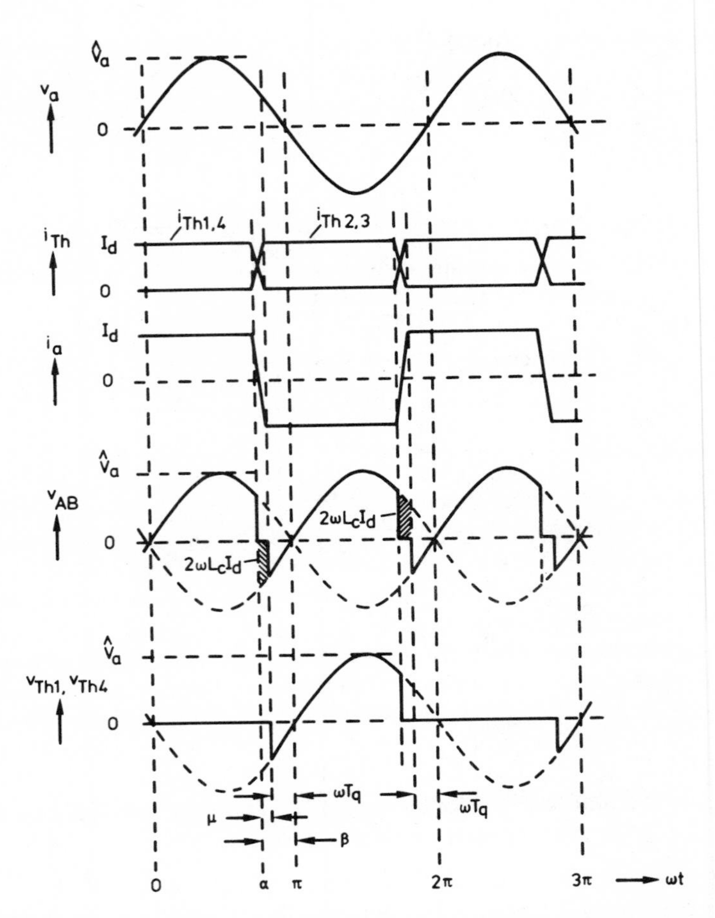

Figure 5.44 Waveforms in a tuned parallel inverter (Figure 5.42) with a significant turn-off interval.

The thyristor firing point must now be further advanced in order to allow an adequate turn-off interval T_q between the end of the overlap period and the end of the voltage half-cycle. In relation to inverter operation, the firing angle is normally expressed as an angle of advance $\beta\,(=\pi-\alpha)$, and, if the overlap angle is μ, $\beta=\mu+\omega T_q$

and
$$\cos\omega T_q-\cos\beta=\frac{2\omega L_c I_d}{\hat{V}_a}\ \text{(c.f. (2.94))} \tag{5.31}$$

Rearranging (5.30) and replacing α by β,

$$V_a=\frac{\pi}{2\sqrt{2}\cos\beta}(V_d-4fL_cI_d) \tag{5.32}$$

More usefully, the voltage equation may be written in terms of ωT_q, as the angle of primary importance, rather than β: given the symmetry of the voltage waveforms in Figure 5.44, a similar process of derivation (see Chapter 2) gives the expression

$$V_d = \frac{2\sqrt{2}}{\pi} V_a \cos\omega T_q - 4fL_c I_d$$

or

$$V_a = \frac{\pi}{2\sqrt{2}\cos\omega T_q}(V_d + 4fL_c I_d) \tag{5.33}$$

The angle of advance required to obtain a given turn-off interval can be calculated by equating the two expressions for V_a:

$$\cos\beta = \cos\omega T_q \left(\frac{V_d - 4fL_c I_d}{V_d + 4fL_c I_d} \right) \tag{5.34}$$

Operation in practice

Operation in the manner described above, with an adequate turn-off interval secured by an appropriate angle of advance, implies that the output current leads the output voltage by a particular phase angle determined by ωT_q and β, and this in turn implies, for the kind of load in question, a particular operating frequency somewhat above the resonant frequency.

If the output current waveform is considered to be approximately trapezoidal, the load phase angle (in fundamental-frequency terms) may be taken approximately as $(\omega T_q + \beta)/2$. Alternatively, equating the input and output power,

$$V_d I_d = V_a I_{a1} \cos\phi$$

where I_{a1} is the fundamental component of I_a, and may justifiably be taken as equal to $2\sqrt{2} I_d/\pi$ with normal overlap angles.

Hence,

$$V_a = \frac{\pi V_d}{2\sqrt{2}\cos\phi} \tag{5.35}$$

and from (5.32) and (5.33)

$$\cos\phi = \cos\beta \left(\frac{V_d}{V_d - 4fL_c I_d} \right) = \cos\omega T_q \left(\frac{V_d}{V_d + 4fL_c I_d} \right) \tag{5.36}$$

The power obtainable within the ratings of a particular inverter is ultimately dependent upon $\cos\phi$, and the inverter has generally to be controlled, therefore, in such a way as to keep β at a minimum (i.e. to operate as close to resonance as possible) consistent with reliable commutation. The means of achieving this is beyond the scope of this discussion. For a given turn-off interval, ωT_q increases in proportion to the output frequency and hence also does ϕ, reducing the available output power.

This type of inverter embodies no practical facility for the control of its output power (unless over a very limited range, by varying β) and with heating loads it is normally supplied from a d.c. source of variable voltage, such as a controlled rectifier. An additional requirement in practice, normally, is a starting system: it should be emphasized that the discussion above relates to steady-state conditions, and does not take account of transient conditions that may arise when the circuit is first energized. The variation of ϕ with frequency as determined by the load characteristics is elucidated further in Appendix 5(i).

Untuned parallel inverters

The current-fed inverter with a resonant load circuit has been shown above to behave essentially as a naturally commutating converter. At the same time, since it is able to invert into a passive load by virtue of a commutating process which depends upon the presence of the capacitor as a parallel branch of the load, and operates ideally with no more than the minimum adequate turn-off interval, it is possible to consider it as a variety of forced-commutation inverter, or at least as a borderline case, with some attractions in relation to other, untuned, loads.

With a resistive load, as in Figure 5.45 (the parallel capacitor now being regarded as part of the inverter), assuming as before that the d.c. inductor is large enough for the input current to be virtually smooth, the waveform of each half-cycle of output voltage is an exponential, the instantaneous voltage approaching the asymptotic value $I_d R$ with a time constant CR. When, as is usual at full load, the half-period is long in comparison with CR (say at least five times as long), the instantaneous output voltage may be taken to reach the asymptotic value by the end of the half-cycle, and the voltage waveform is generally as drawn in Figure 5.46. Suppose Th_1 and Th_4 are fired at $t = 0$; the voltage on the capacitor is applied in reverse across the previously conducting thyristors Th_2 and Th_3 (Y positive with respect to X) and the latter are turned off, provided that the turn-off interval is adequate. The turn-off interval ends when the instantaneous output voltage passes through zero, after a time T_q:

$$I_d R = 2I_d R e^{-T_q/CR}$$

Whence

$$T_q = -\log_e 0.5CR \approx 0.7CR \tag{5.37}$$

The function of the d.c. inductor here is to support the difference between V_d and v_{AB} without permitting the input current to rise excessively during the turn-off

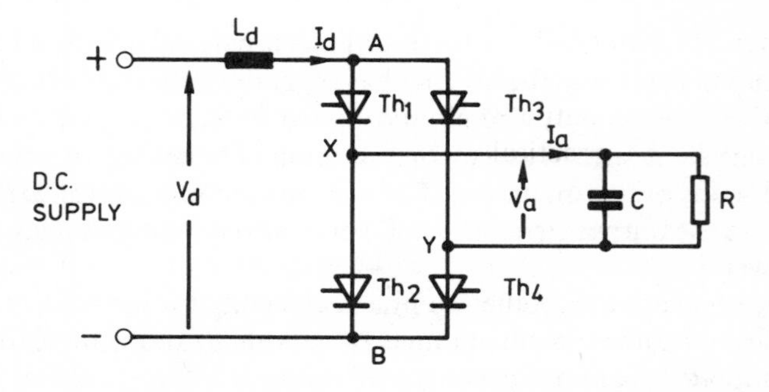

Figure 5.45 Parallel inverter with resistive load.

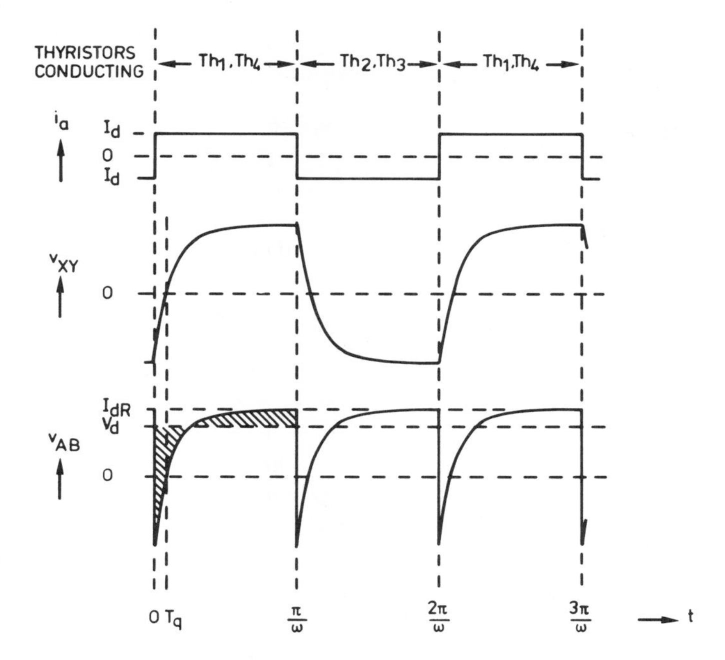

Figure 5.46 Waveforms in a parallel inverter with resistive load (Figure 5.45) ($CR/(\pi/\omega)$ small).

period, which would lead to a reduction of T_q. Under steady-state conditions, the mean voltage across the inductor must be zero, and the positive voltage (i.e tending to increase the input current) which exists from the instant of firing the thyristors up to the time that v_{AB} rises to the level of V_d must therefore be balanced by a negative voltage during the latter part of the half-cycle, such that $\int_0^{\pi/\omega}(V_d - v_{AB})\,dt$ is zero, and the two hatched areas of the waveform of v_{AB} in Figure 5.46 are equal. The peak output voltage, equal to $\hat{V}_{AB}$, must thus exceed V_d in some degree.

Variation of output voltage with load

As the ratio of the half-period π/ω to the time constant CR of the output circuit approaches infinity, the output voltage waveform tends to a square shape, and the r.m.s. output voltage is practically equal to I_dR and to the input voltage V_d. If, on the other hand, CR represents a substantial proportion of the half-period, the asymptotic voltage I_dR is not nearly reached, and the output voltage waveform is more like that shown in Figure 5.47. From inspection of this waveform it is clear that the mean value of v_{AB}—that is, V_d—is considerably reduced relative to the output voltage; in other words, the output voltage is considerably increased.

Although the output waveforms are not sinusoidal, there is in fact little error, so long as $\cos\phi_1$ is less than about 0.8, in applying to this circuit the relationship

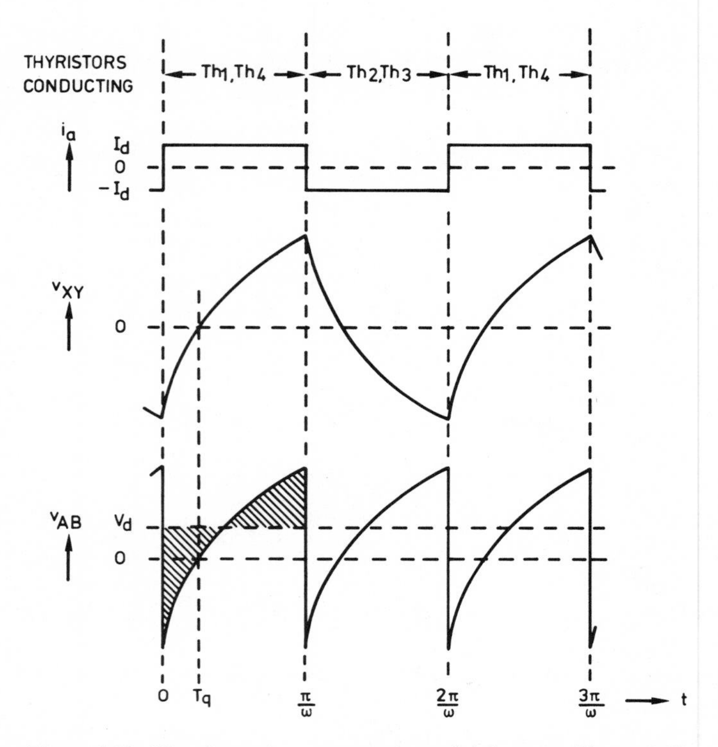

Figure 5.47 Waveforms in an untuned parallel inverter (Figure 5.45) ($CR/(\pi/\omega)$ large).

previously deduced (5.35) in connection with the tuned parallel inverter:

$$V_a = \frac{\pi V_d}{2\sqrt{2}\cos\phi_1} \tag{5.38}$$

where ϕ_1 is the phase angle of the load at the fundamental frequency. (For a justification of this approximation, see Appendix 5 (iv).)

The variation in output voltage, relative to the input voltage, implied by (5.38) has several unfortunate practical consequences. Firstly, since with a given value of $C \cos\phi_1$ varies with R, the regulation with load is poor: secondly, if a reactive load is connected, the output voltage varies with load phase angle: thirdly, operation with no load is impossible, since $\cos\phi_1$ then becomes zero and V_a tends to infinity. These disadvantages severely limit the usefulness of this simple arrangement.

Reactive feedback in current-fed inverters

To make the inverter tolerant of widely varying loads, it is necessary to introduce the reactive feedback diodes referred to previously in connection with the voltage-fed inverter. The diodes cannot be connected directly across the thyristors in this case, however, firstly because the connection would result in a low-impedance discharge path for the commutating capacitor at each firing of the thyristors, and secondly because the inductance on the d.c. side of the inverter bridge would prevent the rapid reversal of the supply current which is essential to the action of the reactive feedback diodes with an inductive load (see Figure 5.8).

The d.c. terminals of the diode bridge must therefore be connected to the (low-impedance) supply on the input side of the d.c. inductor (which may be split, to preserve the symmetry of the circuit) as in Figure 5.48. Operation with varying and reactive loads is then possible, in principle, broadly in accordance with the description previously given in relation to voltage-fed inverters. The direct connection of the diodes between the output terminals and the d.c. supply introduces a further problem, however, in that it suppresses the negataive voltage–time integral across the d.c. inductor, which, as explained above, is necessary to balance the positive voltage–time integral associated with each commutation. The result is a progressive increase in the inductor current with succeeding commutations until the consequent reduction in the turn-off interval due to the increased rate at which the capacitor is charged causes the inverter to fail. During the period before the failure point is reached, the inductor current exceeds the load current, and the difference current flows back to the supply, representing a circulating current through the conducting thyristors and diodes. (In certain practical cases the voltage drops in the thyristors and diodes may provide a sufficient negative voltage-time integral across the inductor to permit continuous operation.)

The increasing output voltage of the capacitively loaded inverter of Figure 5.45 and the increasing inductor current in the arrangement of Figure 5.48 may be seen as alternative manifestations of the same fundamental problem—namely that each commutation entails an input of energy from the supply which, if it cannot be equated to a corresponding output into the load, must progressively augment the energy stored in the reactive components. To enable the inverter to operate satisfactorily, provision must be made for absorbing the surplus energy in losses or returning it to the supply.

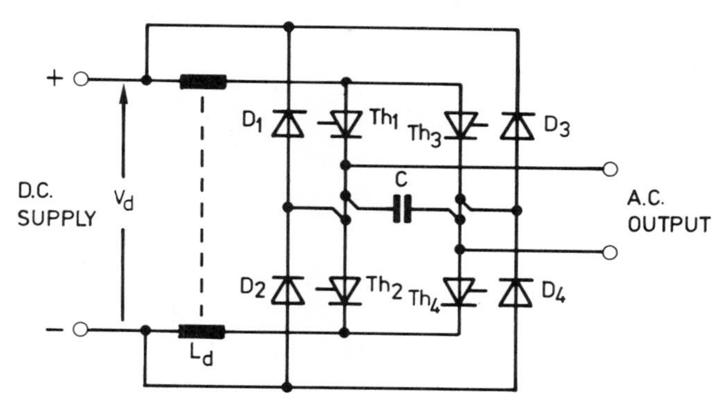

Figure 5.48 Basic (unsatisfactory) connection of reactive feedback diodes in a parallel inverter.

Control of circulating current

The circulating-current problem referred to above is overcome by introducing suitable additional voltages into the feedback paths, whereby the d.c. inductor may be reset after each commutation. This is illustrated in principle in Figure 5.49. Typically, the additional voltages are of such a value that with full resistive load the instantaneous output voltage does not reach the level of voltage across the d.c. terminals of the feedback diode bridge, kV_d, and the diodes do not conduct. As the load is reduced, or made more capacitive, the diodes prevent the instantaneous output voltage from rising above kV_d, and an equilibrium is established such that the inductor current and the turn-off interval remain roughly constant. Alternatively, if k is increased, the no-load circulating current will be reduced at the expense of greater regulation.

With no load, the capacitor is charged linearly by the d.c. inductor current (assumed constant) until the voltage across it is prevented from rising further by conduction of the diodes: the instantaneous output voltage is then clamped at kV_d while the inductor current is fed back to the supply for the remainder of the half-cycle. The resulting waveforms are shown in Figure 5.50.

Since, in the condition of equilibrium, the nett voltage–time integral across the inductor must be zero in any half-cycle, the two shaded areas bcd and defg in Figure 5.50 are equal, and hence the areas ace and abgf are also equal. Therefore

$$(k-1)V_d\frac{\pi}{\omega} = 2kV_dT_q$$

or

$$T_q = \frac{\pi}{2\omega}\left(\frac{k-1}{k}\right) \tag{5.39}$$

But

$$T_q = \frac{kV_dC}{I_{Ld}}$$

$$\therefore I_{Ld} = \frac{2k^2V_d\omega C}{(k-1)\pi} \tag{5.40}$$

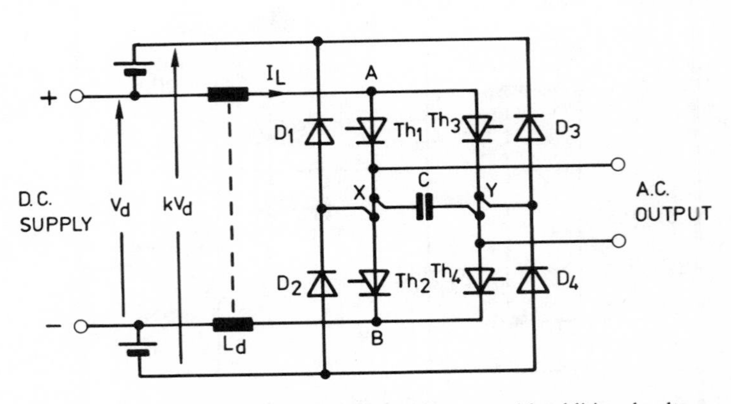

Figure 5.49 Reactive feedback in a parallel inverter with additional voltages introduced to limit circulating current.

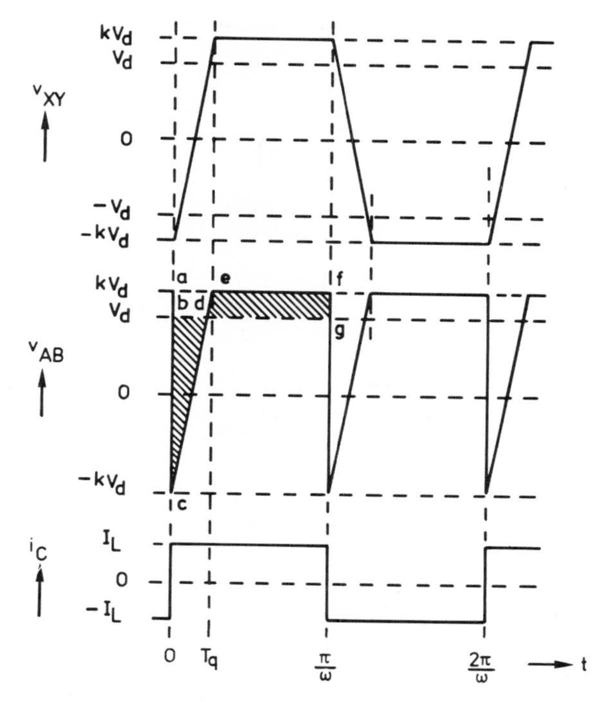

Figure 5.50 Waveforms in the inverter of Figure 5.49 with no load.

Thus for a given capacitance and frequency, both the turn-off interval and the inductor current are effectively defined by the geometry of the waveform diagram, and the factor k may be chosen to limit the circulating current at an appropriate level.

Practical methods employed to introduce the additional feedback voltages include connecting resistors in series with the feedback paths, as in Figure 5.51,

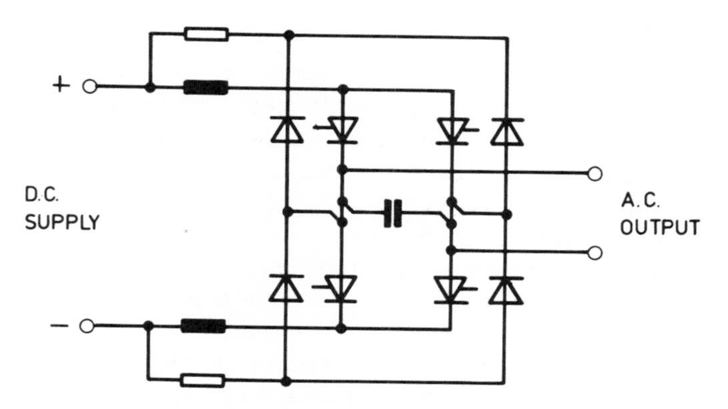

Figure 5.51 Parallel inverter with resistors in the feedback paths.

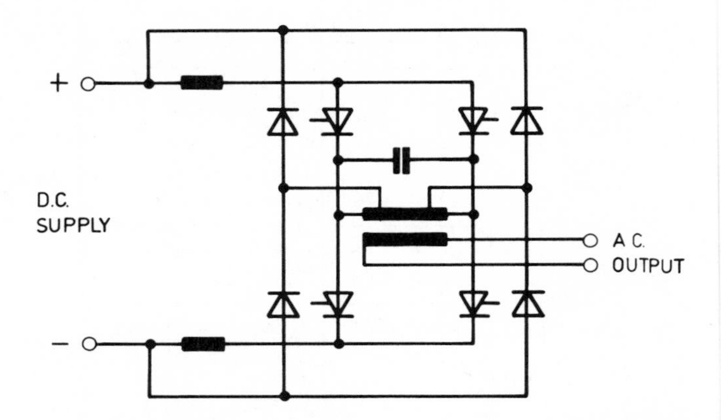

Figure 5.52 Parallel inverter with feedback from output transformer taps.

which results in a loss of power on no-load which may be calculated from a knowledge of I_{Ld}, k, V_d and the time intervals involved. For greater efficiency, the kind of circuit shown in Figure 5.52 is used, where the effect of a total feedback voltage kV_d is obtained by reducing the voltage at the input to the diode bridge by a factor k relative to the inverter output voltage, by means of appropriate transformer taps.

While the circulating current in such an arrangement as that of Figure 5.52 does not represent power drawn from the supply, it does lead to increased losses, and design usually entails, therefore, a compromise between regulation and efficiency.

The operating relationships deduced above for the parallel inverter are modified to some extent by intermediate load conditions, and by the use of a finite value for L_d; detailed consideration of these factors leads to a degree of complication which the results do not justify in an introductory work, and the enquiring reader is referred to more specialized treatments.

Further current-fed inverter configurations

Considered as a forced-commutation inverter, the parallel inverter is one of a class in which the commutating capacitor is enabled to discharge at a controlled rate by virtue of an inductor (or inductors) in the main load-current path, in the absence of which it would discharge so quickly that a significant turn-off interval would not be obtained. In this it is distinct from the McMurray inverter of Figure 5.14, in which the discharge of the capacitor is controlled by generally smaller inductors which carry the capacitor current only.

There are a number of other inverter arrangements based on load-current-carrying inductors, some with, and some without, separate commutating thyristors; Figures 5.53 and 5.55 show two examples. In Figure 5.53, half the inductor winding is in series with each thyristor of the half-bridge, so that the output current flows alternately in the two halves, but with the same magnetic effect. C_1

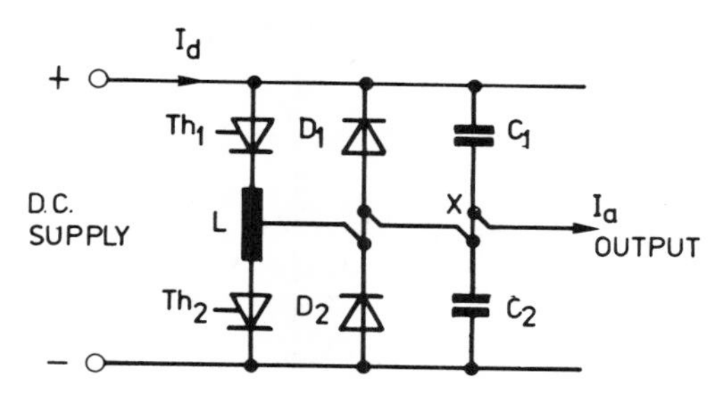

Figure 5.53 Inverter with centre-tapped commutating inductor.

and C_2 are commutating capacitors, and can be regarded as being effectively connected in parallel through the zero source impedance. If Th_1, for example, is conducting at the end of one half-cycle, point X is at the positive supply potential (or at least close to it, since the voltage across L is small except during the commutating process). When Th_2 is fired, the potential at X is maintained initially by the capacitors, and the full supply voltage is impressed upon the lower half of L; a similar voltage is induced in the upper half (the two halves are closely coupled) and as a result Th_1 is reverse-biassed by a voltage V_d, as illustrated in the equivalent diagram of Figure 5.54(a). The capacitors are now discharged by i_a and the inductor current, while the voltage across C_2 is supported by the lower half of L. The turn-off interval ends when the reverse voltage across Th_1 falls to zero, at the instant when the potential of X has fallen by $V_d/2$, as in Figure 5.54(b). (The problem of accumulating inductor energy arises in this inverter, as in that of Figure 5.48, but is ignored here for the sake of simplicity.)

Figure 5.55 shows a bridge inverter with a separate input commutating curcuit. The three-phase bridge comprises the six main thyristors Th_3–Th_8, associated with which are the reactive feedback diodes D_1–D_6. The commutating system consists of the commutating capacitor C_3 and two auxiliary commutating thyristors Th_1 and Th_2, together with the d.c. inductors L_1 and L_2; C_1 and C_2 are much larger in capacitance than C_3, and effectively provide a centre-tap to the supply. The commutating system operates alternately on the positive and negative

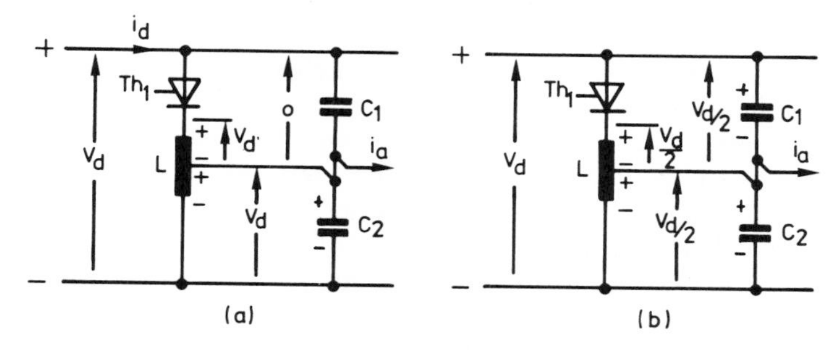

Figure 5.54 Conditions in Figure 5.53: (a) immediately after the firing of Th_2 to turn off Th_1; (b) at the end of the turn-off interval.

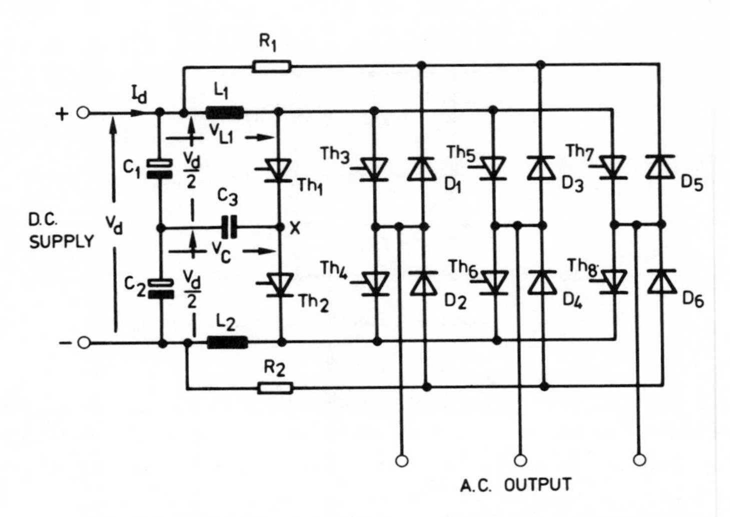

Figure 5.55 Three-phase bridge inverter with input commutation.

inputs to the inverter bridge in the following manner. Suppose that thyristors Th_3, Th_6 and Th_7 are conducting in the bridge, and that it is required to turn off Th_3 to permit the firing of Th_4. As a result of previous operation, point X may be assumed to be at a considerable negative potential $-V_c$ with respect to the negative supply terminal. At the appropriate instant, Th_1 is fired, and the anodes of all the upper-row thyristors, Th_3, Th_5 and Th_7, are brought to the potential of point X, while their cathodes are clamped approximately to the negative supply potential by D_2, D_4 and D_6, so that they all experience a reverse voltage, and will all be turned off, if conducting (provided this condition is of sufficient duration).

Following the firing of Th_1, the potential of point X (v_X) rises at a rate determined by i_{L1}, the current in L_1, as shown in Figure 5.56. At t_2, v_X reaches V_d and continues to rise; firing pulses are meanwhile witheld from all the upper-row thyristors, and any load currents circulate in the lower-row thyristors or diodes and L_2. At t_3, when v_X reaches a value $V_d + V_c$, Th_7 is re-fired, closing a path round L_1 through Th_7, D_5 and R_1, and the anode potential of Th_1 is reduced; Th_1 is thus turned off, and C_3 is left charged in readiness for the next operation of the commutating circuit, when Th_2 is fired to turn off the lower-row thyristors. The surplus energy in L_1, corresponding to the increased current in it at t_3, is subsequently dissipated in R_1.

Referring again to Figure 5.56, the turn-off interval ends at t_1, when both the anodes and the cathodes of the upper-row thyristors are at the potential of the negative supply terminal. This is the instant when $v_{L1} = V_d$, an expression which also marks the end of the turn-off interval in the parallel inverter of Figure 5.45 or 5.49, and in the half-bridge of Figure 5.53. This provides a common basis for calculating the performance of all inverter commutating systems in this category, and a method of calculation which can be applied to practical circuits with finite inductance is given in Appendix 5(v).

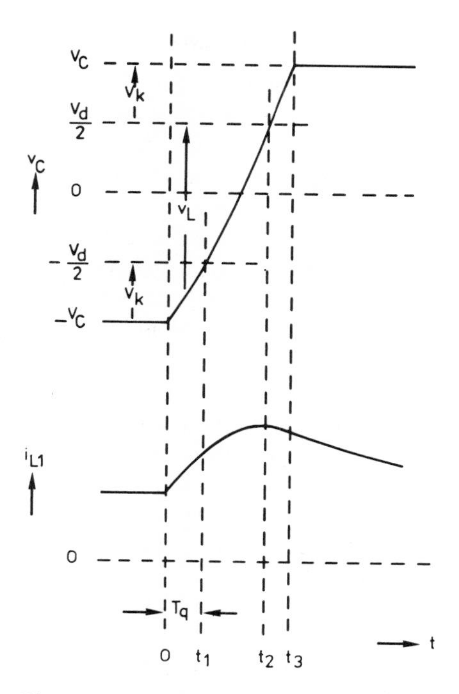

Figure 5.56 Waveforms during commutation in the inverter of Figure 5.55.

Without analysing the circuit of Figure 5.55 in complete detail, the effect of varying L_{d1} and L_{d2} can be understood by considering the extremes of an infinite inductance and one so small that it permits a flow of current into the capacitor which is much greater than the load current. In the first case, the capacitor discharge is linear, and the turn-off interval is given by

$$CV_k = I_1 T_q$$

or, if

$$\frac{V_d + V_k}{V_d} = k$$

$$T_q = \frac{(k-1)\,V_d C}{I_1} \tag{5.41}$$

In the second case, the effect of I_1 may be neglected, and the capacitor voltage follows a cosinusoidal waveform which is independent of the load on the inverter. Then,

$$T_q = \frac{1}{\omega_0}\cos^{-1}\frac{1}{k} \tag{5.42}$$

where

$$\omega_0 = \frac{1}{\sqrt{L_{d1}C}}$$

COMPARISON OF INVERTER TECHNIQUES

It should be apparent from the foregoing discussion that there are cases where the choice between a voltage-fed and a current-fed inverter is determined unequivocally by the nature of the load, once that is established. Clear examples are provided by induction heating or other loads which consist essentially of damped resonant circuits: as explained above, the series-tuned circuit, having a high impedance to harmonic currents and able to support any waveform of voltage impressed upon it, is best supplied by a voltage-fed inverter, while the current-fed inverter is appropriate for a parallel-tuned load, which has a low impedance to harmonics, and will develop a near-sinusoidal voltage waveform with almost any waveform of current.

There may, however be some freedom of choice as to whether a load shall be series- or parallel-tuned, and there are loads—an induction motor is an example—that can be supplied satisfactorily with a predetermined waveform of either voltage or current. Moreover, it will be observed that by the addition of reactive feedback diodes and a means of recovering surplus energy from the commutation system the current-fed inverter (particularly if it is designed with low inductance and an appropriately large capacitance) develops into a hybrid forced-commutation inverter with characteristics very similar to those of the true voltage-fed type. It is useful, therefore, to note some general properties which distinguish the two basic approaches to design from a practical point of view, or criteria upon which a choice may be made.

(i) *Efficiency*. The naturally commutating inverters, voltage- or current-fed, are inherently efficient, having virtually no losses other than those in the semiconductor devices. Voltage-fed inverters with forced-commutation systems such as the McMurray circuit tend to be more efficient than equivalent types derived from the current-fed inverter, in spite of the use of energy-recovery systems in the latter.

(ii) *Frequency limitations*. Losses in inverters generally increase with frequency, and the order of efficiencies noted above is thus also the order of upper frequency limits.

(iii) *Load matching*. With resonant loads, as for induction heating, a series circuit, with a voltage-fed inverter, may be considered less convenient than a parallel one because of the high voltages developed across the load components, relative to the supply voltage.

(iv) *Input ripple*. The large ripple current drawn from the d.c. supply by the voltage-fed inverter is generally an inconvenience, necessitating the use of highly rated capacitors. The true current-fed inverter does not suffer from this disadvantage, although the concomitant lack of a simple voltage-control facility is an offsetting drawback.

(v) *Thyristor operating conditions*. The voltage-fed inverter, with its thyristors connected directly across the low-impedance d.c. supply, tends to present more severe problems in connection with fault protection, $\mathrm{d}v/\mathrm{d}t$ and possibly $\mathrm{d}i/\mathrm{d}t$.

(vi) *Ease of control.* Voltage-fed inverters usually offer more convenient control facilities. In contrast to the tuned parallel inverter, for example, the voltage-fed inverter requires no anticipatory control of firing angle, can be switched off very rapidly as a means of overload protection without any difficulty arising from stored energy, and presents no starting problems.

It will be appreciated that these observations are but generalizations, and may need some qualification in the light of particular circumstances which it is not possible to discuss fully here.

APPENDIX 5(i)

CHARACTERISTICS OF RESONANT LOADS

Series resonant circuits (Voltage-fed inverter)

If the load can be represented by resistance R, inductance L and capacitance C in series, the (leading) phase angle is given by

$$\tan\phi = \frac{1}{R}\left(\frac{1}{\omega C} - \omega L\right) \tag{5.43}$$

where ω is the inverter output frequency, less than the resonant frequency $\omega_0 = 1/\sqrt{(LC)}$. Then if Q is defined as $\omega_0 L/R$

$$\tan\phi = \frac{Q}{\omega_0 L}\left(\frac{1}{\omega C} - \omega L\right) = Q\left(\frac{\omega_0}{\omega} - \frac{\omega}{\omega_0}\right) \tag{5.44}$$

The impedance of the load is $Z = R/\cos\phi$, and the power delivered to the load, ignoring harmonics, is

$$P = V_{a1} I_a \cos\phi = \frac{V_{a1}^2}{Z}\cos\phi$$

$$= \frac{V_{a1}^2}{R}\cos^2\phi$$

$$= \frac{8V_d^2}{\pi^2 R}\left(\frac{1}{1+\tan^2\phi}\right) \tag{5.45}$$

(a bridge inverter is assumed).

Evaluating (5.45) indicates that with loads of reasonably high Q a fairly small range of frequency variation affords a considerable degree of power control. From (5.44), the required frequency, given $\tan\phi$, is given by

$$\frac{\omega}{\omega_0} = \frac{-(\tan\phi)/Q + \sqrt{[(\tan^2\phi)/Q^2 + 4]}}{2} \approx 1 - \frac{\tan\phi}{2Q} \tag{5.46}$$

Parallel resonant circuits (Current-fed inverter)

The load being represented by resistance R, inductance L and capacitance C in parallel, the (leading) phase angle is given by

$$\tan\phi = R\left(\omega C - \frac{1}{\omega L}\right) \tag{5.47}$$

where the inverter output frequency ω is above the resonant frequency $\omega_0 = 1\sqrt{(LC)}$. Then if Q is defined as $R/\omega_0 L$,

$$\tan\phi = Q\omega_0 L\left(\omega C - \frac{1}{\omega L}\right) = Q\left(\frac{\omega}{\omega_0} - \frac{\omega_0}{\omega}\right) \tag{5.48}$$

From (5.48),

$$\frac{\omega}{\omega_0} = \frac{(\tan\phi)/Q + \sqrt{[(\tan^2\phi)/Q^2 + 4]}}{2} \approx 1 + \frac{\tan\phi}{2Q} \tag{5.49}$$

APPENDIX 5(ii)

McMURRAY COMMUTATION CIRCUIT

Figure 5.16 shows waveforms for a complete commutation process. The commutating thyristor Th_1 (Figure 5.15) is fired at $t = 0$, and the resulting capacitor current i_C rises to the level of the inverter output current i_a at t_1. During the period 0–t_1, the current in Th_3 falls from i_a to zero, and the turn-off interval T_q starts at t_1 and ends at t_3, when i_C falls again to equal i_a (i_a is assumed to be constant for the duration of the commutation process); meanwhile the capacitor voltage falls from its initial value V_C to a level V_1 in the reverse direction, $-V_1$ being less than V_C, since less than one half-cycle of oscillation has elapsed. At t_3, assuming Th_3 to have regained its blocking condition, the potential of the cathode of Th_3, relative to the negative supply potential, falls in a step from V_d to zero as ($i_C - i_a$) reverses direction and transfers from D_1 to the only available path, through D_2, and this step of voltage is applied to the resonant circuit L–C in a direction opposing the decline of current in it. The charging of the capacitor is completed under the changed conditions in the resonant loop, and i_C falls finally to zero at t_4, leaving the capacitor with a reversed charge at a voltage V_2, again in excess of the supply voltage, in preparation for a similar operation when Th_2 (Figure 5.14) is fired to turn off Th_4.

If the load is not in fact inductive, or if operation of the commutating circuit is desired with no load current, the step of voltage at t_2 may be induced by firing Th_4 (so long as Th_3 is in a blocking state).

If operation is repetitive and stable, the final capacitor voltage V_2 must be equal to $-V_C$. That V_C cannot be less than V_d can be shown as follows. Immediately after t_3, the nett voltage acting in the resonant loop is $V_d + V_1$; if this is positive (i.e. $V_d > -V_1$) the final nett voltage ($V_d + V_2$) must be negative, because the resonant circuit is underdamped, and $-V_2 > V_d$. If, on the other hand, ($V_d + V_1$) is negative (i.e. $-V_1 > V_d$), any current that flows in Th_1 after t_3 can only increase $-v_c$ further, and again $-V_2 > V_d$.

The ratio of V_C to V_d under given conditions can be determined by means of simultaneous equations representing the state of the circuit at t_4. The capacitor current can be deduced from an equivalent circuit shown in Figure 5.57(a), in which the resonant circuit, initially completely de-energized, is subjected to two consecutive step functions of voltage—the first, at t_0, of amplitude V_C and the second, at t_3, of amplitude V_d, as in Figure 5.57(b). The current is calculated by superposing the separate components due to these two step functions.

In Figure 5.57(b), $\phi = \omega_0 t_3$ and $\psi = \omega_0 t_4$, where ω_0 is the resonant frequency of the loop. The inductor is assumed to have a finite Q factor, and strictly

$$\omega_0 = \sqrt{\frac{1}{LC(1 + 1/4Q^2)}}$$

but for practical purposes it is sufficiently accurate to take ω_0 as equal to $1/\sqrt{(LC)}$ (assuming Q to be greater than about four).

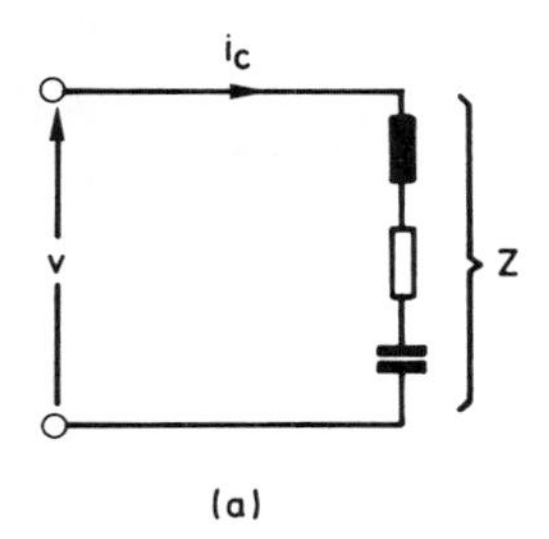

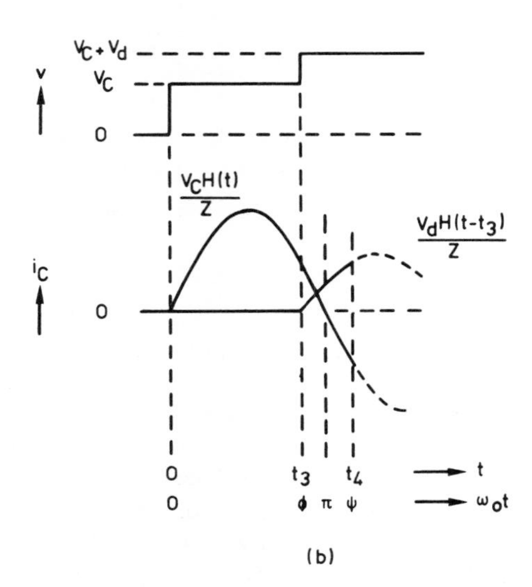

Figure 5.57 Representation of McMurray commutation circuit (Figure 5.14): (a) equivalent circuit; (b) voltage function applied to the equivalent circuit and the resulting components of i_C.

According to the normal response of an underdamped series resonant circuit, the current after the second voltage step at t_3 is

$$i_C = \omega_0 C\{V_C \mathrm{e}^{-\omega_0 t/(2Q)} \sin \omega_0 t + V_\mathrm{d} \mathrm{e}^{-(\omega_0 t - \phi)/(2Q)} \sin(\omega_0 t - \phi)\} \tag{5.50}$$

Therefore, when $i_C = 0$ at t_4,

$$V_C \mathrm{e}^{-\psi/(2Q)} \sin \psi + V_\mathrm{d} \mathrm{e}^{-(\psi - \phi)/(2Q)} \sin(\psi - \phi) = 0 \tag{5.51}$$

Whence

$$\frac{V_C}{V_\mathrm{d}} = -\frac{\mathrm{e}^{\phi/(2Q)} \sin(\psi - \phi)}{\sin \psi} \tag{5.52}$$

The total change in capacitor voltage from t_0 to t_4 is $2V_C$, and is related to i_C by

$$2V_C = \frac{1}{C} \int_0^{\psi/\omega_0} i_C \mathrm{d}t \tag{5.53}$$

Integrating (5.50), with a slight approximation,

$$\begin{aligned} 2V_C = {} & V_C\left\{1 - \mathrm{e}^{-\psi/(2Q)}\left(\cos\psi + \frac{\sin\psi}{2Q}\right)\right\} \\ & + V_\mathrm{d}\left\{1 - \mathrm{e}^{-(\psi - \phi)/(2Q)}\left(\cos(\psi - \phi) + \frac{\sin(\psi - \phi)}{2Q}\right)\right\} \end{aligned} \tag{5.54}$$

or

$$\frac{V_C}{V_\mathrm{d}} = \frac{1 - \mathrm{e}^{-(\psi - \phi)/(2Q)}\left\{\cos(\psi - \phi) + \dfrac{\sin(\psi - \phi)}{2Q}\right\}}{1 + \mathrm{e}^{-\psi/(2Q)}\left\{\cos\psi + \dfrac{\sin\psi}{2Q}\right\}} \tag{5.55}$$

Equations (5.52) and (5.55) can be solved simultaneously by taking successive values of ϕ and finding by an iterative procedure the value of ψ that makes the two expressions equal, yielding a curve such as that shown in Figure 5.58 for the particular parameter $Q = 10$.

If the voltage step at t_3 occurs automatically when i_C falls to the level of i_a, ϕ is itself dependent on V_C and i_a, since at t_3

$$i_\mathrm{a} = i_C = \omega_0 C V_C \mathrm{e}^{-\phi/(2Q)} \sin \phi \tag{5.56}$$

This results in an increase of capacitor voltage with increasing load current, which can be seen as a virtue of the circuit, in that the maximum losses in the commutation process occur only when a high capacitor voltage is actually required by the load on the inverter. The usefulness of this feature is however somewhat reduced by the fact that the increase of voltage follows the increase of current, and provides effective compensation only if the rate of increase of current is limited. The capacitor voltage produced by a given set of conditions in this mode of operation can be estimated by solving simultaneously the three equations (5.52), (5.55) and (5.56).

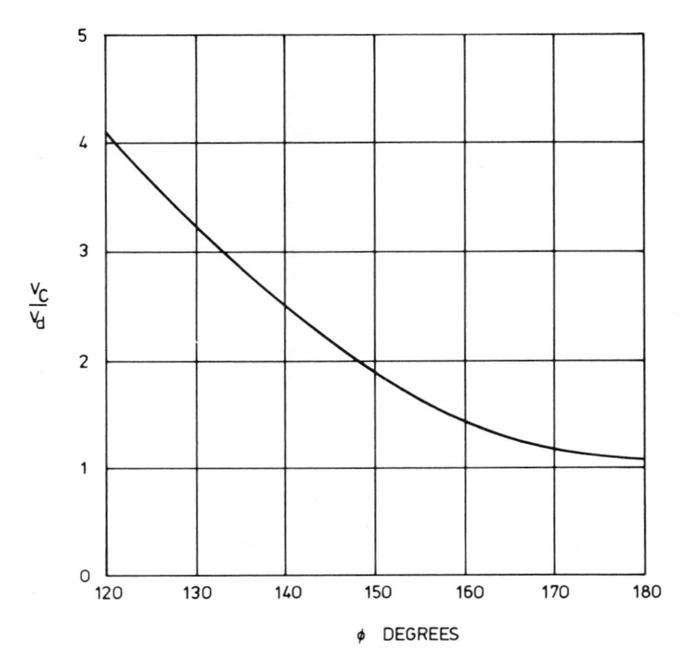

Figure 5.58 Variation of capacitor voltage in the McMurray commutation circuit with $Q = 10$.

Under no-load conditions, the voltage step at t_3 has to be produced by firing Th_4, and ϕ may reach the maximum value π. The operation of the circuit may then be considered in two separate parts, as illustrated in Figure 5.59. Since each part constitutes a complete half-cycle of oscillation,

$$V_1 = -V_C e^{-\pi/(2Q)} \tag{5.57}$$

and

$$V_d - V_C = -(V_1 + V_d)e^{-\pi/(2Q)}$$

$$= V_C e^{-\pi/Q} - V_d e^{-\pi/(2Q)} \tag{5.58}$$

$$\therefore \frac{V_C}{V_d} = \frac{1 + e^{-\pi/(2Q)}}{1 + e^{-\pi/Q}} \tag{5.59}$$

(Equation (5.59) also follows from (5.55) with $\phi = \pi$ and $\psi = 2\pi$).

There is a loss of energy at each operation of the commutating circuit approximately equal to the loss that would occur in a half-cycle of natural oscillation starting with the capacitor charged to V_C:

$$-\delta E = \frac{C}{2}\{V_C^2 - (V_C e^{-\pi/(2Q)})^2\} = \frac{CV_C^2}{2}(1 - e^{-\pi/Q}) \tag{5.60}$$

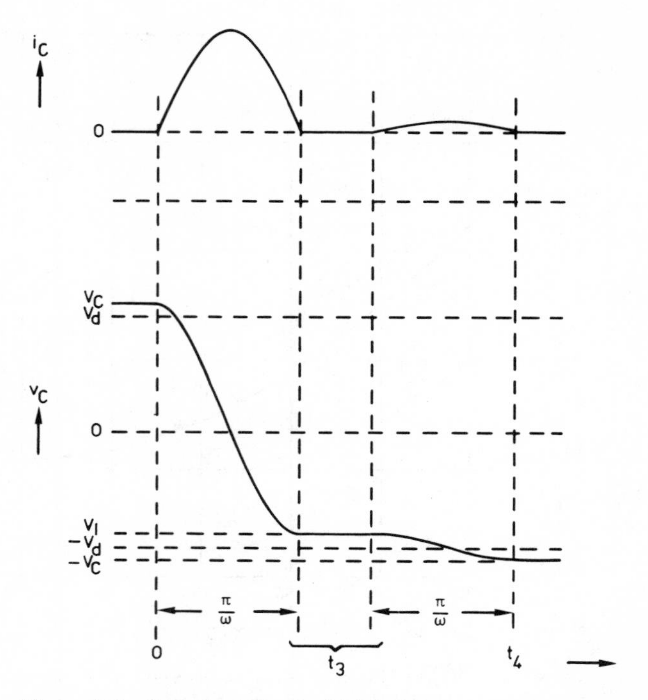

Figure 5.59 Waveforms in the McMurray commutation circuit under no-load conditions.

APPENDIX 5(iii)

FOURIER ANALYSIS OF RECTILINEAR WAVEFORMS

Figure 5.60 shows the basic rectilinear waveform, as of the output voltage of a voltage-fed single-phase bridge inverter, with angle of delay α and pulse duration λ/ω, arranged for convenience so that the harmonic analysis yields only (odd) sine terms. If the waveform, of unit amplitude, is represented as

$$f(\omega t) = a_1 \sin \omega t + a_3 \sin 3\omega t + a_5 \sin 5\omega t + \dots + a_n \sin n\omega t + \dots$$

then

$$a_n = \frac{2}{\pi} \int_0^{\pi} f(\omega t) \sin n\omega t \, d\omega t \tag{5.61}$$

$$= \frac{2}{\pi} \int_{\alpha/2}^{\pi - \alpha/2} \sin n\omega t \, d\omega t$$

$$= \frac{2}{\pi} \left[-\frac{\cos n\omega t}{n} \right]_{\alpha/2}^{\pi - \alpha/2} = \frac{4}{n\pi} \cos \frac{n\alpha}{2} \tag{5.62}$$

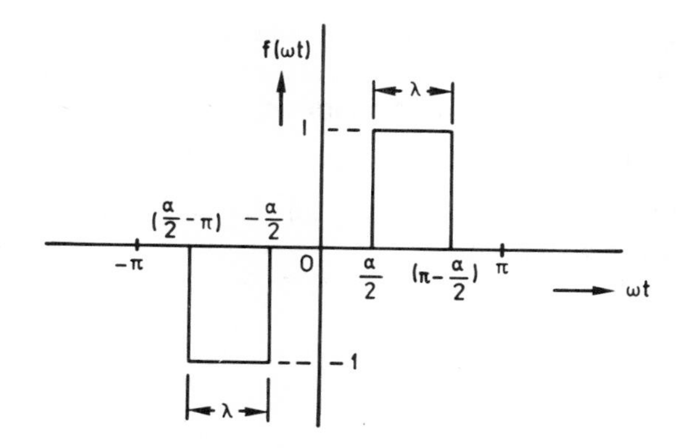

Figure 5.60 Generalized rectilinear waveform.

Alternatively, since $\lambda = \pi - \alpha$,

$$a_n = \frac{4}{n\pi}\cos\left(\frac{n\pi}{2} - \frac{n\alpha}{2}\right) = \frac{4}{n\pi}\sin\frac{n\alpha}{2} \tag{5.63}$$

APPENDIX 5(iv)

OUTPUT VOLTAGE OF SIMPLE CURRENT-FED INVERTER WITH *CR* LOAD

In the inverter of Figure 5.45 with a capacitive and resistive load and a d.c. inductor sufficiently large to make the input current substantially smooth, the output current I_a may be taken to be of square waveform. The fundamental component of output current is therefore

$$I_{a1} = \frac{2\sqrt{2}}{\pi} I_d \tag{5.64}$$

and the fundamental-frequency output voltage is

$$V_{a1} = I_{a1} R \cos\phi_1 = \frac{2\sqrt{2}}{\pi} I_d R \cos\phi_1 \tag{5.65}$$

where

$$\cos\phi_1 = \frac{1/\omega_1 C}{\sqrt{(R^2 + 1/\omega_1^2 C^2)}}$$

Equating the input and output powers,

$$\frac{V_a^2}{R} = V_d I_d = \frac{V_d V_{a1}\pi}{2\sqrt{2}R\cos\phi_1} \tag{5.66}$$

$$\therefore \frac{V_a}{V_d} = \left(\frac{V_{a1}}{V_a}\right)\frac{\pi}{2\sqrt{2}\cos\phi_1} \tag{5.67}$$

The r.m.s. output voltage V_a is the root-sum-square of the fundamental and harmonic components, which are the products of the harmonic components of I_a and the impedance of the load at the harmonic frequencies. Thus

$$V_a = \sqrt{[(I_{a1} R \cos \phi_1)^2 + (I_{a3} R \cos \phi_3)^2 + (I_{a5} R \cos \phi_5)^2 + \ldots} \quad (5.68)$$

If C is negligible, so that $\cos \phi_n \approx 1$ for all significant harmonics, the output voltage is of square waveform

and
$$\frac{V_{a1}}{V_a} = \frac{2\sqrt{2}}{\pi}$$

Hence
$$\frac{V_a}{V_d} = \frac{1}{\cos \phi_1} = 1 \quad (5.69)$$

If $\cos \phi_1$, is substantially less than one, however, the significance of the terms in I_{a3}^2, I_{a5}^2 etc. is reduced, since $\cos \phi_n$ decreases with frequency, and if $\cos \phi_1$ is less than about 0.8 all except the fundamental term can be ignored without introducing an error of more than 2% in V_{a1}/V_a. Thus for $0 < \cos \phi_1 < 0{\cdot}8$,

$$\frac{V_a}{V_d} \approx \frac{\pi}{2\sqrt{2} \cos \phi_1} \quad (5.70)$$

The reciprocal function V_d/V_a is plotted against $\cos \phi_1$ in Figure 5.61.

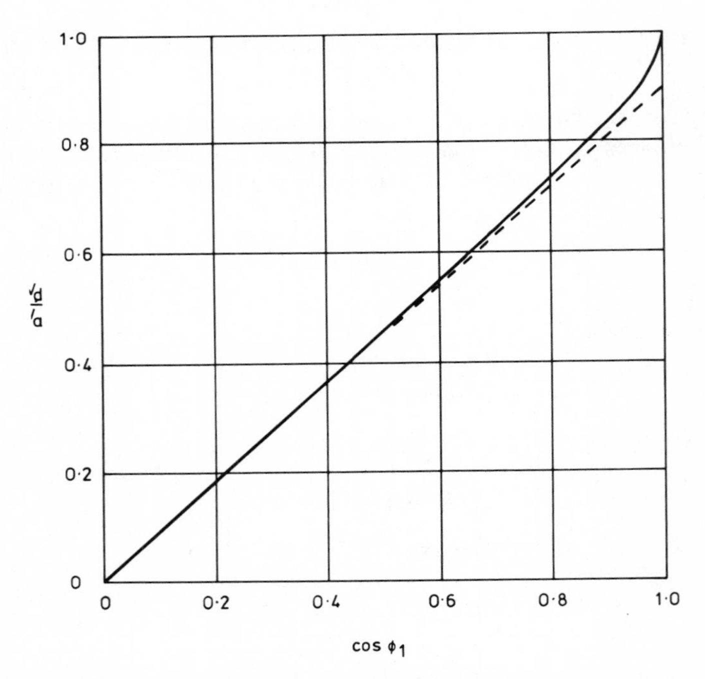

Figure 5.61 Variation of V_d/V_a with (leading) load power factor in a simple current-fed inverter.

APPENDIX 5(v)

COMMUTATION IN CURRENT-FED INVERTERS WITH REACTIVE FEEDBACK

The commutating circuits of inverters of the general type represented by Figures 5.49, 5.53 and 5.55 can be analysed in terms of the equivalent circuit of Figure 5.62, wherein V_1 and I_1 are the initial voltage and current acting in the L–C loop at $t = 0$ when the commutating thyristor (or the next thyristor or thyristors in the sequence) is fired. Referring to the inverter of Figure 5.55, $V_1 = V_d/2 + V_C$, while $I_1 = i_d$ unless it is augmented by any circulating current still flowing at the instant of commutation.

Ignoring losses, the current in the loop is

$$i = I_1 \cos \omega_0 t + \frac{V_1}{Z_0} \sin \omega_0 t \tag{5.71}$$

where $\omega_0 = 1/\sqrt{(LC)}$ and $Z_0 = \sqrt{(L/C)} = \omega_0 L = 1/(\omega_0 C)$

This may be re-written:

$$i = \sqrt{(I_1^2 + V_1^2/Z_0^2)} \sin (\omega_0 t + \alpha) \tag{5.72}$$

where $$\alpha = \tan^{-1} \frac{I_1 Z_0}{V_1}$$

The voltage across the inductor is

$$\begin{aligned} v_L = L\frac{\mathrm{d}i}{\mathrm{d}t} &= \omega_0 L\sqrt{(I_1^2 + V_1^2/Z_0^2)}\cos(\omega_0 t + \alpha) \\ &= \sqrt{(V_1^2 + I_1^2 Z_0^2)} \cos (\omega_0 t + \alpha) \\ &= \frac{V_1 \cos (\omega_0 t + \alpha)}{\cos\alpha} \end{aligned} \tag{5.73}$$

As observed previously, the turn-off interval ends when $v_L = V_d$. If, therefore, V_1/V_d is defined as k, a ratio which expresses the necessary voltage rating of the inverter thyristors relative to the supply voltage,

$$\cos (\omega_0 T_q + \alpha) = \frac{V_d}{V_1} \cos \alpha = \frac{\cos\alpha}{k} \tag{5.74}$$

Equation (5.73) is illustrated graphically in Figure 5.63.

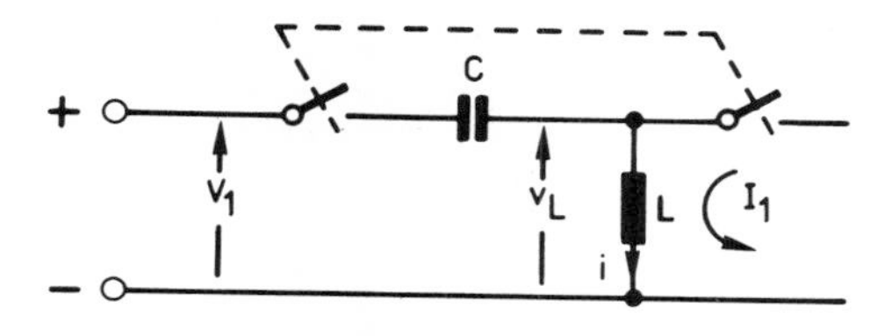

Figure 5.62 Equivalent of commutation circuit in a current-fed inverter.

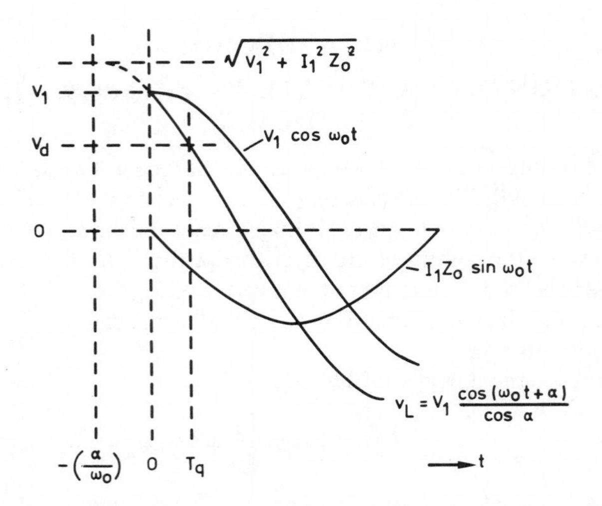

Figure 5.63 Waveforms in the circuit of Figure 5.62.

For a given required value of T_q and a given ratio V_1/V_d, putting any arbitrary value of α (between 0 and $\pi/2$) into (5.74) gives a corresponding value of ω_0, while the definition of tan α leads to a value for Z_0. From these a particular combination of $C\,(= 1/\omega_0 Z_0)$ and $L\,(= Z_0/\omega_0)$ may be obtained. The results can be presented in a general form by defining two notional time constants, $\tau_C = C V_d/I_1$ and $\tau_L = L I_1/V_d$: then

$$\left.\begin{aligned} \tau_L \tau_C &= LC = \frac{1}{\omega_0^2} \\ \frac{\tau_L}{\tau_C} &= \frac{L I_1^2}{C V_d^2} = k^2 \tan^2 \alpha \\ \frac{\tau_L}{\tau_q} &= \frac{k \tan \alpha}{\omega_0 T_q} \\ \frac{T_C}{T_q} &= \frac{1}{\omega_0 k \tan \alpha \, T_q} \end{aligned}\right\} \tag{5.75}$$

Based on these equations, τ_C/T_q is plotted against τ_L/T_q for two values of k in Figure 5.64. The resulting curves are approximately hyperbolic in shape, asymptotic to the value of τ_C/T_q required with infinite inductance, which is given by

$$I_1 T_q = C V_k = C V_d (k-1)$$

$$\therefore \frac{\tau_C}{T_q} = \frac{C V_d}{T_q I_1} = \frac{1}{k-1} \tag{5.76}$$

Given the relationship of C to L, an optimum combination may be chosen on the basis of cost, the variation of T_q with load, or other criteria.

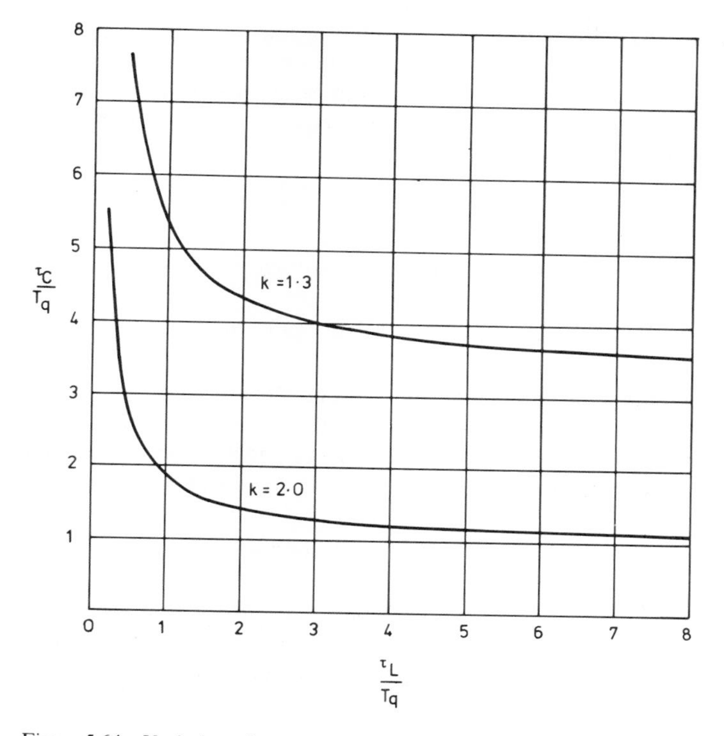

Figure 5.64 Variation of commutating capacitance with d.c. inductance in a current-fed inverter with reactive feedback.

The same approach may be applied to the bridge and half-bridge circuits (Figures 5.49 and 5.53). In the bridge, $V_1 \approx 2V_d$ $(k \approx 2)$, $I_1 \approx 2i_d$ and L is the total input inductance. In the half-bridge, calculation is more conveniently in terms of $V_d/2$ and the inductance of one half-winding of the inductor; then $V_1 = V_d/2 + V_C \approx V_d$ $(k \approx 2)$, $I_1 \approx 2i_d$ and $C = C_1 + C_2$.

APPENDIX 5(vi)

HARMONICS GENERATED BY SINUSOIDAL P.W.M. IN A BRIDGE INVERTER

If the pulse repetition frequency in a pulse-width-modulation inverter producing the kind of waveform shown in Figure 5.39 is high in comparison with the modulating (fundamental) frequency, the effect may be considered as that of a switching regulator operating with a switching ratio γ (see Chapter 4) which varies through the half-cycle in proportion to the instantaneous value of the modulating signal: i.e.,

$$\gamma(t) = m \sin \omega_1 t \qquad (5.77)$$

where m is the modulation index, which is the ratio of the amplitude of the fundamental output voltage to the d.c. supply voltage.

At any point in the half-cycle, this produces a series of harmonics of the pulse repetition frequency in accordance with equation (4.10), with an r.m.s. value (measured over a short period)

$$V_{cn}(t) = V_d \frac{\sqrt{2}\sin n\gamma\pi}{n\pi} \tag{5.78}$$

Related to the peak fundamental output voltage $\hat{V}_{o1} = mV_d$, this becomes

$$V_{cn}(t) = \frac{\sqrt{2}\,\hat{V}_{o1}}{nm\pi} \sin(nm\pi \sin\omega_1 t) \tag{5.79}$$

This represents a carrier, at the pulse repetition frequency, and its harmonics, modulated by a function of the fundamental modulating signal, producing the groups of sidebands referred to earlier. The most important components are those centred about the carrier frequency itself—i.e., those produced by

$$V_c(t) = \frac{\sqrt{2}\,\hat{V}_{o1}}{m\pi} \sin(m\pi \sin\omega_1 t) \tag{5.80}$$

The modulating function of (5.80) has a waveform that varies with m but is generally of the form shown in Figure 5.65, and is found, by numerical Fourier analysis, to contain significant components at the fundamental frequency (f_1) and at the third and fifth harmonics. Each of these components gives rise to a product

$$v = \hat{V}_{o1} \times a_n \sin n\,\omega_1 t \sin\omega_c t \tag{5.81}$$

where a_n is the Fourier coefficient of the nth-harmonic sine term obtained from the modulating waveform (Figure 5.65). Hence,

$$v = a_n \hat{V}_{o1} \frac{\sin(\omega_c + n\omega_1)t + \sin(\omega_c - n\omega_1)t}{2} \tag{5.82}$$

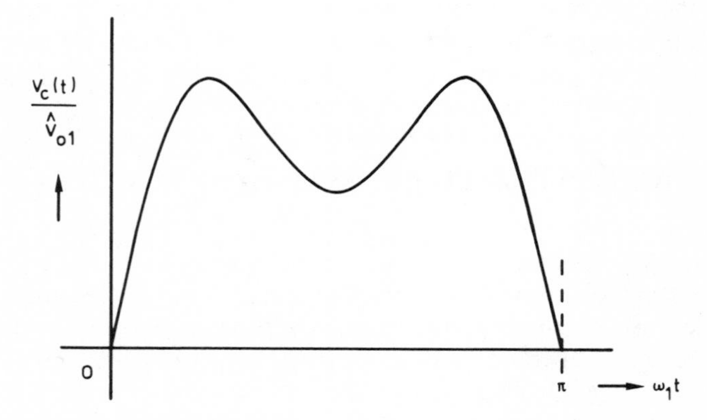

Figure 5.65 Typical waveform of carrier modulation in a pulse-width modulation bridge inverter ($m = 0.8$).

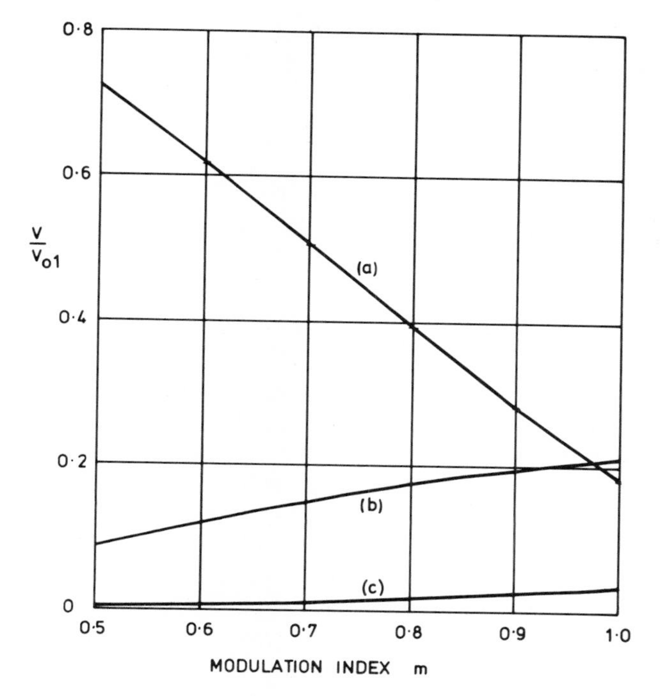

Figure 5.66 Variation with modulation index of the principal harmonics in the output voltage of a pulse-width modulation bridge inverter: (a) $f_c \pm f_1$; (b) $f_c \pm 3f_1$; (c) $f_c \pm 5f_1$.

and the r.m.s. modulation products are $(a_n/2)V_{o1}$ at $(f_c + nf_1)$ and $(a_n/2)V_{o1}$ at $(f_c - nf_1)$.

The significant harmonics are thus at frequencies $(f_c \pm f_1)$, $(f_c \pm 3f_1)$, and $(f_c \pm 5f_1)$, with respective magnitudes $a_1/2, a_3/2$, and $a_5/2$ relative to the fundamental. These quantities are plotted for a range of modulation indices in Figure 5.66; they are quite accurate so long as the carrier frequency is not less than about fifteen times the modulating frequency. Similar methods may be applied to other inverter configurations. For lower frequency ratios, it is usual to employ more specific computer analysis techniques.

CHAPTER 6

Application techniques

COOLING OF SEMICONDUCTOR DEVICES

The heat produced by losses in a semiconductor device has to be dissipated sufficiently effectively to enable the device to operate within the upper limit of temperature imposed by its ratings, its characteristics, or both. Dissipation entails the transmission of heat to whatever may be considered as the ultimate heat-sink—i.e., the repository of heat whose temperature does not change significantly as a result of the dissipation; this is normally either atmospheric air or water. The semiconductor device is usually mounted (unless it is very small) on a cooling device, which is essentially a means of coupling the device thermally to the ultimate heat-sink; the cooling device itself may, however, constitute a heat-sink for limited periods of time, by virtue of its thermal capacity, and is frequently referred to as such.

The complete thermal path from the source of heat (loosely identified as 'the junction') includes at least three links:

(i) junction to cell base or stud;
(ii) cell base or stud to cooling device;
(iii) cooling device to ultimate heat-sink

This system may be represented, under steady-state conditions, by a chain of thermal resistances analogous to an electrical resistance chain, with temperature corresponding to electrical potential and heat, expressed as power, corresponding to current.

Figure 6.1 represents the case of a cooling fin dissipating to atmosphere, or alternatively a liquid cooler carring mains water. W represents the generated heat (measured in watts), R_{jb}, R_{mtg} and R_c are respectively the thermal resistances from junction to base, from base to cooler and from cooler to air or water (measured in °C/watt), and θ_j, θ_b, θ_c and $\theta_{a(w)}$ are the temperatures of the junction, the base of the cell, the cooler at the mounting location of the cell and the ambient air (water). Then

$$\theta_j = \theta_{a(w)} + W(R_{jb} + R_{mtg} + R_c) \tag{6.1}$$

In this equation, R_{jb} and R_{mtg} are linear resistances, albeit subject to manufacturing tolerances, while R_c may be linear or somewhat non-linear. All three values are simplified representations of the actual elements, since these consist of three-dimensional components, and it is necessary to specify points at which θ_b, θ_c and $\theta_{a(w)}$ are taken to exist if measurements are to be made with any consistency.

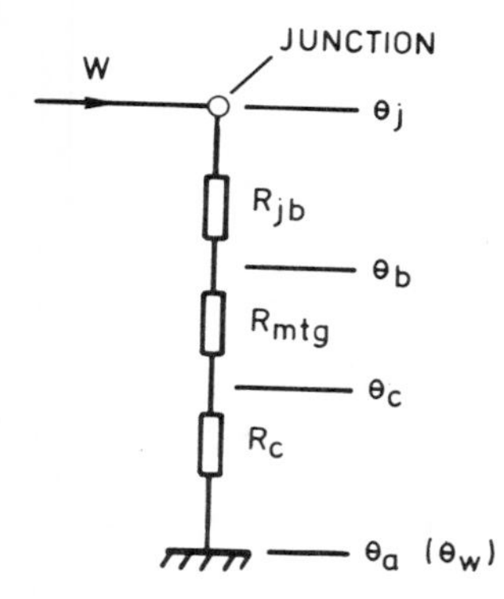

Figure 6.1 Thermal resistance network representing a semiconductor device in the steady state.

The steady-state ratings of diodes and thyristors are normally related to base temperature rather than junction temperature, and in many cases, therefore, calculation will be restricted accordingly:

$$\theta_b = \theta_{a(w)} + W(R_{mtg} + R_c) \tag{6.2}$$

Mounting of semiconductor devices

Most single-ended power semiconductor devices have either a parallel-threaded stud, to be screwed into a tapped hole in the cooler or secured with a nut, or a flat base to be clamped to a flat surface. (Less usual methods of mounting include taper threads and soft-soldering.) Capsule-type cells are clamped either to single coolers, or, for more effective cooling, between pairs of coolers.

The thermal resistance of the interface between the base of the cell and the cooler, R_{mtg}, may be a substantial fraction of the internal thermal resistance of the cell, and it is generally important to minimize it by accurate machining of the mating surfaces and tapping of threads, the removal of burrs, careful assembly and the use of recommended jointing compounds.

In most cases the cooler serves additionally as the means of making an electrical connection to the base of the cell, but it is sometimes preferred for high-power equipment incorporating flat-based cells to provide the base with a lug for direct connection in order to reduce the voltage drop and to avoid the possibility of electrolytic corrosion at the junction of the dissimilar materials of the cell base and the cooler. Capsule cells have an advantage from the electrical point of view, because of the higher mounting pressures used with them.

In some cases the equipment design concept requires the semiconductor devices to be electrically insulated from their coolers. Ordinary insulating materials can be used for this purpose, but much better performance can be obtained with one of the several materials available with good electrical strength combined with exceptionally high thermal conductivity. Beryllia (beryllium oxide) has a thermal conductivity about equal to that of aluminium at moderate temperatures; that of alumina (aluminium oxide) is about an order lower. Boron nitride lies between beryllia and alumina in regard to thermal conductivity, but has very different mechanical properties which make it more attractive, for many purposes, than either. (It is also free from the toxicity which often makes beryllia unacceptable.)

Very small devices mounted by their leads may be adequately cooled by dissipation from their bodies, but often the ratings are dependent on heat conduction through the leads to their points of attachments.

Air coolers

The material of an air-cooling device (usually aluminium) serves firstly to present an exposed surface from which the heat is dissipated by convection and radiation, and secondly to conduct heat from the semiconductor device to the surface. The conflicting requirements of a large surface area and a low thermal resistance between the surface and the heat source make the optimum design of coolers a matter of specialized skill. Accuracy in design is in any case difficult to achieve and the principles referred to below are best regarded as a basis for an empirical approach.

Cooling fins intended for limited dissipation may be constructed from sheet metal, but their small cross-sectional area makes it difficult to obtain a total thermal resistance much below about 1°C/W. More effective profiles can be employed in castings or extruded sections; even so, there is a practical limit to the performance that can be achieved, almost regardless of size. 0.25°C/W represents about the best that can be obtained with natural convection: forced convection enables this to be reduced to around 0.1 °C/W (these figures refer to single-sided cooling). Because a much greater heat flow is possible with forced than with natural convection, the optimum relationships of surface to cross-sectional area are different for the two modes, and a particular cooler design cannot be ideal for both, although it may be used for both as a matter of convenience.

Natural air cooling

For practical purposes, a naturally cooled fin dissipates heat by two processes—convection and radiation. Expressions relating the rate of heat loss to the temperature of the surface are given in Appendix 6(i), but it is important to note that these are basic relationships which apply to ideal situations and cannot be expected in themselves to indicate accurately the characteristics of practical fins: firstly, the overall characteristics are considerably affected by the thermal resistance between the cell and the surface of the fin, and secondly, the heat lost by radiation is normally much reduced by radiation emitted or reflected from neighbouring components, including adjacent fins, and by the re-entrant shapes of any but the simplest structures, which prevent much of the surface from radiating to the surrounding space. The diminished significance of radiation also means that the performance of a fin is less affected by the surface finish than might be supposed, particularly with complicated profiles, and varies according to the environment.

The natural dissipating characteristic shows a non-linearity, which is represented nearly enough by $W \propto (\theta_c - \theta_a)^x$; in the ideal case of natural convection,

$x = 1.25$, but in practice it normally lies between 1.1 and 1.2 because of the addition of linear terms. there is thus no true, fixed thermal resistance, but a notional value which varies with temperature rise, typically according to the power of from -0.1 to -0.2.

Forced air cooling

Large semiconductor devices, especially thyristors, with their relatively low maximum junction temperatures, cannot operate near their full current ratings with natural air cooling. Even a modest degree of forced air circulation reduces the thermal resistance of the cooler considerably, and greater rates of air flow, up to a normal limit of around ten or fifteen metres a second air velocity, yield a performance about three times better than is obtainable with natural air cooling, assuming the cooler to be appropriately designed. Under these conditions the effect of radiation becomes negligible, and because the rate of air flow is virtually independent of the cooler temperature the thermal resistance becomes practically constant, so that the whole thermal system can be assumed to be linear.

The estimation of heat loss from a surface at uniform temperature by forced convection is complicated by the question of whether the air flow is laminar or tubulent: if laminar, dissipation is proportional to the square root of the air velocity—if turbulent, to velocity to the power of 0.8. The flow in fact changes

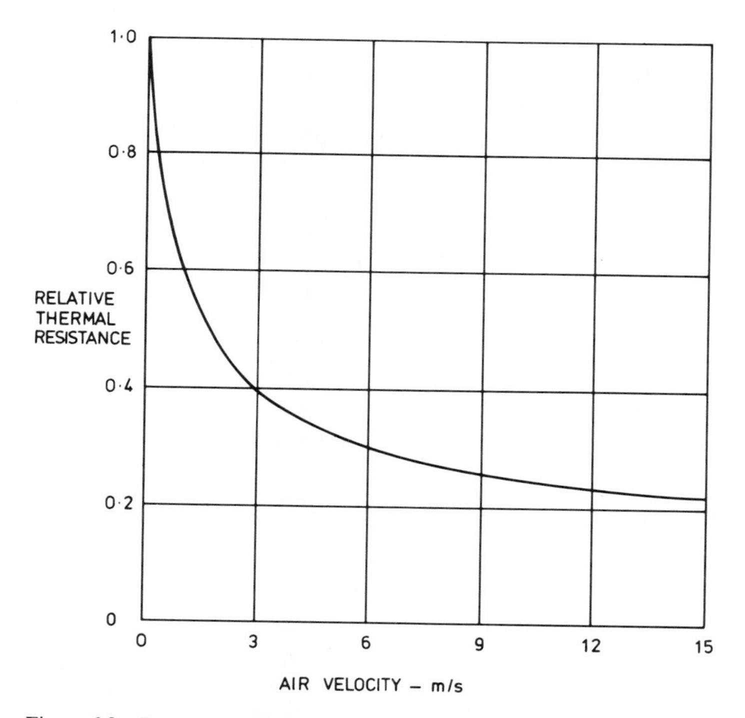

Figure 6.2 Typical variation with air velocity of the thermal resistance of a forced-convection cooler.

from laminar to turbulent when the air velocity is increased beyond a critical value which depends on the design of the cooler: in practice it is likely to be turbulent in most of the range of velocities used. Figure 6.2 illustrates a typical relationship between air velocity and measured thermal resistance, and idealized relationships in terms of a uniform surface temperature are quoted in Appendix 6(i).

Thermal time response

Apart from its steady-state dissipation characteristics, a cooler has appreciable thermal mass, which gives it a finite response time which is of interest when an equipment is required to have a short-term or overload rating in excess of its continuous rating. Commonly, albeit not quite accurately, this is expressed as an exponential time constant

$$\tau_{th} = R_{th}C_{th} \tag{6.3}$$

While the thermal resistance of a given cooler is reduced by forced air circulation, its thermal capacity is not changed, and its thermal time constant therefore becomes shorter. Any sizeable cooler, however, has a thermal time constant very much greater than those within the semiconductor device mounted on it, and the two components can be treated separately in questions of transient analysis.

Multiple cooler assemblies

The design of equipment for natural air cooling or of forced air cooling systems often entails mounting coolers one above the other, so that air passes over them in sequence. In such a case allowance has to be made for the fact that the downstream coolers receive air at a higher temperature than those upstream, and in an optimum design they may for that reason be considerably larger. In some designs for natural air cooling the problem is avoided by setting the fins at an angle so that each cooler draws fresh air from the side rather than pre-heated air from the cooler below.

Liquid cooling

In certain classes of industrial equipment, semiconductor devices on suitable fins are cooled by immersion in an oil tank, commonly shared with other components; more generally applicable forms of liquid cooling involve either hollow liquid-carrying bus-bars on which a series of parallel-connected semiconductor devices, most conveniently in flat-based or capsule form, are mounted, or separate coolers associated with individual cells and connected, for liquid flow, by electrically insulating pipes. Liquid cooling may be preferred to air, in spite of practical problems associated with it, because it can provide much more effective cooling and permits a very compact semiconductor assembly; it also facilitates total enclosure of the equipment, and enables the waste heat to be dissipated remotely.

A typical individual liquid cooler for capsule thyristors or diodes carrying water at the rate of about 0.07 litres/second, provides a thermal resistance of the order of 0.05°C/W at each surface, or 0.025°C/W total for double-sided cooling, including the mounting thermal resistance. Using oil instead of water increases these figures by a factor of two or three.

THE ESTIMATION OF INSTANTANEOUS TEMPERATURE IN SEMICONDUCTOR DEVICES

While under normal steady operating conditions the mean junction temperature of a semiconductor device can be estimated directly as the product of average dissipation and thermal resistance (generally an unnecessary calculation for diodes or thyristors, since the steady-state ratings of these devices are given in terms of base temperature), the instantaneous junction temperature in pulse or transient operation is a complex function of time as well as dissipation and a special technique is necessary to estimate it when required.

Knowledge of instantaneous junction temperature is frequently necessary for such purposes as determining pulse or intermittent current ratings—that is, estimating the extent to which advantage can be taken of brevity of conduction, given that the rated junction temperature is not to be exceeded—or ascertaining whether or not a thyristor will retain its forward blocking characteristic under certain surge conditions (see 'Automatic current limiting').

The time response of a semiconductor device to a varying input of heat is determined physically by the thermal resistances and capacities of its various constituent parts. The overall thermal circuit generally takes the form shown in Figure 6.3: the earth symbol represents whatever may be regarded, in the time scale applicable, as the ultimate receptacle of heat, at constant temperature. Calculation of the thermal response from first principles is difficult and unreliable, and it is therefore usually determined by direct experiment, and published as a curve of transient thermal impedance, which represents the time response of the junction temperature to a unit step function of heat input, as illustrated in Figure 6.4. (It is important to note that the published data for thyristors refers to cells in the fully conducting condition, and cannot be used directly in connection with transient heating due to loss during the turn-on period, when the transient

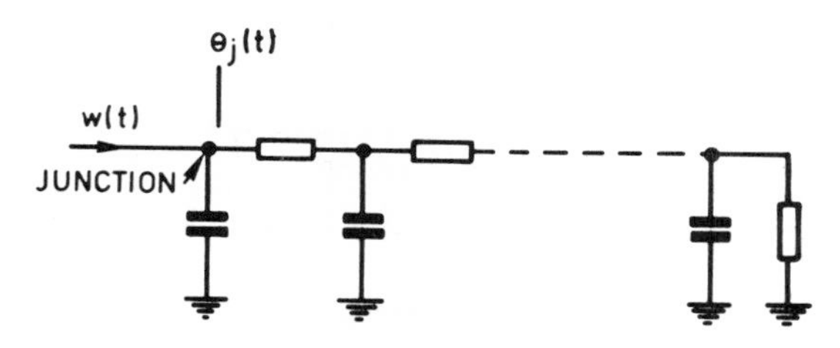

Figure 6.3 Representation of the transient thermal impedance of a semiconductor device.

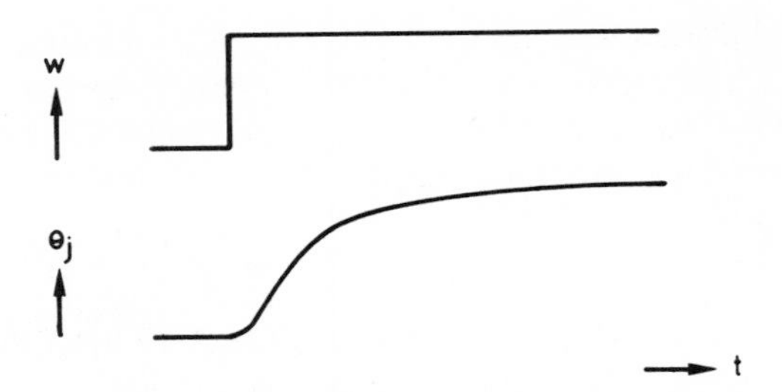

Figure 6.4 Response of junction temperature to a step function of dissipation.

thermal impedance, as well as the loss, varies with time as conduction spreads.) Figure 6.5 illustrates the typical form of such a curve.

If, in a practical application, the dissipation, calculated from the current and the limit voltage-drop characteristic, is in the form of a simple step function, the transient thermal impedance curve may be used directly. Suppose, for example, that a thyristor in a forced-commutation circuit carries a large constant current under a particular fault condition, and that it is required to know whether, at a particular time subsequent to the fault, the junction temperature will have reached such a level as to prevent the thyristor from being turned off: the effective thermal resistance is read off the curve for the time-interval concerned; if this is Z_t for a time interval t, and the dissipation during the interval is W, the junction temperature rise at the end of the interval is WZ_t.

Since the transient thermal impedance curve is no more than a statement of the step response under specific conditions, it cannot be modified to embrace other

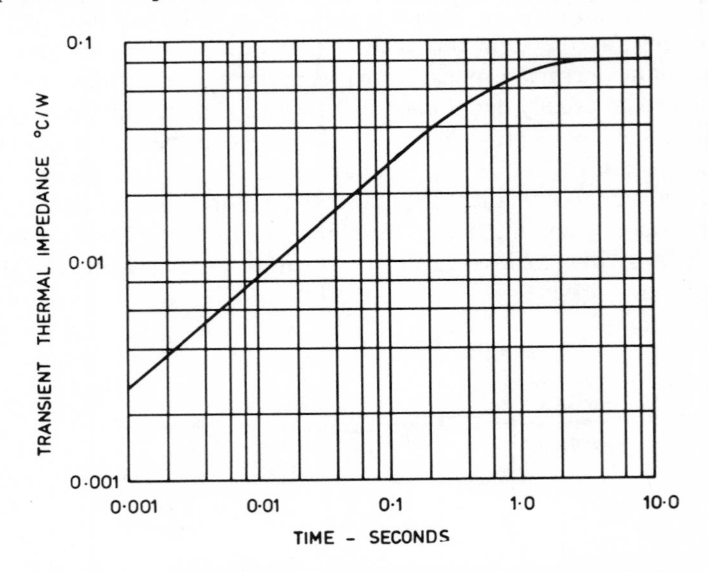

Figure 6.5 Typical transient thermal impedance characteristic of a thyristor.

conditions (the case of a cell mounted on a fin of very little thermal mass, for example, has to be dealt with by means of a separate curve for the particular combination of cell and fin if accurate results are needed over a wide range of time) and it can only be applied directly to dissipation waveforms that can be expressed in terms of step functions.

The method is applied to more complex waveforms by recourse to the principle of superposition, combined in many cases with judicious averaging or other approximation. The procedures available may be illustrated by cronsidering cases in an ascending order of complexity.

Single rectangular pulse

If the dissipation is in the form shown in Figure 6.6(a), and it is required to know the instantaneous junction temperature rise at $t = 0$, the pulse is represented as the sum of two step functions, $+W$ occurring at t_2 and $-W$ at t_1, as in Figure 6.6(b). The junction temperature rise at $t = 0$ is then the sum of the separate effects of the two step functions: i.e.,

$$\delta\theta_{\mathrm{j}} = WZ_{t2} - WZ_{t1} = W(Z_{t2} - Z_{t1}) \tag{6.4}$$

Repeated rectangular pulse

The junction temperature rise at $t = 0$ due to a succession of pulses, as in Figure 6.7, is, by a simple extension of the previous method,

$$\delta\theta_{\mathrm{j}} = W_1(Z_{t2} - Z_{t1}) + W_2(Z_{t4} - Z_{t3}) + W_3(Z_{t6} - Z_{t5}) + \cdots \tag{6.5}$$

or if all the pulses have the same amplitude W,

$$\delta\theta_{\mathrm{j}} = W\{(Z_{t2} - Z_{t1}) + (Z_{t4} - Z_{t3}) + (Z_{t6} - Z_{t5}) + \cdots\} \tag{6.6}$$

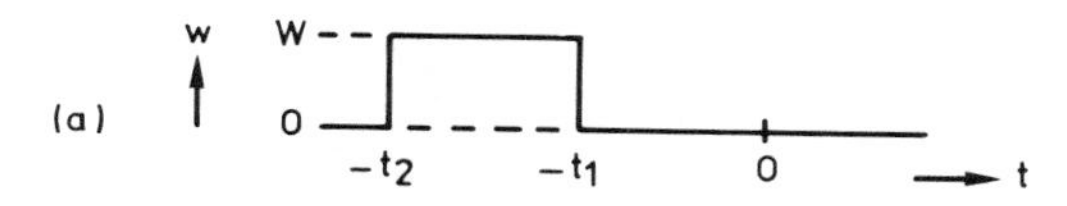

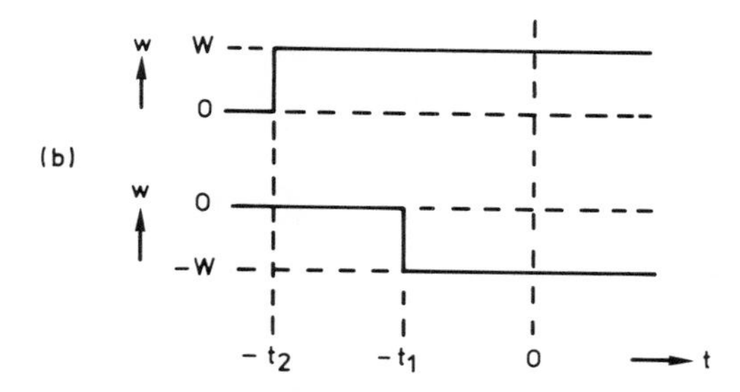

Figure 6.6 Representation of a rectangular pulse by superposed step functions.

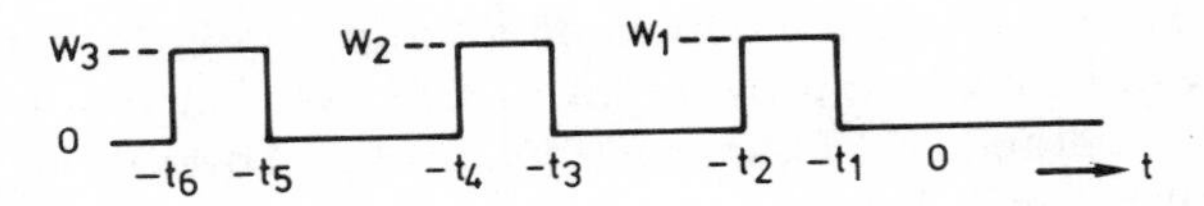

Figure 6.7 Repeated rectangular power pulses.

Extended train of similar rectangular pulses

Only the last few pulses make substantial individual contributions to the final junction temperature rise; this does not mean that the earlier pulses can be totally ignored, since their cumulative effect may be considerable, but it does mean that they can be averaged without introducing an excessive error. The junction temperature resulting from a train of pulses typically varies in the manner illustrated in Figure 6.8. If the average dissipation $WT_a/(T_a + T_b)$ is treated as persisting until the end of a pulse, as shown in Figure 6.9(a)—that is, at the point where the junction temperature is near its maximum value—the result is in some degree optimistic, whereas if it is treated as terminating at the beginning of a pulse, as in Figure 6.9(b), it is pessimistic; the latter is obviously a safer basis of calculation. In any case, working backwards pulse by pulse shows when the incremental effect of an individual pulse is small enough to be ignored.

The junction temperature rise at $t = 0$ calculated on the basis of two discrete pulses, as in Figure 6.9(b), is

$$\delta\theta_j = W\left(\frac{T_a}{T_a + T_b}\right)(Z_\infty - Z_{t4}) + W(Z_{t4} - Z_{t3}) + W(Z_{t2} - Z_{t1}) \qquad (6.7)$$

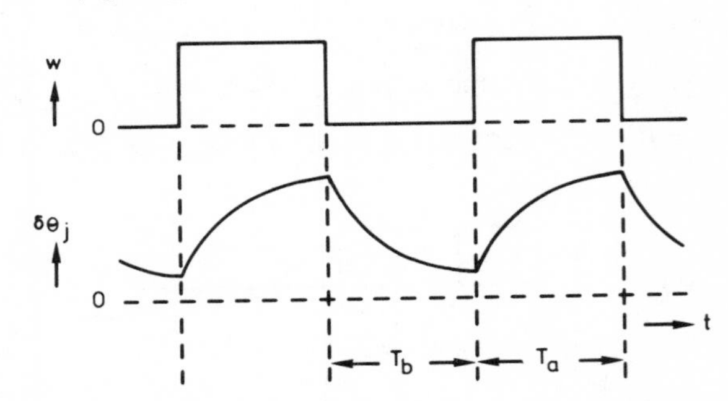

Figure 6.8 Variation of thyristor junction temperature with repeated rectangular power pulses.

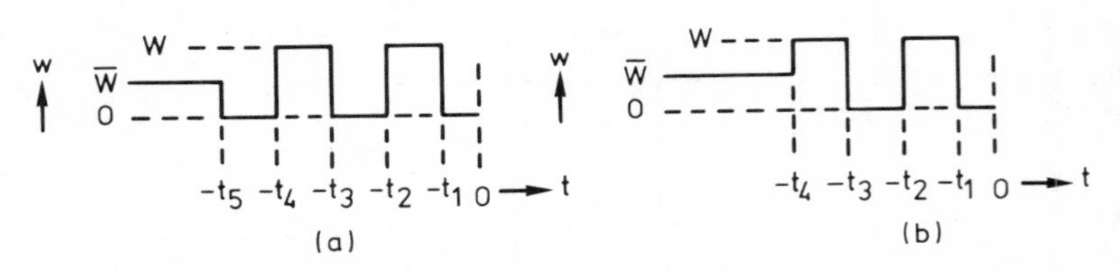

Figure 6.9 Alternative representations of repeated power pulses by average dissipation.

Single pulse of other than rectangular waveform

A pulse of any waveform can be represented by step approximation with little error so long as each step is not of such duration as to encompass an excessive variation of instantaneous power coupled with a large variation of Z_t. Thus the pulses shown in Figure 6.10(a) and (b) are represented with considerable accuracy by ten rectangular pulses of equal duration, the amplitude of each pulse being equal to the average amplitude of the actual pulse over the same period, although in the (unusual) case of a pulse such as that illustrated in Figure 6.10(c) the calculated temperature rise may be appreciably in error on the high side unless the final steep fall is itself divided into a number of steps, as shown. Errors are most likely to be significant when the temperature rise is to be estimated at or shortly after the end of the pulse. The approximately equivalent stepped pulse, illustrated for a general case in Figure 6.11, may be dealt with in several ways: regarding it as a succession of step functions gives the solution in the form

$$\delta\theta_j = W_{10}Z_{t11} + (W_9 - W_{10})Z_{t10} + (W_8 - W_9)Z_{t9} + \cdots + (W_1 - W_2)Z_{t2} - W_1 Z_{t1} = \sum(W_r - W_{r+1})Z_{t(r+1)} \tag{6.8}$$

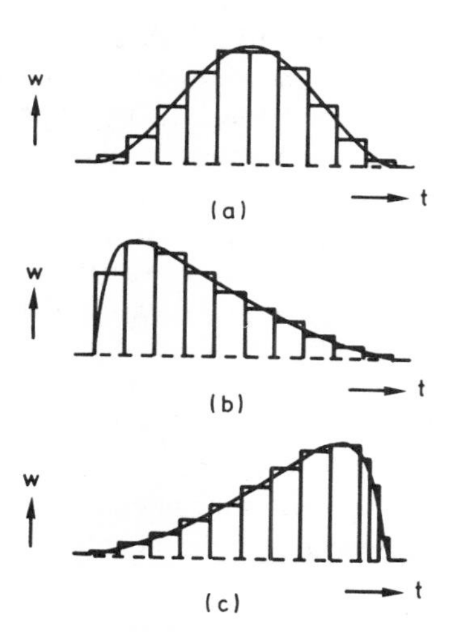

Figure 6.10 Step approximations to power pulses.

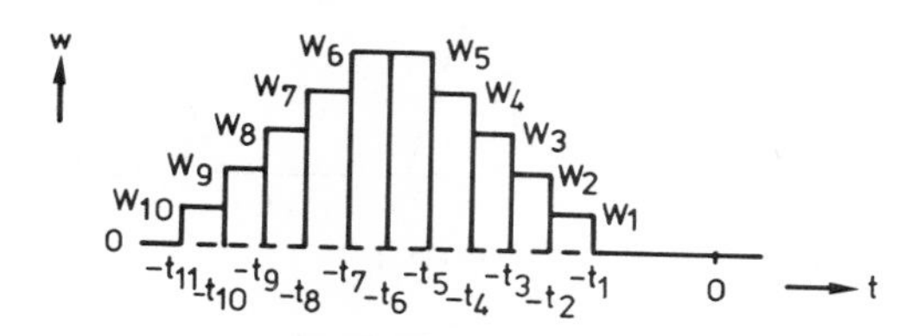

Figure 6.11 Step representation of power pulse.

Extended train of pulses of other than rectangular waveform

This case can be treated in the same way as a single pulse, but the tedium of such a procedure can be considerably reduced by the following methods:

(i) Assuming all the pulses to be similar, by evaluating a transient thermal resistance function representative of each elemental pulse repeated at the period of the main pulse, as illustrated in Figure 6.12. Thus the function for a particular elemental pulse is

$$Z_r = \{Z_{t(r+1)} - Z_{tr}\} + \{Z_{t(r+1)+T_p} - Z_{tr+T_p}\} + \{Z_{t(r+1)+2T_p} - Z_{tr+2T_p}\} + \dots \tag{6.9}$$

where T_p is the repetition period of the main pulse. For an extended train of pulses the averaging technique described above can be applied up to the last two or three pulses, in which case the complete function for the rth element might be

$$Z_r = \frac{t_{r+1} - t_r}{T_p}\{Z_\infty - Z_{t(r+1)+T_p}\} + \{Z_{t(r+1)+T_p} - Z_{tr+T_p}\} + \{Z_{t(r+1)} - Z_{tr}\} \tag{6.10}$$

and for the whole series of pulses,

$$\delta\theta_j = \sum W_r Z_r \tag{6.11}$$

The advantage of this approach is realised when results are required for a large number of different waveforms at one particular frequency, e.g. mains frequency.

(ii) Replacing pulses, other than the last, by approximately equivalent rectangular pulses. A very closely equivalent rectangular pulse for this purpose cannot be determined without performing the detailed calculation which the substitution is designed to avoid, but given, as previously noted, that the errors resulting from judicious approximations decrease as the time interval between the end of the pulse and the reference point increases relative to the duration of the pulse, sufficiently accurate results are normally obtained by choosing a rectangular pulse of the same energy content as the original, and of the same peak amplitude (the pulse duration consequently being reduced) as illustrated in Figure 6.13. Since the substitute pulse is so placed as to terminate at the same point in time as the original pulse, the general shift of energy towards the reference point leads to an error which, if appreciable, is on the safe side. Figure 6.14 illustrates at (b) the application of this principle to an extended train of pulses (a) in combination with the averaging approach described above.

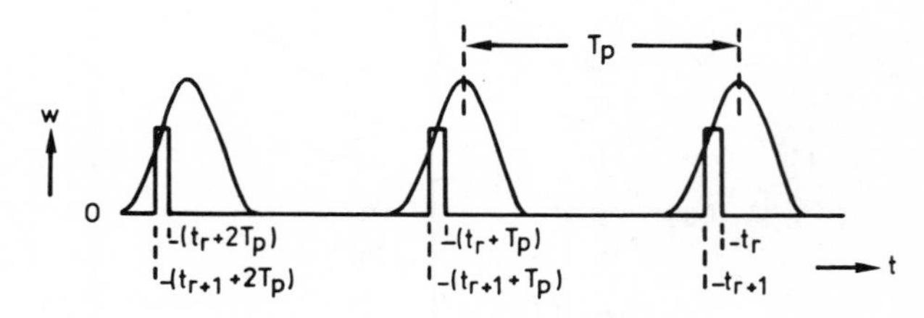

Figure 6.12 Element of a step approximation to a repeated power pulse.

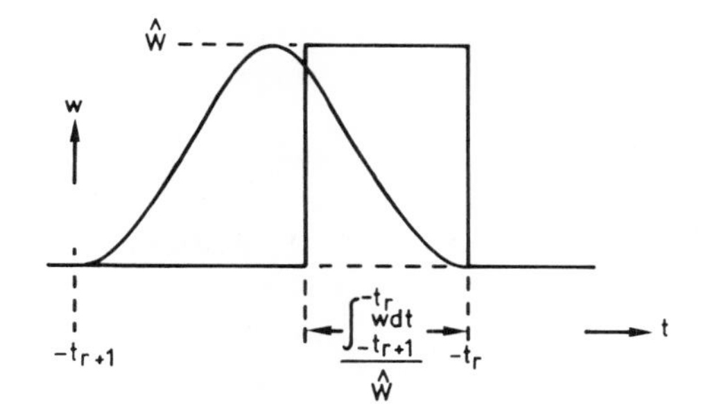

Figure 6.13 Representation of power pulse by approximately equivalent rectangular pulse.

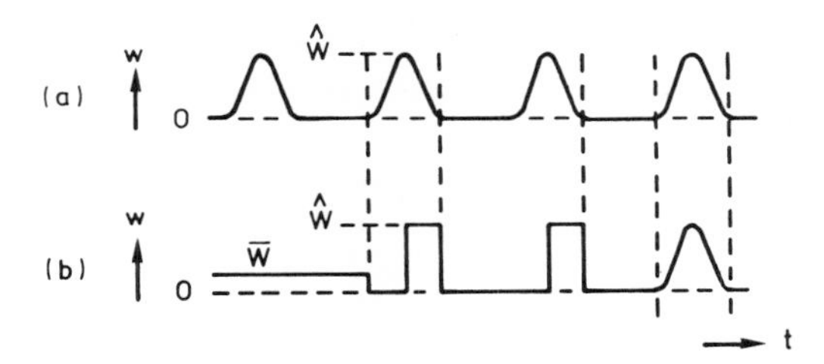

Figure 6.14 Partial representation of repeated power pulses by average dissipation and approximately equivalent rectangular pulses.

OVERCURRENT PROTECTION IN THYRISTOR AND DIODE CIRCUITS

This section is written in terms of thyristors, but where it is applicable it refers equally to diodes.

A distinction should be made between an overload, in the sense of a momentary or intermittent forward current in excess of the normal continuous rated current, and a fault current, which may represent a condition only just short of possible immediate failure of the thyristor, and is generally tolerable only in exceptional circumstances and on a limited number of occasions. In the absence of more precise information, it may be assumed that a safe intermittent loading is one that does not at any time lead to an instantaneous junction temperature higher than the maximum continuous rated value, and this is in general a valid basis for a continuously repeated condition; if an overload is relatively infrequent, however, and provided it does not prevent satisfactory operation of the circuit, it may well be possible (subject to the thyristor manufacturer's agreement) to exceed instantaneously the normal rated junction temperature without detriment to the life of the thyristor, with a consequent improvement in the overload rating.

The limitations upon the surge current ratings (this term may be used generally to refer to overload or fault ratings) of a thyristor are basically thermal, and arise from the effects of an abnormal rise in junction temperature of which the most significant are as follows, in ascending order of temperature:

(i) The forward break-over voltage V_{BO} falls below the specified value, and may approach zero. This is not necessarily destructive in itself, but may cause maloperation of the circuit and initiate a destructive condition, or may lead to a di/dt failure. In the terms of the ratings the user is not entitled to expect any forward blocking capability above the rated junction temperature, but in practice some margin is always present, and the variation of V_{BO} with junction temperature is typically as shown in Figure 6.15. Such a variation permits some selection to be made by the manufacturer if blocking at abnormally high temperatures is essential.

(ii) The reverse leakage current increases very greatly so that effectively the cell loses its reverse blocking capability and is liable to be destroyed by excessive dissipation on the application of reverse voltage. The cell normally fails to a short-circuit.

(iii) The cell fails through excessive forward dissipation alone, usually to a short circuit, but possibly to an open circuit if the fault current is extremely high.

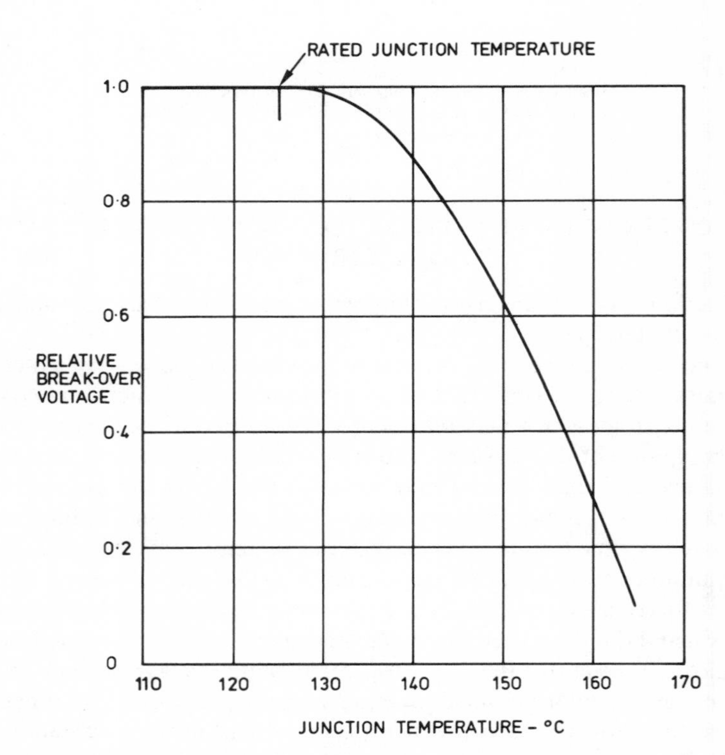

Figure 6.15 Typical variation of thyristor break-over voltage with junction temperature.

From the operational point of view the distinction between overloads and faults may be drawn on the basis that an overload is a relatively normal and frequently occurring condition which has to be catered for in such a way as to involve no expense and little inconvenience, while a fault is an unexpected and infrequent occurrence which generally results in an interruption of service. Partly for this reason and partly because fault currents are generally higher than overloads, overload protection is usually provided by circuit breakers or by automatic gate control of the thyristors, while fault protection generally depends on fuses.

Overcurrent protection by fuses

A fuse protecting a thyristor is required to carry continuously as high a proportion as possible of the normal current which the thyristor is rated to carry, but to open the circuit in the event of a fault before the thyristor is damaged. Difficulties arise because the thermal and electrical behaviour of a fuse is different from that of a thyristor, and because, in most cases, the restriction is imposed upon the characteristics of the fuse that in blowing it must not produce an excessive arc voltage.

The precise demands made upon the fuse vary according to the nature of the equipment in which it is used. Fuses may in fact be used for either or both of two distinct purposes:

(i) to protect thyristors from the effects of short-circuits or severe faults, either external to or within the equipment.
(ii) to effect the harmless disconnection of a faulty cell from a parallel group, to enable the equipment to remain in operation.

Figure 6.16 illustrates three examples of the first class of duty: in (a) the fuse protects either thyristor in the event of a short-circuit across the output of the regulator; in (b) FS_1 protects Th_1 in the event either of a short-circuit across the output of the rectifier or of a failure of D_1; in (c) the fuse protects both thyristors in the inverter if it presents a short-circuit to the supply by failing to commutate, either because of an excessive overload or through some malfunction. Figure 6.17 shows a rectifier bridge with four parallel-connected diodes in each arm, one diode in each group being redundant; if diode D_8, for example, fails to a short-circuit, the resulting fault current, flowing through the arm comprising D_{10}, D_{12}, D_{14}, D_{16}, and their associated fuses, disconnects D_8 by blowing fuse FS_8.

The case of Figure 6.17 represents a much easier task for the fuse than those of Figure 6.16 (a) and (b) ((c) is not strictly comparable) in that whereas in the latter cases a fuse carrying the current in one cell is required to prevent damage to that same cell in the event of a fault, in Figure 6.17 each fuse carries the current of one cell but protects a group of cells carrying the fault current in parallel. The fault current in Figure 6.17 is likewise shared by the corresponding group of fuses, so that, in the instance cited, FS_{10}, FS_{12}, FS_{14} and FS_{16} will survive intact while FS_6

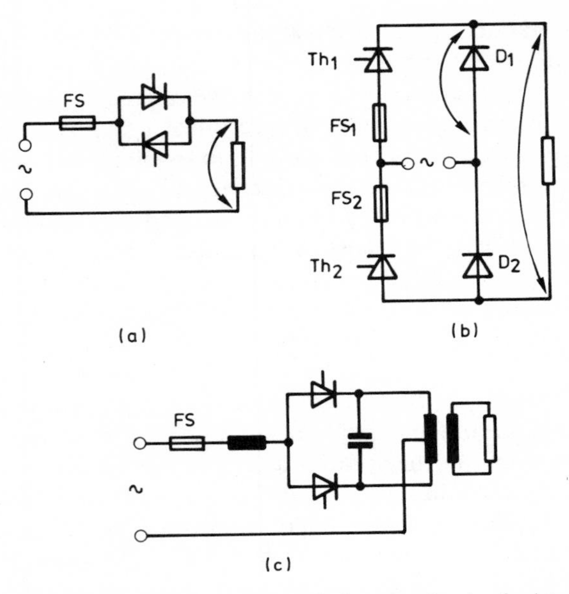

Figure 6.16 Fuses applied to fault protection in thyristor circuits.

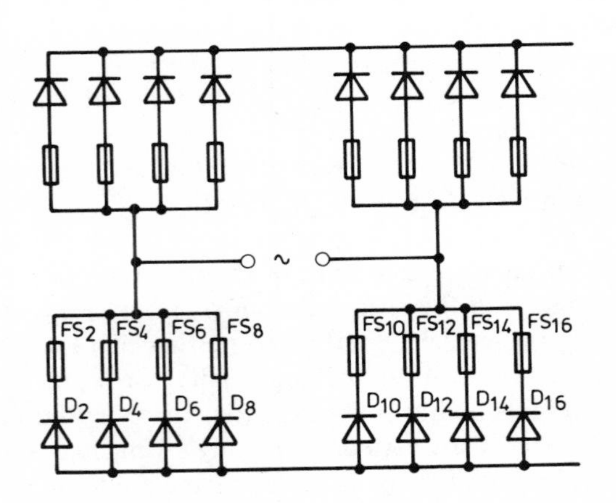

Figure 6.17 Fuses applied to disconnection of faulty cells in a rectifier bridge.

blows, assuming that the system is designed to achieve the necessary fuse discrimination (see below).

Fuses, like semiconductor devices, have both ratings, which determine the conditions under which they will serve without deterioration for a satisfactory length of time, and characteristics, which determine the manner in which they operate when required. The basic ratings are the continuous r.m.s. current, which may be affected by the ambient temperature or the duty cycle, and the working voltage, which is the maximum peak a.c., or maximum d.c., voltage at which the fuse can be relied upon to clear.

Fuse characteristics are of two kinds. The first mode of operation occurs with an a.c. supply and a moderate fault current, such that it takes at least one half-cycle for the fuse to melt (a.c. fuse characteristics are normally given for a 50 Hz supply, or 60 Hz in the U.S.A.) In these circumstances the arc which forms in the fuse is extinguished at the first current zero following the melting, and the characteristic is adequately represented as a graph of melting, or pre-arcing, time against prospective fault current (i.e. the current that would flow in the absence of the fuse). The general form of the current/time characteristic of a high-speed semiconductor fuse is shown in Figure 6.18. At some current only moderately in excess of the rated current the melting time becomes indefinitely long; the ratio of this current to the rated current is the fusing factor, typically about 1.4 for this type of fuse.

The second mode applies when the prospective fault current is so high that the circuit must be broken in less than one half-cycle of the a.c. supply, or when the supply is d.c. In this case the current is forced to zero, after the fuse has melted, by virtue of an arc voltage across the fuse exceeding the instantaneous supply voltage (hence the limitation upon the working voltage of any particular fuse). For most practical purposes the maximum current is reached nearly enough at the instant when the fuse starts to melt, and thereafter falls in a manner which depends on the circumstances; fuses used to protect semiconductor devices are for this reason sometimes referred to as current-limiting fuses. Figure 6.19 illustrates typical waveforms for a large prospective fault current.

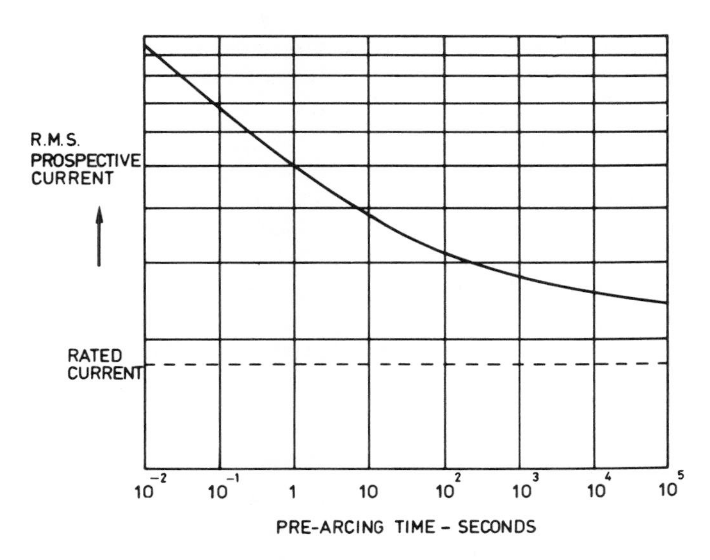

Figure 6.18 General form of the current–time characteristic of a high-speed semiconductor fuse.

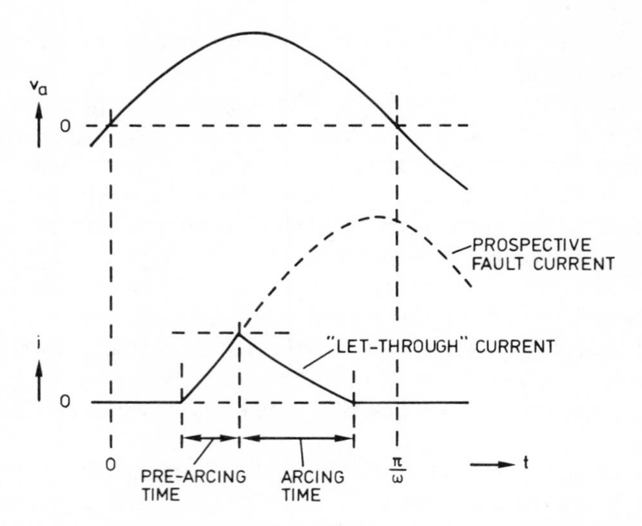

Figure 6.19 Typical waveform of current in a current-limiting fuse with an asymmetrical fault.

Before it melts, the fuse is affected only by the current flowing in it; the maximum, or cut-off, current, and the pre-arcing time are therefore given simply as functions of prospective current. As an alternative to a pre-arcing time curve, a single value may be quoted for $\int i^2\,dt$ (commonly referred to as I^2t) for the pre-arcing period, this being a sufficiently constant parameter over the range of times concerned. After the fuse has started to melt, it is exposed to the influence of the various voltages acting in the circuit, and the characteristics in this region are given in terms of supply voltage as well as prospective current; these characteristics are the arc voltage, which generally increases with increasing supply voltage, and the total $\int i^2\,dt$ (pre-arcing plus arcing) which also increases with increasing supply voltage and with prospective current.

Since a large fault current in an a.c. circuit is usually limited by a predominantly inductive impedance, it is likely to be highly asymmetrical, as illustrated in Figure 6.19. The inductance also reduces the rate at which the current can be forced down, and generally makes conditions more severe as regards both arc voltage and total $\int i^2\,dt$. Fuse characteristics are therefore given on the basis of an arbitrary degree of asymmetry which is considered to represent the worst case normally encountered: typically a peak asymmetrical prospective current 1.6 times the peak symmetrical value is assumed, corresponding to an R/X ratio of 0.15.

Matching of thyristors and fuses in a.c. circuits

In general terms, given that a fuse in circuit with a thyristor is adequately rated to carry the thyristor current, it protects the thyristor to the extent that its 'let through' characteristic lies below the fault current rating of the thyristor. To make the necessary comparison, it is necessary to express characteristics and ratings in similar terms; this generally entails some manipulation of manufacturers' data. To ensure complete overall protection the correlation has to be examined separately for long-term and sub-cycle surges.

The thyristor rating curve for long-term surges is drawn in terms of the permissible amplitude of a succession of half-sine-wave current pulses, representing the common case of a thyristor carrying fault current in alternate half-cycles of a 50 Hz a.c. supply. Figure 6.20 shows the usual form of the curve. For a large number of cycles, it is sufficiently accurate to convert from the peak current to the r.m.s. value simply by dividing by two to permit a direct comparison with the fuse characteristic, but for short periods this leads to a considerable error, since the only points of interest on the time scale are those at which the fuse has an opportunity of clearing, namely the current zeros at the ends of the conduction half-cycles, where the number of conduction periods exceeds the number of non-conducting periods by one. If the number of surge half-cycles is n, the effective r.m.s. current is

$$I = \frac{\hat{I}}{\sqrt{2}} \sqrt{\left(\frac{\text{number of conducting half cycles}}{\text{total number of half-cycles}}\right)}$$

$$= \hat{I} \sqrt{\left(\frac{n}{2(2n-1)}\right)} \qquad (6.12)$$

The thyristor rating curve being thus modified, it is typically found to compare with the current/time characteristic of a correspondingly rated fuse in the manner

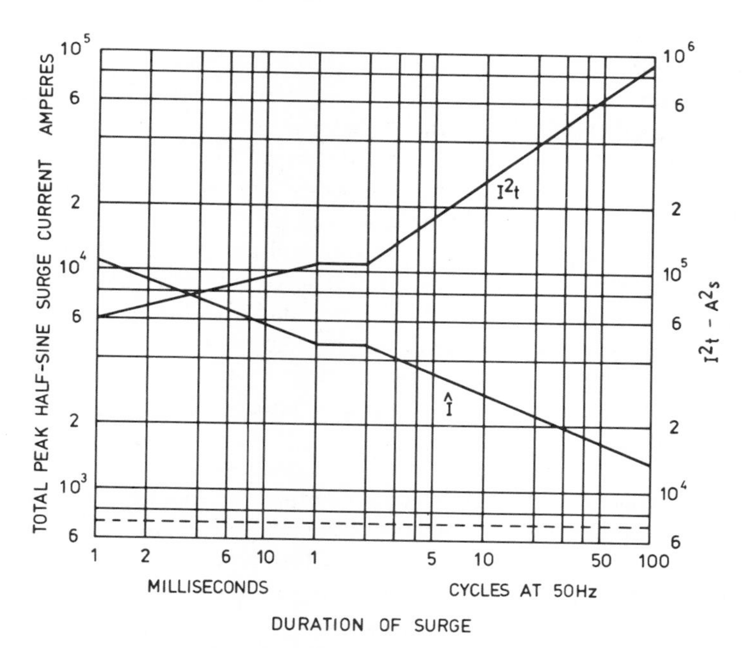

Figure 6.20 Typical surge current rating curve for a thyristor. - - - - maximum continuous mean current rating.

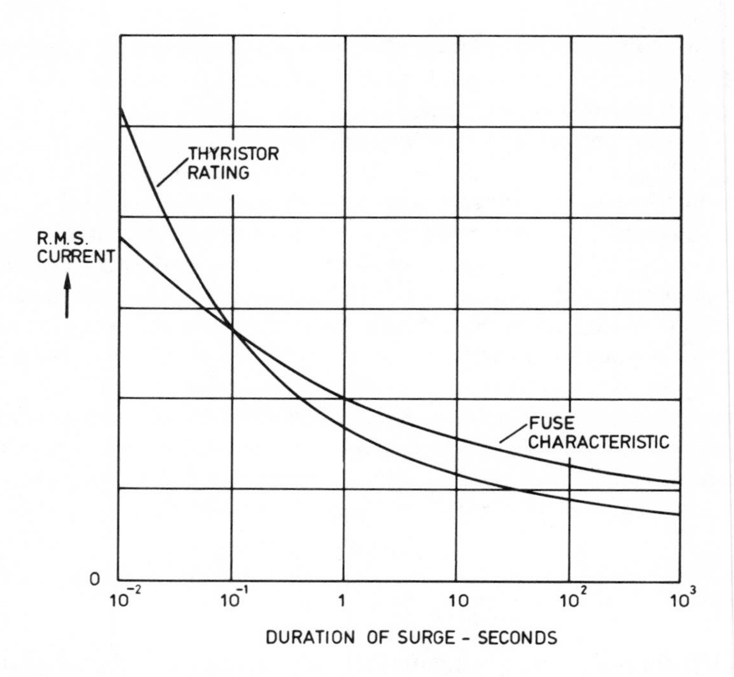

Figure 6.21 Typical relationship of thyristor surge rating and fuse characteristic.

illustrated in Figure 6.21; a fuse rated to carry the maximum rated current of the thyristor will not usually protect it for periods of more than a few seconds, while conversely a fuse that will protect the thyristor over the whole range of time will not permit it to be operated at its full rated current. To guard against moderate overcurrents therefore, it is normally necessary to employ a circuit-breaker or an automatic current-limiting system as well as fuses.

Since, as observed previously, a thyristor will survive at a higher junction temperature if no reverse voltage is applied, it is possible to have alternative surge rating curves according to whether reverse voltage is present during the non-conducting periods or not; circuits in which reverse voltage cannot appear are in a minority, and a single curve, for applied reverse voltage, may be considered sufficient for practical purposes. Alternative curves may also be given for different base temperatures.

In the case of sub-cycle surges, there is no simple parameter common to fuse and thyristor which may be used to compare the two on a general basis. Given even the fixed assumptions on which the fuse characteristics are based, the cut-off current, and commonly the total $\int i^2\,dt$, of the fuse vary with the prospective fault current, which may itself be a variable, while the rated $\int i^2\,dt$ for the thyristor varies appreciably with the surge duration, and implies a peak current rating which is dependent on both surge duration and waveform. Correlation thus depends on some reasonable assumption regarding the surge waveform and the way in which it varies with surge duration; given this, it is possible to make a simplified comparison on the basis of an $\int i^2\,dt$ rating for the thyristor, which should exceed the total $\int i^2\,dt$ for the fuse, coupled with a peak surge current rating, which should exceed the fuse cut-off current, for a given level of prospective current.

Fuse protection frequently has a restricting influence on the usable current rating of a thyristor, particularly when the continuous rating of the thyristor can be increased by very efficient cooling, as in the case of double-sided capsule cells, and when the supply impedance is exceptionally low. Sometimes impedance is deliberately introduced into the supply lines in order to ease protection problems, although improvements in fuse characteristics have made this less common than formerly.

The significance of the arc voltage generated by the fuse, from the point of view of the associated circuit, can be seen by reference to Figures 6.16 and 6.17. In Figure 6.17, the arc voltage appears directly across D2, D4, and D6 in reverse, in the condition postulated, while in Figure 6.16 (b) it appears in reverse across Th_2, and the cell voltage ratings must be adequate to allow for this. In general this is not an embarrassment to the circuit designer, since an allowance for transient overvoltages is usually necessary for other reasons.

Fuses in d.c. circuits

Fuse protection in d.c. circuits presents greater problems than in a.c. circuits. Since natural periodic current zeros do not occur, there is no equivalent to the long-term current/time characteristics for fuses in a.c. circuits, and moderate or slowly rising fault currents can produce a dangerous condition by allowing the fuse to arc continuously without breaking the current. The use of fuses in d.c. circuits is therefore subject to important restrictions regarding the minimum prospective fault current, the maximum circuit time constant (L/R) under fault conditions, and the supply voltage.

Fuses in series and parallel

Improved protection can sometimes be obtained by connecting fuses in parallel or in series. Series connection of fuses is not, however, a safe means of working at a voltage in excess of the rated voltage of one fuse, except in special circumstances, because of the tendency of one fuse, in arcing, to prevent the melting of the other.

Automatic current limiting

Automatic current-limiting may be employed in thyristor circuits (a) to limit the conditions of operation from the point of view of the load, —e.g., to limit the accelerating current in a motor or charging current in a battery, (b) to protect the thyristors or other parts of the equipment under overload or fault conditions, or (c) to prevent the current from rising to a level that causes maloperation of the circuit, possibly itself leading to a damaging condition—e.g. to limit the output current of a chopper regulator to a level at which commutation can be maintained.

Generally speaking, automatic current limiting may be achieved in controlled rectifiers, or other thyristor systems which embody a means of controlling their output over a wide range, by means of a current feedback which overrides the normal mode of control when the desired limiting value of output current is reached, or by virtue of a control system which operates primarily in terms of current under all conditions, with a direct limitation on the current demand signal. Current sensing for this purpose can be effected by current transformers, shunts or any other appropriate means.

In naturally commutating a.c. systems such as controlled rectifiers and a.c. regulators, apart from the stability problems which are always liable to arise in feedback systems with variable gain and possibly with relatively large amounts of ripple in the feedback signals, the main limitation upon performance—that is upon the degree of overload for which the current-limiting action is effective—lies in the delay which can occur before any limiting action can begin, and the possibility, which has to be avoided, that the rise in the thyristor junction temperature consequent upon the unrestricted increase in current may so reduce the breakover voltage of the thyristor that control of the output by the gate is lost. To meet this difficulty, margins of temperature have to be allowed in normal operation, the necessary extent of which can be calculated by estimating the dissipation that occurs during the delay period and applying the techniques described in the previous section.

In forced-commutation circuits the situation is different, since it is not necessary to wait for a naturally-occurring commutation, but only to ensure that the current-limiting action can be put into effect before the current has risen beyond the capacity of the commutation circuit.

d*v*/d*t* SUPPRESSION

In most thyristor circuits protection is necessary against the effects of excessive rate of rise of forward voltage (dv/dt) across the thyristors, which can otherwise cause unintended breakover, leading to a malfunction of the circuit and possibly to failure of the thyristors.

The tendency to excessive dv/dt may arise from external causes such as the closing of a main supply contactor, or from the operation of the circuit itself.

dv/dt suppression is achieved by means of a capacitive network in parallel with each thyristor, or sometimes in parallel with a circuit or sub-circuit, which defines the rate of rise of voltage in conjunction with a series inductance. Thus, while the details of the actual circuit may vary according to the application, virtually all cases are represented by a series LCR circuit with a step function of voltage applied to it as illustrated in Figure 6.22. The resistor R is included firstly to limit the discharge current from the capacitor into the thyristor when the latter is fired and secondly to damp the oscillatory circuit formed by L and C, and so to limit the overshoot of voltage across the thyristor.

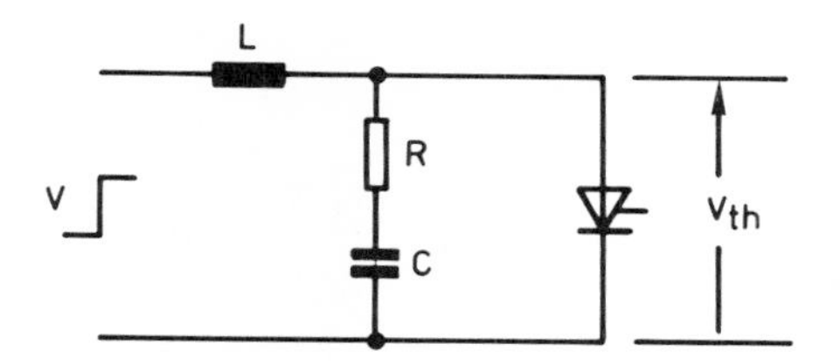

Figure 6.22 Basic dv/dt suppression circuit.

Calculation of the performance of the *LCR* suppression circuit is in terms of the peak output voltage $\hat{V}_{Th}$ across R and C, and the maximum instantaneous value of dv_{Th}/dt, for an input step function of amplitude V and ideally, at least, the design is a question of compromise between these two quantities.

The appropriate expressions are derived in Appendix 6 (ii). From (6.50), (6.53), and (6.54), $\hat{V}_{Th}$ and dv_{Th}/dt max are plotted in non-dimensional form in Figure 6.23 against the damping factor $\delta = R/2\sqrt{(L/C)}$.

Since the waveform of voltage across the thyristor is neither linear nor exponential, the value of dv/dt arrived at in the above way does not relate exactly to either of the standard methods of assessing the dv/dt capability of thyristors. The linear convention in this context affords a useful margin of safety for all waveforms, i.e. for any damping factor.

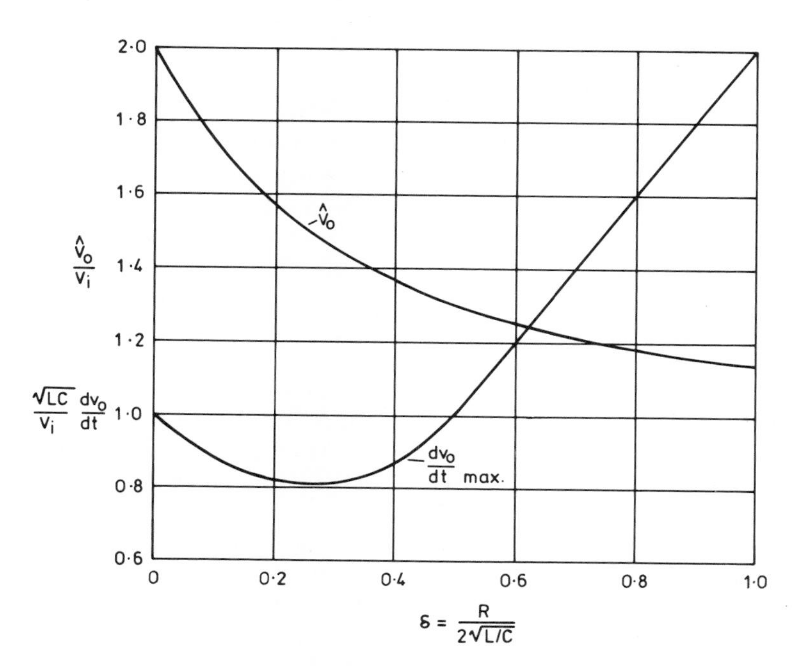

Figure 6.23 Variation of $\hat{V}_o(\hat{V}_{Th})$ and maximum $dv_o/dt\,(dv_{Th}/dt)$ with damping factor in an *LCR* circuit.

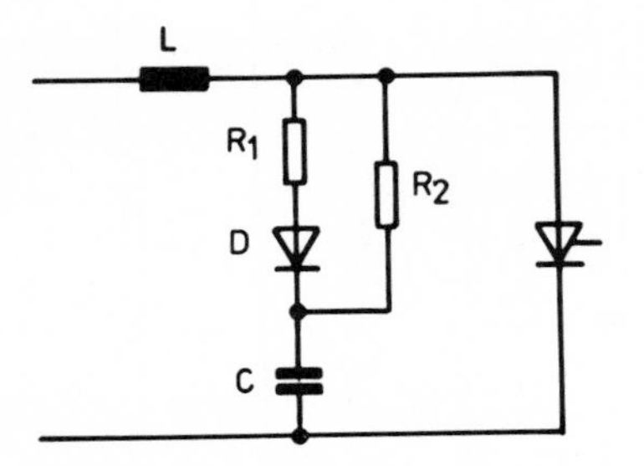

Figure 6.24 Diode–capacitor dv/dt suppression circuit.

The discharge of the capacitor through the thyristor constitutes a special aspect of rating, in that nothing is included specifically to limit di/dt, and is normally covered by maximum limits on the value of C, or on the stored energy, and the peak discharge current. L is frequently provided, generally in excess of the minimum value required, by supply or transformer inductance (assuming that the benefit of this inductance is not nullified by the presence of RC surge-suppression circuits) but has in many cases to be provided in the form of an air-, iron- or ferrite-cored inductor; difficulty may then be met in limiting dv/dt to an acceptable level if technical or economic considerations place a constraint upon L, particularly in high-power equipment where tolerable inductance values tend to be lower but the permissible values of C and R are not correspondingly scaled.

In the modified suppression circuit illustrated in Figure 6.24, the addition of a diode D to limit the capacitor discharge current enables C to be considerably increased and R ($R1$) to be reduced, so that the design of the circuit is relatively unconstrained, and more effective suppression can be achieved. $R2$ is provided to enable the capacitor to discharge within the shortest conduction period of the thyristor.

dv/dt suppression circuits may introduce considerable power losses at high operating frequencies.

OVERVOLTAGE PROTECTION

It is almost always necessary in power semiconductor equipment to provide a means of limiting transient overvoltages which could otherwise overstress the semiconductor devices. Overvoltages can be classified broadly according to their sources—i.e., those arising repetitively from the normal operation of the circuit, those which arise from occasional switching in or associated with the equipment, and those produced by events external to the equipment, such as switching in other equipment connected to the same supply.

Most overvoltages are produced either by the interruption of current in an inductance or by overshoot in an oscillatory circuit with an applied step function of voltage, although it is sometimes necessary to take account of other possible sources, such as the capacitive coupling between the windings of a supply transformer with a high step-down ratio. The common cases are illustrated by the following consideration of surges produced by switching in an equipment supplied from an a.c. system through a transformer, although the basic principles apply also to other circumstances.

Transformer switching surges

Overvoltages are generated in transformers mainly in three ways:

(a) through the interruption of the primary magnetizing current with the secondary winding unloaded (Figure 6.25(a));
(b) through the interruption of load current flowing in the leakage inductance (Figure 6.25 (b));
(c) through overshoot, in the oscillatory circuit formed by the leakage inductance and connected or stray shunt capacitance, when the supply is connected (Figure 6.25 (c)).

Suppression of the surge produced by switching off the supply entails absorbing the energy stored in the magnetizing inductance of the transformer without an excessive rise in voltage. A satisfactory method is to connect a capacitor across either the primary or the secondary winding—whichever is more convenient—as in Figure 6.26. (In practice a damping resistor is connected in series with the capacitor, but it can be ignored in this particular context.)

A value for the shunt capacitor can be calculated in terms of the magnetizing characteristic if various complicating factors are ignored. Idealized waveforms are

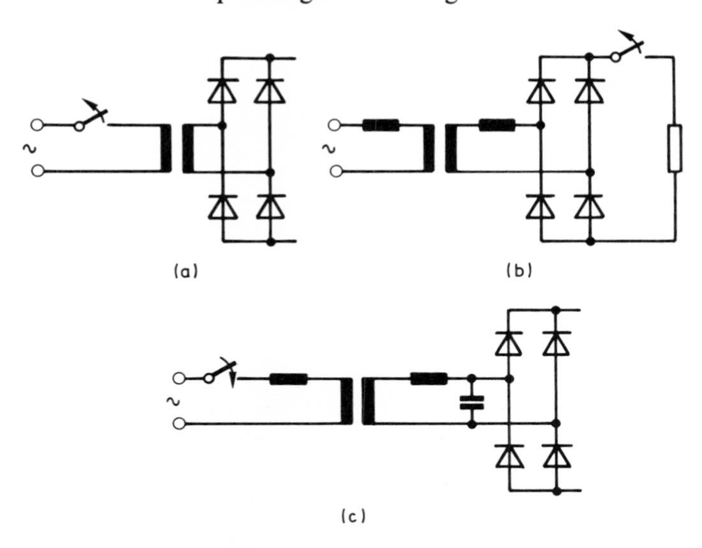

Figure 6.25 Generation of transient voltages by switching in reactive circuits.

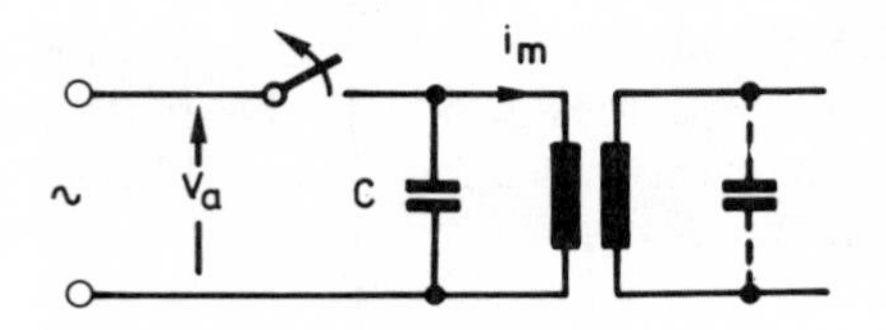

Figure 6.26 Suppression of transient voltage due to the interruption of transformer magnetizing current.

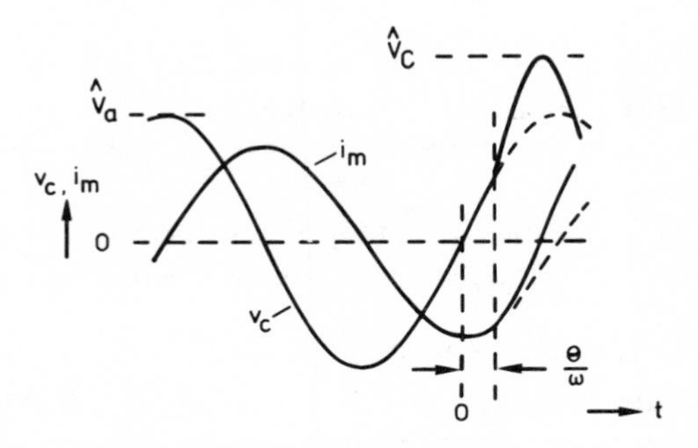

Figure 6.27 Transient overvoltage due to interruption of transformer magnetizing current with shunt capacitor connected.

shown in Figure 6.27. If the supply voltage is $v_a = \hat{V}_a \sin\theta$ and the magnetizing industance is L_m,

$$i_m = -\frac{\hat{V}_a}{\omega L_m}\cos\theta \tag{6.13}$$

Then when the supply is disconnected at $t = 0$ the capacitor voltage (ignoring damping) is

$$v_c = \hat{V}_a \sin\theta\cos\omega_0 t + \frac{\hat{V}_a}{\omega L_m}\cos\theta\sqrt{\left(\frac{L_m}{C}\right)}\sin\omega_0 t \tag{6.14}$$

$$= \hat{V}_a\sqrt{\left(\sin^2\theta + \frac{\cos^2\theta}{\omega^2 L_m C}\right)}\sin\{\omega_0 t + \tan^{-1}(\omega\sqrt{(L_m C)}\tan\theta)\}$$

$$= \hat{V}_a\sqrt{\left\{1 + \cos^2\theta\left(\frac{\omega_0{}^2}{\omega^2} - 1\right)\right\}}\sin\left\{\omega_0 t + \tan^{-1}\left(\frac{\omega}{\omega_0}\tan\theta\right)\right\} \tag{6.15}$$

where $\omega_0 = 1/\sqrt{(L_m C)}$.

If $\omega_0 < \omega$, expression (6.14) has its maximum value when $\cos\theta = 0$, so that $\hat{V}_c = \hat{V}_a$. In the more realistic case when $\omega_0 > \omega$, the maximum occurs when $\cos\theta = 1$: then

$$\frac{\hat{V}_c}{\hat{V}_a} = k_V = \frac{\omega_0}{\omega} \tag{6.16}$$

From which

$$C = \frac{I_m}{\omega k_V{}^2 V_a} \tag{6.17}$$

This expression may be applied equally to a capacitor connected across a secondary winding by using referred values of I_m and V_a.

In practice, the value of C cannot be calculated accurately by such a simple approach. The energy released from the magnetizing inductance is greatly reduced by hysteresis and eddy currents in the transformer core; on the other hand it may be considerably increased if the transformer is switched off during a period of asymmetrical magnetization shortly after it has been switched on. Expression (6.17) is therefore to be regarded as an approximate indication subject to possible correction factors based on experience. Insofar as it is valid, or suitably corrected, it can be applied to individual phases of a three-phase transformer, although advantage may be taken of the non-coincidence of the peak stored energies in the phases to reduce the total capacitance required, depending on the connections.

With a shunt capacitor connected, the situation on switching on at the peak of the supply voltage is that of a damped oscillatory circuit, as shown in Figure 6.28, where L represents the supply inductance, plus the transformer leakage inductance if the capacitor is on the secondary side of the transformer. The maximum instantaneous output voltage depends on the damping factor $R/2\sqrt{(L/C)}$ in the manner shown for the essentially similar dv/dt suppression circuit of Figure 6.22, with a theoretical maximum of $2\hat{V}_a$ when $R = 0$. The degree of damping is not critical, a damping factor in the region of 0.5 being generally satisfactory; an unnecessarily high resistance is avoided since it results in excessive dissipation and may reduce the effectiveness of the capacitor for other purposes (see below).

The curve of $(\sqrt{(LC)}/V)\,dv_{Th}/dt$ in Figure 6.23 also applies to the situation of Figure 6.28; the surge-suppression capacitor frequently serves usefully to limit dv/dt when the supply is switched on.

The situation represented by Figure 6.25(b) becomes as shown in Figure 6.29 in the worst case when the peak input current $\hat{I}_a$ is interrupted coincidentally with

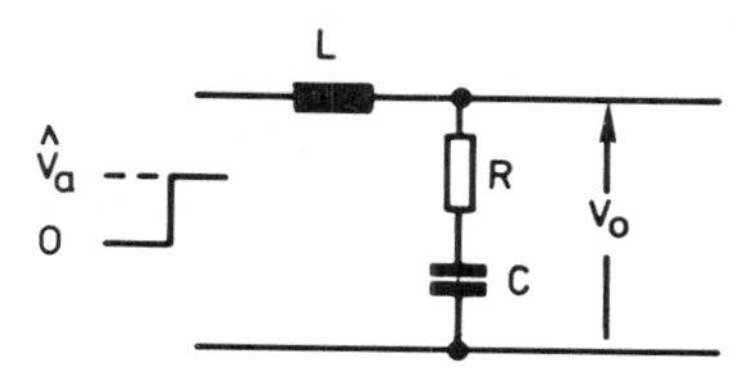

Figure 6.28 Equivalent circuit of transformer with secondary shunt capacitor on connection to a supply.

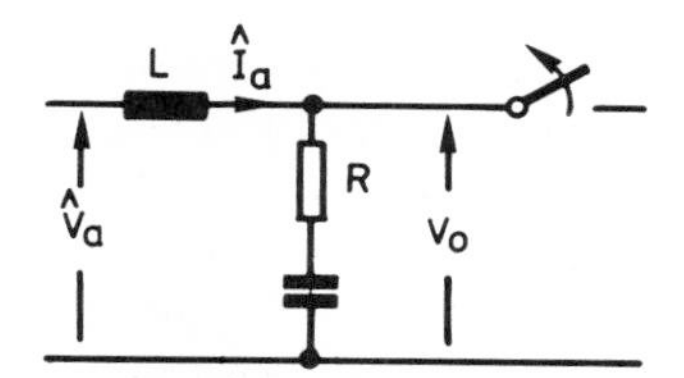

Figure 6.29 Equivalent circuit of transformer with secondary shunt RC network on disconnection of load.

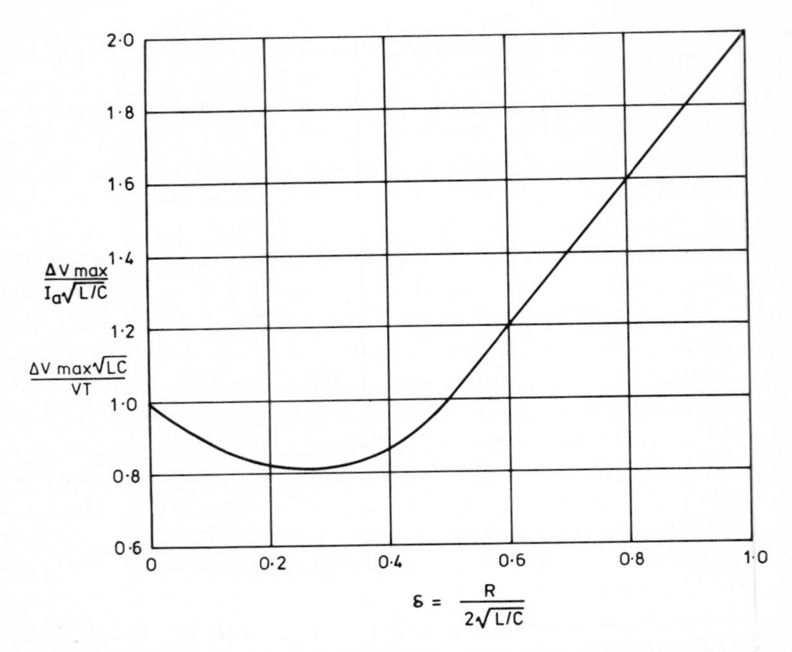

Figure 6.30 Peak voltage generated by interruption of current in an inductance with *RC* suppression, or by the application of an impulse to a damped *LCR* circuit.

the peak supply voltage. The increment in voltage across the capacitor and resistor as a function of time is then given by expression (6.58) in Appendix 6(ii), and its peak value is plotted against δ in Figure 6.30.

The stored energy in the leakage inductance of a transformer is generally of the same order as that in the magnetizing inductance under normal conditions, and can be much greater under overload or fault conditions; the use of a high-speed circuit-breaker on the secondary side of a transformer can thus give rise to severe surge-suppression problems.

Supply-borne transients

Transient overvoltages on the a.c. supply can be attenuated by a shunt capacitor on the secondary side of the transformer in conjunction with the transformer leakage inductance, or on the primary side with the benefit of the supply inductance only, provided that they are of short duration in comparison with the half-period of oscillation of the resonant circuit so formed.

The response to a short pulse in terms of the applied voltage–time integral is derived in Appendix 6(ii), equation (6.57), and is represented by the curve of Figure 6.30. For satisfactory performance δ should not greatly exceed 0.5.

In the absence of full information on the magnitude and duration of supply-borne transients, surge suppression based on the stored energy associated with the magnetizing current of the transformer is commonly found adequate, even if it is connected on the primary side of the trnasformer. This does not necessarily apply to high-energy surges such as may be produced by switching off a large lightly loaded transformer from which the supply is derived; in such a case the means of suppression must be commensurate with the source of the surge.

Rectifier-fed suppression capacitors

Connecting the *RC* network to the a.c. supply through a rectifier, as shown in Figure 6.31 for single-phase and three-phase systems respectively, has the advantage that there is virtually no restriction on the capacitance that may be used, since it draws virtually no reactive current from the supply, and more effective suppression may thus be achieved, particularly if electrolytic capacitors are used. R_1 is the normal damping resistor, and R_2 is a relatively high-value resistor provided to allow the capacitor to discharge to approximately the peak supply voltage following a surge. Where the semiconductor equipment to be protected includes an uncontrolled rectifier, the *RC* network would normally be connected across it rather than through a separate rectifier.

The estimation of maximum voltage described above for directly connected *RC* networks is equally valid for rectifier-fed networks except that in (6.17) k_V^2 has to be replaced by $(k_V^2 - 1)$, giving

$$C = \frac{I_{\mathrm{m}}}{\omega (k_V^2 - 1) V_{\mathrm{a}}} \tag{6.18}$$

since the capacitor is initially charged to $\hat{V}_{\mathrm{a}}$.

A slight disadvantage of rectifier-fed suppression networks is that their effectiveness is reduced if surges occur in rapid succession unless the discharge resistance is of low value, in which case the dissipation in it may be inconvenient.

Surge voltages due to carrier storage

Transient overvoltages are frequently produced by the rapid cessation of reverse recovery current when a forward-conducting diode or thyristor, with

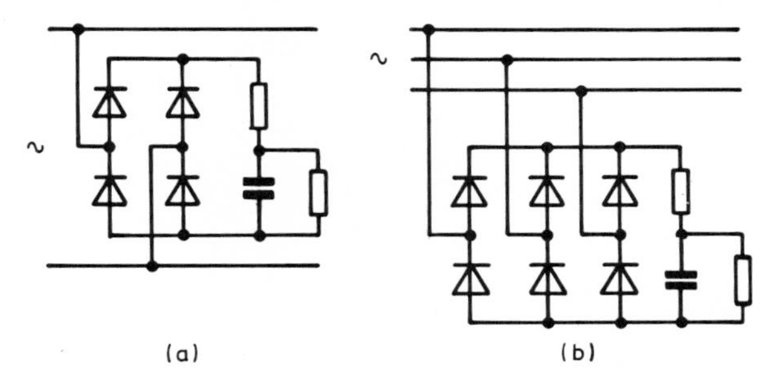

Figure 6.31 Transient suppression networks connected to a.c. circuits through rectifiers: (a) single-phase; (b) three-phase.

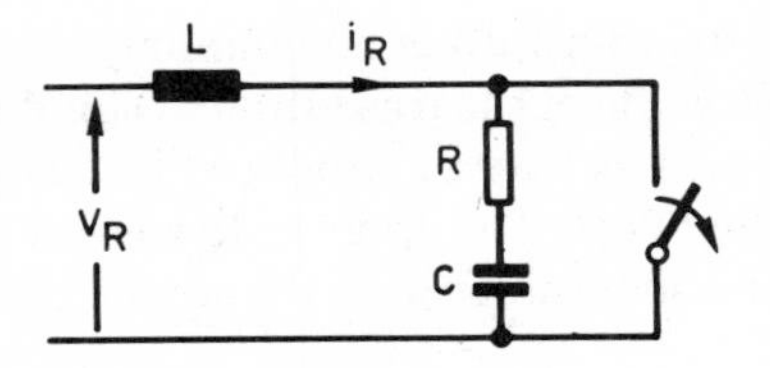

Figure 6.32 Equivalent circuit representing an *RC* network used to limit transient overvoltage due to carrier storage in a diode or thyristor.

inductance in series, is subjected to a reverse voltage, as in the process of commutation. The equivalent circuit is that of Figure 6.32, wherein L represents the total operative inductance, R and C are transient suppression components, and the blocking of the diode or thyristor is represented by the opening of S. If the reverse current is assumed to cease instantaneously as it reaches its peak, the peak voltage developed is given by equation (6.58) in Appendix 6 (ii), and the curve of Figure 6.30.

In practice this calculation is somewhat pessimistic, since the reverse current does not actually fall instantaneously to zero, but decreases at a rate depending on the recovery characteristics of the cell, some of the stored energy in L being thereby dissipated. It is in any case difficult to carry out owing to the lack of the information necessary to determine $\hat{I}_R$. In diode rectifier circuits, suppression networks designed to limit transformer switching surges or supply-borne transients are normally more than adequate to cope with carrier-storage spikes (assuming that the capacitors have satisfactory high-frequency characteristics) so long as they are connected directly across the rectifier terminals without any interposed inductance. In thyristor circuits such direct connection of relatively large capacitive networks may not be permissible because of the resulting high $\mathrm{d}i/\mathrm{d}t$; acceptable spike suppression is normally achieved in such a case by means of networks similar to those used for $\mathrm{d}v/\mathrm{d}t$ suppression.

Non-linear surge suppressors

Non-linear devices having the kind of characteristic illustrated in Figure 6.33 are frequently used in place of, or in conjunction with, reactive networks as a means of limiting voltage surges, being so selected that they pass an acceptably small current at the normal working voltage, but a sufficiently large current at higher voltages to limit surges at a tolerable level.

Devices of this kind are:

(a) Voltage-dependent resistors. These have been made for many years with a thyrite material, which provides a current variation in proportion to the fourth or fifth power of the voltage; they are not normally satisfactory as the sole means of protection for semiconductor circuits, because surge voltages cannot be limited

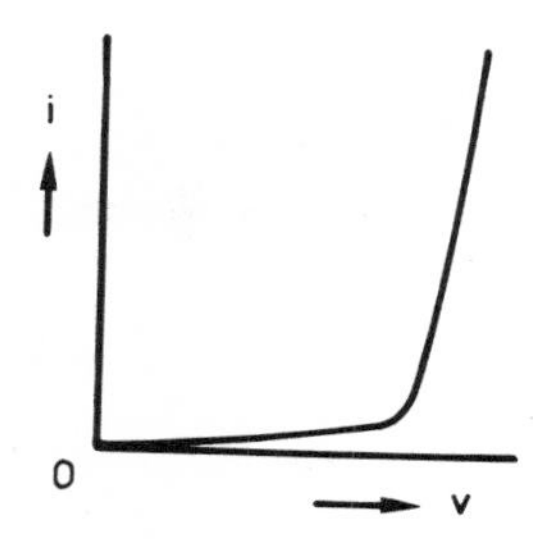

Figure 6.33 General form of the characteristic of a non-linear surge-suppressor.

adequately without excessive dissipation under normal conditions. New types—so-called metal oxide varistors—are now available with a very much better characteristic, corresponding to a thirteenth, or higher, power law, and can be used very effectively within their limitations of size.

(b) Selenium surge suppressors. These are reverse-biassed selenium rectifiers specially processed to produce a stable reverse characteristic which can be sufficiently non-linear to limit surges to well under twice the steady-state peak voltage (sometimes with the qualification that limiting may be less effective for very rapidly rising transient voltages); suppressors can be built to virtually any size, and have considerable energy-absorption capabilities.

(c) 'Avalanche' devices—i.e. thyristors or, more commonly, silicon diodes, which are capable of conducting relatively large reverse currents in the avalanche mode without deterioration. 'Controlled avalanche' devices are those for which the upper limit of avalanche voltage is specified as well as a lower limit. Such devices may be incorporated into the main circuits of rectifiers, converters etc., or used in conjunction with ordinary cells, purely as suppressors; in either case their virtually square reverse characteristics afford very positive protection against overvoltages, but in general their reverse dissipation ratings tend to be somewhat restrictive, so that their suitability for use as the sole means of suppression is dependent on individual circumstances.

Non-linear devices have the advantages, in comparison with capacitive networks, firstly that they are free from certain inconvenient tendencies associated with stored energy in reactive suppression circuits and secondly that in the event of a surge of unexpectedly high magnitude they will normally fail to a short-circuit rather than permit a damaging voltage to be applied to the more expensive equipment to which they are connected.

Crowbar circuits

A positive and rapid means of preventing excessive voltages in circumstances where the amount of energy involved may be too great to permit the use of any of

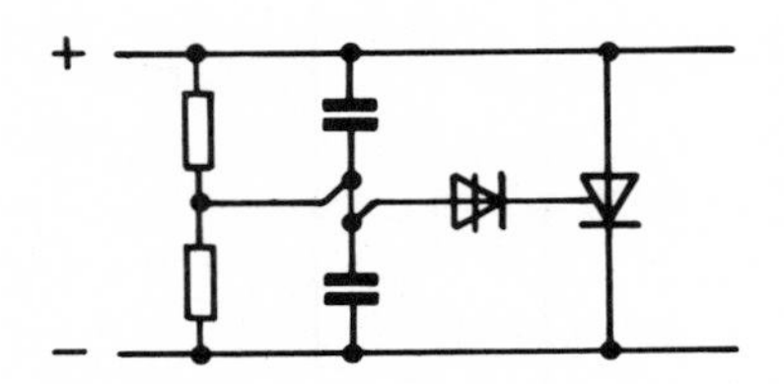

Figure 6.34 Thyristor crowbar with trigger diode control.

the suppression devices discussed above is an electronic crowbar, consisting of a forward-biassed thyristor with a voltage-responsive firing circuit, whereby a virtual short-circuit is placed across the circuit to be protected if the voltage across it tends to exceed a defined level. A simple arrangement using an *RC* voltage divider and a trigger diode to define the operating voltage is shown in Figure 6.34. Very high current levels can be sustained in this way, the ultimate limit being set by the fault current rating of the crowbar thyristor. In certain cases, with an a.c. supply, for example, it is possible to make the crowbar automatically self-resetting, but in most cases some other means is employed to limit and interrupt the current when the crowbar has operated.

SERIES OPERATION OF DIODES AND THYRISTORS

The use of series-connected strings of diodes and thyristors to obtain voltage ratings beyond the capabilities of single cells is less common than formerly as a result of the development of cells with ratings adequate, without such expedients, for the voltage levels involved in the majority of applications. A number of applications remain, however, in which series operation is still necessary, either because the operating voltage is exceptionally high, or because some special requirement, such as a short turn-off time in a thyristor, restricts the cell voltage rating obtainable.

Cells connected in series tend to share unequally the total voltage across the string in the blocking condition because of the variations in steady-state leakage current and in transient behaviour. (At very high voltages, stray capacitance to earth may also have a significant effect on the voltage distribution; such specialized problems are beyond the scope of this book.)

To ensure that individual cells are not subjected to voltages in excess of their ratings, it is necessary in most cases to add to a series string a voltage-sharing network. For both technical and economic reasons such a network cannot ensure perfect voltage sharing, and its design is a compromise based on an acceptable degree of mis-sharing, which implies a degree of de-rating, expressed as the ratio of the maximum permissible voltage across the string to the sum of the individual cell ratings. What may loosely be termed 'steady-state' sharing, i.e. that affected by variations in steady-state leakage current, and transient sharing generally demand different treatments, which are combined in the complete network. The following is a slightly simplified analysis.

Steady-state voltage sharing

This is achieved in nearly all cases by connecting a resistor of suitable value in parallel with each cell. Calculation of the necessary resistance value is based on the worst-case situation illustrated generally for n cells in Figure 6.35, where all the cells pass the maximum leakage current except for one, D_1 (the approach applies equally to thyristors blocking reverse or forward voltage), which has the lowest possible leakage; D_1 tends to experience the highest individual cell voltage, which is thus the determining factor in the rating of the string, and this tendency is aggravated by the assumption that the corresponding resistor $R1$ is at the upper limit of its tolerance, while all the others are at the lower limit. Referring to Figure 6.35, let

$$I_{R2} = I_{R3} = \ldots = I_{Rn} \quad \text{and} \quad I_{R1} = I_{R2} - \Delta I_R$$

$$R_2 = R_3 = \ldots = R_n = R \quad \text{and} \quad R_1 = R(1+a)$$

V_1 = maximum cell voltage

V_s = total voltage across string

$$k_s = \text{sharing factor} = \frac{V_s}{nV_1}$$

Then
$$V_1 = V_s\left(\frac{1+a}{n+a}\right) + \Delta I_R R\frac{(1+a)(n-1)}{(n+a)} \tag{6.19}$$

Putting $V_s = nV_1k_s$,

$$R \ngtr \frac{V_1}{\Delta I_R(n-1)}\left(\frac{n+a}{1+a} - nk_s\right) \tag{6.20}$$

A figure for ΔI is not, as a rule, explicitly available. A safe solution is given by assuming that ΔI could be as high as the maximum instantaneous leakage current for the type of cell; on the other hand if sufficient information is available on the cell characteristics, it may be possible to arrive at a higher value of resistance, or to use a higher value of k_s, by taking advantage of such possible factors as a working junction temperature below the permitted maximum.

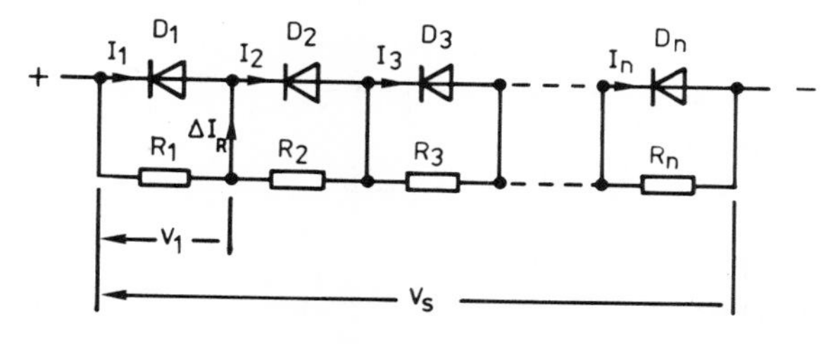

Figure 6.35 Series diode string with steady-state voltage-sharing resistors.

Transient voltage sharing

As distinct from the mis-sharing of zero-frequency or low-frequency voltages, which can normally be adequately limited by steady-state sharing resistors, transient mis-sharing can arise from three factors.

(i) variations in cell capacitance and stray capacitance, causing mis-sharing of sharply rising voltages and (in thyristors) dv/dt;
(ii) variations in reverse recovery charge;
(iii) variations in turn-on time (in thyristors).

The first factor, although important, is in general less significant than the others, and is adequately catered for in the measures adopted to deal with the more serious ones.

Reverse recovery

When a string of diodes or thyristors is switched rapidly from the forward conducting to the reverse blocking state the delay which occurs before the cells assume their blocking condition varies from one cell to another according to the variation in stored charge; cells which recover while the rest are yet unable to support voltage tend to experience an excessive proportion of the total voltage applied to the string. In forced-commutation circuits, as well as the tendency to cause damage to individual cells through excessive voltage, this leads to a further undesirable effect, namely, a variation in the turn-off interval, whereby the cells with the highest stored charge experience the shortest forward-recovery interval; this is aggravated, in the absence of a suitable sharing network, by the fact that the reverse recovery time of the slower cells is further increased by the reduction in reverse current caused by the recovery of the faster cells.

Voltage-sharing under reverse recovery conditions is achieved by a shunt capacitor chain (Figure 6.36). This again is equally applicable to thyristors except that in such cases a resistor is connected in series with each capacitor to limit the discharge current when the thyristors are fired; assuming the resistors to be of suitably low value, they do not significantly affect operation with regard to reverse recovery.

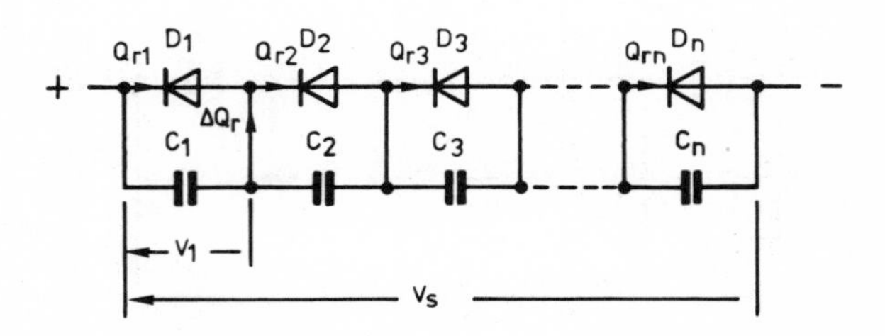

Figure 6.36 Series diode string with transient voltage-sharing capacitors.

The worst-case assumption in Figure 6.36 is that D_1 has the minimum stored charge while all the other cells have the maximum, and that C_1 is at the lower limit of tolerance while all the other capacitors are at the upper limit.

Let

$$Q_{r2} = Q_{r3} = \ldots = Q_{rn} \quad \text{and} \quad Q_{r1} = Q_{r2} - \Delta Q_r$$

$$C1 = C \quad \text{and} \quad C_2 = C_3 = \ldots = C_n = C(1+b)$$

With a step reverse voltage $V_s = nV_1k_s$ applied to the string,

$$V_1 = nkV_s\left(\frac{1+b}{n+b}\right) + \frac{\Delta Q_r}{C}\left(\frac{n-1}{n+b}\right) \tag{6.21}$$

whence

$$C \nless \frac{\Delta Q_r}{V_1}\left\{\frac{n-1}{n+b-nk_s(1+b)}\right\} \tag{6.22}$$

In a diode string, capacitors can effect steady-state voltage-sharing as an alternative to resistors, particularly at relatively high frequencies, in a manner somewhat analogous to the use of inductors for current sharing in parallel-connected groups. The capacitance required can be determined by means of (6.22), with ΔQ_r supplemented by $\int \Delta i_R \, dt$, representing the spread of reverse leakage current integrated over the reverse voltage period, in the case of a square wave, or, approximately, over half the reverse voltage period in the sine-wave case. Capacitors are not necessarily suitable for steady-state sharing in thyristor strings, since thyristors may operate in a completely blocking mode, without any conducting periods in which the capacitor voltages might be reset to zero.

Variations in turn-on time

When a string of series-connected thyristors supporting forward voltage is fired, the slowest cells to turn on tend to experience an excessive forward voltage together with a high dv/dt as the voltage across the faster cells collapses. To avoid possibly damaging break-over, a low-value resistor is usually connected in series with a capacitor across each cell, and an inductance is connected in series with the whole string, as in Figure 6.37. The inductance may be present in the form of supply or transformer inductance, or it may be provided for the purpose, in which case it may be advantageous, in view of the relatively high inductance that can be

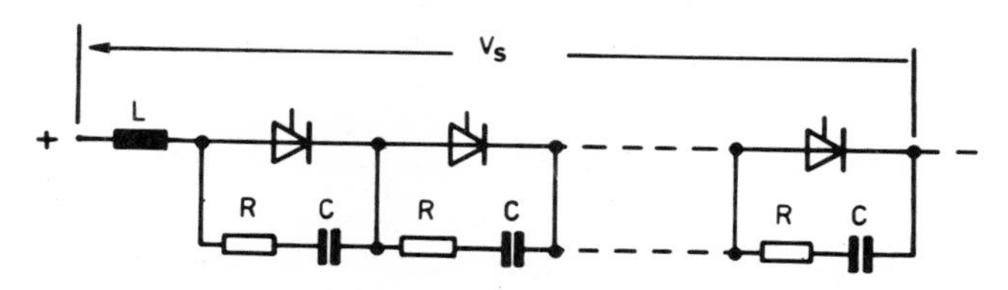

Figure 6.37 Series thyristor string with *LCR* network to accommodate variations in turn-on time.

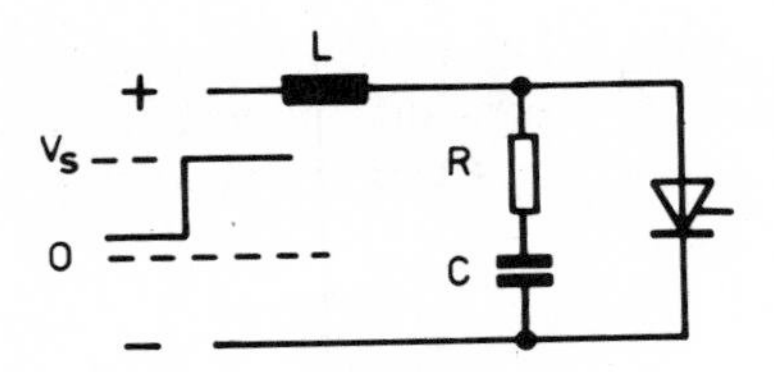

Figure 6.38 Equivalent circuit for the slowest thyristor of Figure 6.37 at turn-on.

required, to use an iron- or ferrite-cored inductor which is allowed to saturate once the string is fully turned on.

The safe basis for estimating the required component values is that all but one of the thyristors turn on simultaneously, leaving the remaining slower cell to support the entire applied voltage in conjunction with the series inductance and one resistor and capacitor (Figure 6.38), until it, too, turns on.

Since it cannot be assumed that the total string voltage is perfectly shared before the string is fired, the initial condition is based on a sharing factor k_{s1}, which represents the degree of mis-sharing permitted by the steady-state sharing network and any effects of reverse stored charge remaining from a preceding turn-off process. During the interval before the slowest thyristor turns on, the voltage across it rises from its initial value V_s/nk_{s1} thus:

$$v_1 = V_s \left\{ 1 - \left(1 - \frac{1}{nk_{s1}} \right) e^{-d\omega_0 t} (\cos \omega_0 t - d \sin \omega_0 t) \right\} \tag{6.23}$$

where

$$d = \frac{R}{2\omega_0 L}$$

(cf. Appendix 6(ii), (6.47)). Then if v_1 reaches V in time Δt, representing the spread of turn-on delay times, and the overall sharing factor is k_{s2},

$$V_1 = V_1 nk_{s2} \left\{ 1 - \left(\frac{nk_{s1} - 1}{nk_{s1}} \right) e^{-d\omega_0 \Delta t} (\cos \omega_0 \Delta t - d \sin \omega_0 \Delta t) \right\}$$

or

$$\frac{k_{s1}(nk_{s2} - 1)}{k_{s2}(nk_{s1} - 1)} = e^{-d\omega_0 \Delta t} (\cos \omega_0 \Delta t - d \sin \omega_0 \Delta t) \tag{6.24}$$

The slow cell experiences a maximum dv/dt approaching $V_s R/L$.

If other factors determine that π/ω_0 is long in comparison with Δt, (6.24) simplifies approximately to

$$\frac{k_{s1} - k_{s2}}{k_{s2}(nk_{s1} - 1)} \approx \frac{R\,\Delta t}{L} \tag{6.25}$$

Conditions during turn-on are clearly improved by making Δt as small as possible, to which end firing pulses of high amplitude and short rise-time are recommended.

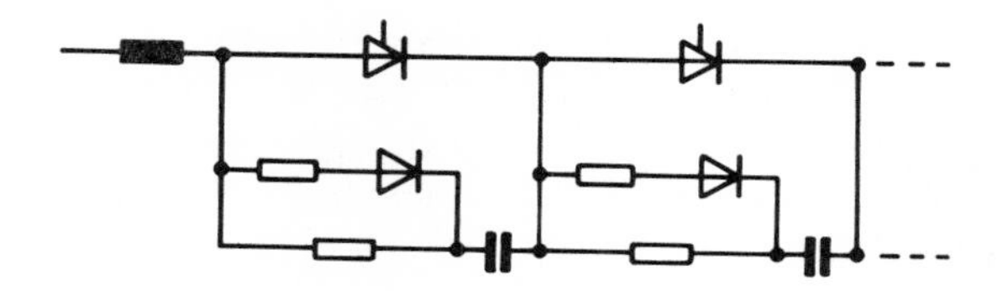

Figure 6.39 Modification of Figure 6.37 to permit the use of larger capacitors.

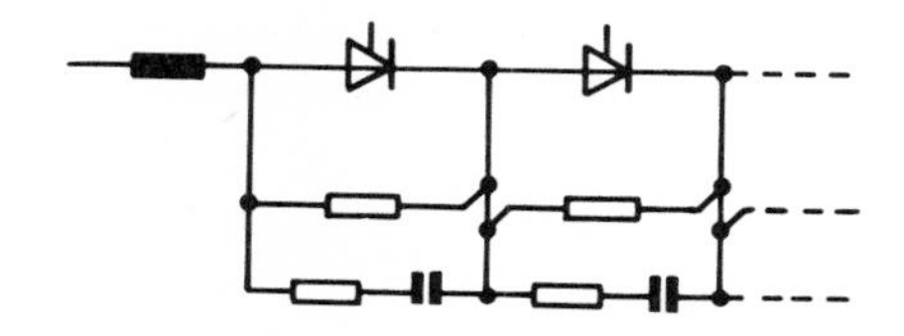

Figure 6.40 Typical complete voltage-sharing network.

Where the desired degree of voltage sharing cannot be obtained with an acceptable value of L and the lowest value of R permitted by considerations of the capacitor discharge currents when the thyristors are fired, the resistors may be shunted by diodes, as in Figure 6.39.

Overall design of sharing networks

A complete network for a string of series-connected thyristors is typically of the form shown in Figure 6.40, including provision for steady-state sharing, reverse recovery and the spread of turn-on times. Networks of this type serve multiple purposes, in that the component values required are generally similar for voltage sharing, dv/dt suppression and voltage spike suppression, while the discharge from the capacitors may serve usefully to provide latching current, and the series inductor possibly to assist current sharing in parallel paths.

Selection of cells for series operation

Since the problems of voltage sharing stem from variations in thyristor or diode characteristics, there is in principle a possibility of improving matters by matching cells, and some benefit may be obtained by selection within a limited band of stored charge. Special selection of cells, and particularly of thyristors, is to be avoided as far as possible, however, in view of the difficulties it causes in production.

PARALLEL OPERATION OF THYRISTORS AND DIODES

To achieve equipment current ratings higher than those permitted by the largest available thyristors or diodes used singly, it is common practice to connect cells in parallel, in virtually any number from two upwards. In general, parallel-connected cells are not used if cells with sufficient ratings to carry the required current without paralleling are available, although parallel operation does offer some advantages: the losses are more distributed, making cooling easier, and redundant cells may be provided in order to increase the operational reliability of the equipment.

The basic technical problems associated with the parallel operation of thyristors are those of ensuring that each cell of a group carries nearly enough its proper share of the total current and that each cell is properly fired despite the possible influence of the other cells in the group. The problem of current sharing is met also with diodes, but is generally less severe, because diode characteristics are usually more consistent than those of thyristors.

Current sharing in diodes and thyristors

Semiconductor devices are normally tested to maximum limits of conducting voltage drop, and unselected cells connected directly in parallel are unlikely to carry the total current in even approximately equal shares. Good current distribution is, however, necessary if the cells are to be used economically, since the rating of the group is the total current that flows when one individual cell is carrying its rated current. The degree of current sharing achieved may be expressed by a sharing or de-rating factor, k_s, which is the ratio of the effective rating of a number of cells in parallel to the rating that would be permitted if sharing were perfect. If the most heavily loaded cell carries more than its ideal share by an amount Δ_I p.u.

$$k_s = \frac{1}{1 + \Delta_I} \qquad (6.26)$$

The means employed to secure adequate sharing in parallel cells are firstly, the selection of cells to a specified degree of matching in terms of voltage drop and, secondly, the use of impedances to reduce the effect of variations between cells within the selection bands. Diodes are normally produced with such consistency of forward characteristics that the selection of cells within a certain range of voltage drop at the continuous rated current is sufficient in itself to ensure a sharing factor in the region of 0.8–0.9 so long as no other factors, such as unbalanced bus-bar impedances, are introduced to disturb the current distribution, and the use of sharing impedances is not normally justified. The cost of selecting thyristors to close limits is considerably greater, and the higher basic cost of thyristors is an incentive to obtain the highest practicable sharing factor; for these reasons sharing impedances are commonly used with parallel-connected thyristors.

In most cases, sharing impedances are inductors, sometimes in the form of centre-tapped or double-wound inductors providing a coupling between two current paths, as in Figure 6.41(a), more often a separate inductor in each path as in Figure 6.41(b). For operation at very low frequency, it may be necessary to use

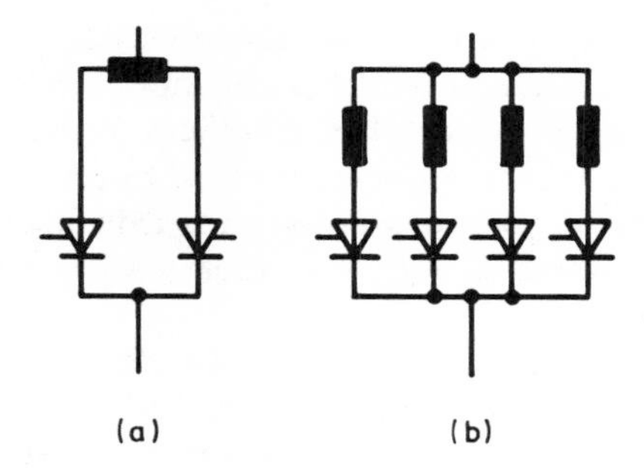

Figure 6.41 Current sharing with (a) centre-tapped inductor; (b) individual inductors.

resistors, although the resultant power loss may be comparable with the thyristor dissipation.

Current sharing with individual inductors

For the purpose of estimating the degree of current-sharing in a parallel-connected group of thyristors, the simplified representation of the forward conducting characteristic of the thyristor as a threshold voltage in series with a linear resistance can be used in all normal circumstances with sufficient accuracy (see Figure 1.4). A group of thyristors with individual sharing inductors is then equivalent to the circuit of Figure 6.42, wherein L, V and R indicate nominal values. Both the threshold voltage and the incremental resistance of the thyristors may be subject to considerable spreads, and there may be a significant tolerance on the inductor value, but in addition the resistance and inductance values in the complete equivalent circuit of the group must include the impedance of any additional components, such as fuses, and also that of the connecting bus-bars or cables, the effect of which may be beneficial or adverse, depending on the degree of balance.

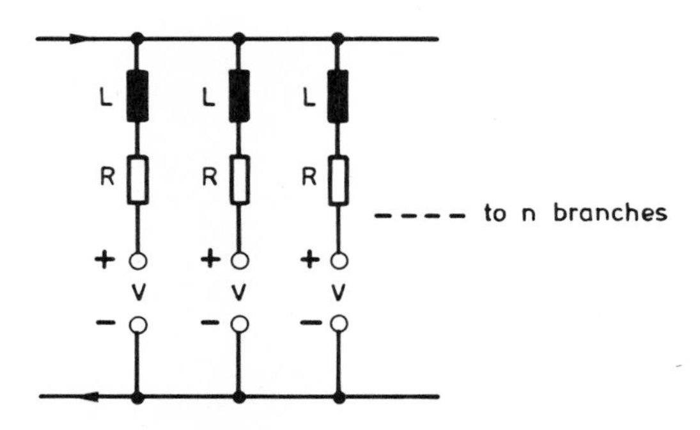

Figure 6.42 Equivalent representation of a group of parallel thyristors with individual sharing inductors.

The degree of current sharing in the worst case may be calculated by assuming that, of a total of n branches, $n-1$ branches are similar while the remaining one is different and carries a higher current than the others, the relationship between the characteristics being the least favourable permitted by the various tolerances. The equivalent circuit from the point of view of the heavily loaded branch is then as shown in Figure 6.43, where $L_1 - R_1 - V_1$ represents one branch and $L_2 - R_2 - V_2$ the remaining $n-1$.

The most generally representative current waveform to be considered for a rectifier or converter is a rectangular one, with a conduction period depending on the supply frequency, the converter configuration and possibly the angle of delay. The distribution of current at the beginning of conduction may be assumed to be determined entirely by the inductances; if the total current has an amplitude I and the instantaneous current i_1 in L_1 has an initial value i_s, then

$$i_s = I \frac{L_2}{L_1 + L_2} \tag{6.27}$$

If the conduction period were infinite, i_1 would reach an ultimate value i_u determined by R_1, R_2, V_1, and V_2:

$$i_u = \frac{IR_2 + (V_2 - V_1)}{R_1 + R_2} \tag{6.28}$$

Thus i_1 starts with the value i_s and approaches the value i_u exponentially with a time constant

$$\tau = \frac{L_1 + L_2}{R_1 + R_2} \tag{6.29}$$

i.e.

$$i_1 = i_u - (i_u - i_s)e^{-t/\tau} \tag{6.30}$$

With n branches, i_1 should ideally be I/n; if Δ_s is the initial, and Δ_u the ultimate, per-unit deviation of i_1 from I/n,

$$\left.\begin{aligned} \Delta_s &= \frac{i_s n}{I} - 1 \\ \text{and} \qquad \Delta_u &= \frac{i_u n}{I} - 1 \end{aligned}\right\} \tag{6.31}$$

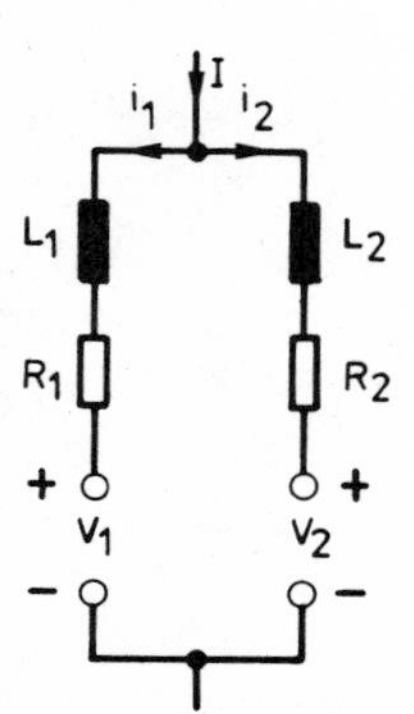

Figure 6.43 Equivalent circuit for the calculation of current sharing with individual sharing inductors.

Suppose R_1, L_1 and V_1 have values R, L and V, while the values for each of the other branches are $R(1+a_R)$, $L(1+a_L)$ and $V(1+a_V)$, where a_R, a_L and a_V represent the total ranges of variation for the three quantities: then

$$\left.\begin{aligned} R_2 &= \frac{R(1+a_R)}{n-1} \\ \text{and} \quad L_2 &= \frac{L(1+a_L)}{n-1} \end{aligned}\right\} \quad (6.32)$$

From the above,

$$\left.\begin{aligned} \Delta_s &= \frac{(n-1)a_L}{n+a_L} \\ \text{and} \quad \Delta_u &= \left(\frac{n-1}{n+a_R}\right)\left(a_R + a_V\frac{V_n}{RI}\right) \end{aligned}\right\} \quad (6.33)$$

If Δ_i is the per-unit instantaneous deviation of i_1 from I/n,

$$\Delta_i = \Delta_u - (\Delta_u - \Delta_s)e^{-t/\tau} \quad (6.34)$$

and if the duration of the conduction period is T, the mean deviation is

$$\overline{\Delta}_I = \frac{1}{T}\int_0^T \{\Delta_u - (\Delta_u - \Delta_s)e^{-t/\tau}\}\,dt \quad (6.35)$$

$$= \Delta_u - \frac{\tau}{T}(\Delta_u - \Delta_s)(1 - e^{-T/\tau}) \quad (6.36)$$

Since all the quantities in (6.33) are known, by putting

$$\frac{\tau}{T} = \frac{L_1 + L_2}{T(R_1 + R_2)} = \frac{L(n+a_L)}{RT(n+a_R)} \quad (6.37)$$

equation (6.36) can be used to determine the degree of mis-sharing in a given situation, or alternatively the value of L required to limit mis-sharing to a given amount.

A particular case of interest is that for $n = \infty$, which gives a result that is on the safe side for any finite number of parallel paths. Equations (6.33) then become

$$\left.\begin{aligned} \Delta_s &\to a_L \\ \Delta_u &\to a_R + a_V\frac{V}{R}\left(\frac{n}{I}\right) \end{aligned}\right\} \quad (6.38)$$

where I/n is the current per path.

In applying the above expressions to a practical case, it is necessary to recognize the constraints which the method of selecting thyristors for parallel operation places upon the possible combinations of a_R and a_V, and to determine the most unfavourable combination for the purpose of the worst-case calculation. If,

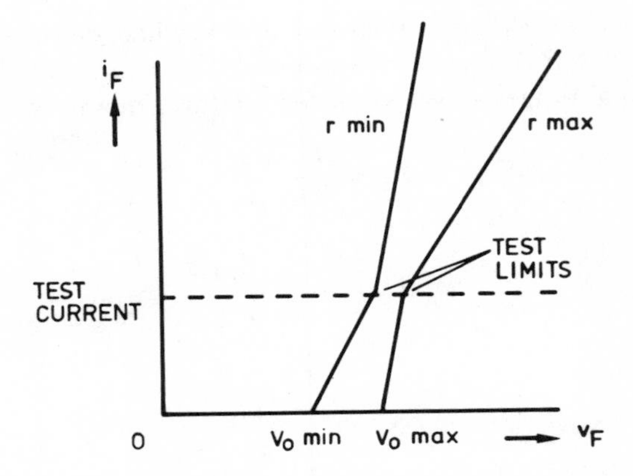

Figure 6.44 Possible spread of forward conducting characteristics in thyristors matched for voltage drop at once current.

for example, the matching is carried out on the basis of a defined range of instantaneous voltage drop at a particular current in the normal working range, the overall spread of characteristics may be somewhat as suggested (in idealized form) in Figure 6.44; clearly in this case the worst combination of threshold voltages and incremental resistances is different according to whether the working current is greater or less than the test current. More elaborate matching procedures can be employed to obtain greater uniformity of characteristics if the cost is justified by savings in the cost of sharing impedances or other components.

Whereas the matching of diodes to achieve satisfactory current sharing without sharing impedances is not difficult, series inductors are normally desirable with parallel-connected thyristors; they often serve more than one purpose, however, being necessary for di/dt limitation or dv/dt suppression.

Firing parallel-connected thyristors

The firing of thyristors connected in parallel calls for special attention to ensure that all the thyristors conduct correctly and that unnecessary damage is not caused by circuit malfunctions.

If a number of thyristors connected directly in parallel, with no sharing impedances, apart, perhaps, from fuses, are supplied with firing pulses only marginally above their I_{GT} values, they are likely to exhibit a considerable spread of turn-on time, with at least two undesirable consequences: firstly, the reduction in anode voltage brought about by the first thyristor to turn on may prevent the slower ones from firing, and secondly the whole of the di/dt in the circuit, initially at least, may be concentrated in the first thyristor. For these reasons, if for no other, firing pulses of high amplitude should be used when thyristors are required to operate in this way, and thyristors intended for such use should be tested for firing current at a lower anode voltage than might otherwise by necessary.

Sharing impedances (inductors more than resistors) prevent a too-rapid collapse of voltage across the thyristor group and limit the di/dt in each thyristor individually; generally the special testing of the thyristors is then unnecessary, apart from matching.

When a centre-tapped sharing inductor is employed, as in Figure 6.41(a), the firing of the slower thyristor tends to be assisted by a momentary increase in the voltage across it caused by the turning-on of the faster one; it is important, however, that the peak voltage should not exceed the repetitive peak rating of the cell, and that the dv/dt produced should not be such as to introduce any possibility that the firing process is not wholly under the control of the gate. These considerations tend to offset the apparent advantages of the arrangement, and make firing currents of high amplitude as desirable in this as in other parallel circuits.

A partial control-circuit failure that results in one thyristor of a group not being fired aggravates the potential hazard of excessive anode voltage or dv/dt with a centre-tapped inductor, and in all cases leads to the possibility of overloading the thyristors that are fired. It is not uncommon, therefore, in large equipment, to provide a means of proving conduction in all the branches of the group, by detecting, for example, pulses of voltage across the (uncoupled) sharing inductors; such a system also serves to indicate non-conduction due to other causes, such as blown fuses.

THE FIRING OF THYRISTORS

The basic necessities for the successful firing of a thyristor are that the current supplied to the gate should:

(a) be of adequate amplitude and sufficiently short rise time:
(b) be of adequate duration:
(c) occur at a time when the main circuit conditions are favourable to conduction.

Gate current amplitude and rise time

The quoted firing current I_{GT} expresses the minimum gate current required to fire all thyristors of a given type at a standard temperature. This is not normally a figure that can be applied directly in practice; a margin is required not only to cover possible errors and uncertainties but to ensure that the turn-on time of the thyristor, which becomes long and indeterminate as the gate current approaches the critical firing level, is acceptably short. A firing current of about $1.4I_{GT}$ is commonly satisfactory so long as the main circuit conditions do not impose special requirements; a considerably higher current is often necessary, however, in order to reduce the turn-on time (usually to reduce the spread of turn-on times in a group of thyristors), to increase the di/dt rating of the thyristor, or to reduce the turn-on switching loss. The gate voltage must exceed the quoted gate firing voltage, V_{GT}, by a corresponding margin.

The gate drive chosen must trigger the thyristor under the most adverse conditions, and, in particular, at the lowest thyristor junction temperature that is likely to occur. Both I_{GT} and V_{GT} exhibit a negative variation with temperature (see Figure 1.26).

A specification of the gate current amplitude is incomplete unless the rise time of the pulse is also specified. To be useful, an adequate level of gate current must be reached before the thyristor has turned on—that is, within the turn-on delay time—and even given that this condition is met, the effectiveness of the pulse still increases somewhat as the rise time is further reduced; on the other hand, there is no particular advantage in a very short rise time if the amplitude is not sufficient to give a commensurately short turn-on time. Hence amplitude and rise time must be considered together in designing a firing circuit in accordance with any stipulated di/dt capability and switching performance.

The design of circuits for very short gate pulse rise times has to allow for the finite response time of the gate–cathode junction itself, which makes it necessary to apply a higher voltage than is apparent from the static gate characteristic. Commonly an open-circuit source voltage of 15–20 volts is specified in order to achieve the highest useful current amplitude with a commensurately short rise time.

Gate pulse duration

Under favourable conditions a thyristor may be fired successfully by a gate pulse of a duration approximately equal to the turn-on time of the cell. Normally, however, a considerably longer pulse duration is desirable for one or more of the following reasons:

(a) a relatively long period may be required for the anode current to rise to the latching current level:
(b) oscillations, reflections or other disturbances may conspire to turn off the thyristor shortly after it is first fired:
(c) there may be uncertainty as to whether or not the anode circuit conditions are favourable to conduction when the firing pulse is initiated.

The relative importance of these factors, and possibly the ease with which they can be modified (by the addition of bleed networks, for example), determine the pulse duration required in a particular application. A certain way of ensuring that the pulse is of adequate duration is to extend it to the whole period for which the thyristor is intended to conduct; this is often done in cases where the moment when the thyristor becomes forward-biassed is more or less unpredictable (as in many types of forced-commutation inverter with reactive load). Alternatively the pulse is made to cover at least the period of uncertainty. Since, however, the generation of a very long pulse, particularly if high amplitude and a short rise time are required, usually entails an unwelcome degree of complexity and cost, pulses of short duration are more often used where possible. An incidental advantage of a short pulse is that if it coincides with a reverse anode voltage the resulting reverse loss in the thyristor is less likely to be significant. In general, a pulse duration of less than about 10 μs requires considerable care in the detailed design of the anode circuit, while a duration of 30–60 μs is usually sufficient to avoid problems so long as the anode circuit conditions are favourable to conduction.

Two alternatives to the use of a gate pulse long enough to cover the region of uncertain firing are:

(a) a control system in which the timing of the gate pulse is synchronized to the thyristor voltage or the current zero, rather than to the supply voltage;
(b) an extended train of short pulses.

The first of these two possibilities is generally more appropriate in relatively simple applications in which isolation is not required between the control system and the thyristor main circuit. The second technique is of more general application, and has the additional advantage that it gives the thyristor repeated opportunities to turn on; however, a disadvantage occurs if the first pulse of the train is not effective, since the actual instant of firing is then somewhat indeterminate.

Pulse waveforms

Even if the anode circuit conditions require a gate pulse of high amplitude, it is not necessary for this to be maintained for the whole duration of the pulse; if the thyristor starts to conduct at the beginning of the pulse, the conditions dictating the high amplitude will have disappeared after the first few microseconds, while if it becomes forward-biassed at some time within the duration of the pulse, at a dv/dt within the capability of the cell, neither a severe di/dt condition nor the necessity for a very short switching time can exist. On the other hand, a pulse of continuing high amplitude has the disadvantages of making a bigger demand on the firing circuit, increasing the mean gate dissipation, and increasing the latching current. It is thus a sound practice to use a pulse waveform with a leading edge of the required amplitude and rise time followed by a tail of no more than the practical minimum amplitude required to fire the thyristor, namely around $1.4I_{GT}$. The duration of the initial high-amplitude pulse may depend on its amplitude, since a high amplitude is associated with fast switching; if it is not a well-defined square pulse, an exponentially decaying pulse with a time constant of about 10 microseconds is generally satisfactory. Figure 6.45 illustrates a pulse waveform of this kind.

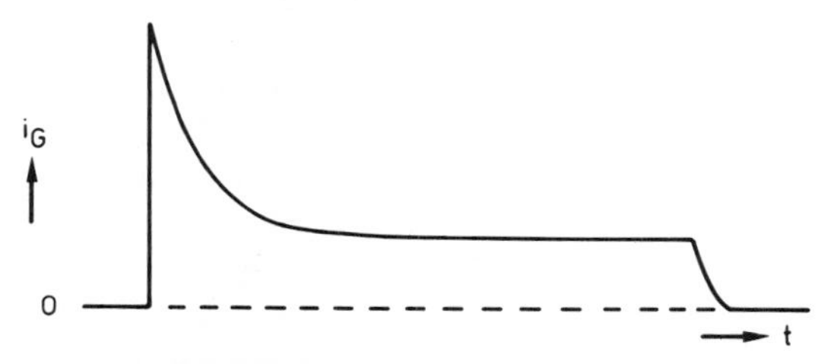

Figure 6.45 Extended gate firing pulse with high initial amplitude.

A negative overshoot at the end of the gate pulse is generally to be avoided, since it may tend in some circumstances to turn off the anode current that has been initiated in the thyristor.

Spurious triggering

Spurious triggering, whether through stray pick-up on the gate or in the firing control circuit or through excessive anode voltage or dv/dt in the main circuit, can result directly in damage to a thyristor, apart from the possibility of indirect damage through a resulting malfunction of the circuit, because the thyristor so triggered may not have the di/dt capability to withstand the anode circuit conditions. It follows that 'self-protecting' circuits depending on the automatic break-over of a thyristor in case of excessive voltage are to be avoided, unless the thyristor is specifically rated for such a duty.

Firing-angle determination in naturally commutating converters

The basic function of the firing-angle control circuit attached to a naturally commutating converter is normally to determine the delay angle, in relation to the a.c. supply voltage, so that the desired voltage is applied to the load, in accordance with a control input—i.e., in most cases, to serve as a voltage-to-delay-angle transducer. The systems principally employed are

(i) variable phase-shift circuits;
(ii) magnetic amplifiers;
(iii) circuits in which the firing angle is determined by the intersection of a repetitive waveform, synchronous with the supply, and a reference or control level;
(iv) digital timing circuits.

While it is not the intention here to discuss the design of firing circuits in detail, the following general observations will give some indication of the considerations entailed.

Variable phase shift

A voltage or current is produced with a controllable phase relationship to the supply voltage, and firing pulses are generated at the zero-crossing points of the phase-shifted waveform. Most circuits of this type are suited less to electrical than to manual control, and do not readily provide the full 180° range that is commonly required.

Magnetic amplifiers

In suitable configurations, magnetic amplifiers can be made to produce output voltage waveforms closely resembling those of a half-controlled bridge rectifier, and their ideally linear relationship of mean output voltage to mean control voltage implies exactly the variation of 'firing angle' required to obtain a similar characteristic in a controlled rectifier; this makes them superficially attractive for thyristor firing circuits. In practice the output from a conventional magnetic amplifier is not generally suitable for firing a thyristor directly, because its rise time is poor and its initial amplitude varies with the firing angle, and even though this can be overcome by the addition of pulse-forming networks, the disadvantages remain, that the characteristic tends to be seriously non-linear in the region of minimum firing angle, and that the virtually immediate response so easily provided by transistor circuits is not obtainable.

Waveform-intersection systems

Figure 6.46 illustrates the basic principle; the firing point is determined by the coincidence of the instantaneous level of a ramp or timing waveform and a triggering level; at (a), the firing point is controlled by varying the slope of the ramp, while (b) illustrates the more common and useful method of control in which the timing waveform is unvarying and control is effected by altering the triggering level. This principle is amenable to relatively easy predictable design employing normal analogue and digital techniques, and represents the great majority of systems in use at present.

The ramp, shown as linear in Figure 6.46, may in some cases be exponential, or can, with advantage for some purposes, be a cosine function, which yields a linear overall control characteristic with a fully or half-controlled converter.

The main problem in designing a circuit of this type is to synchronize the ramp function to the zero-crossings of the a.c. supply voltage in such a way that errors are not caused by transient disturbances or waveform distortion, without introducing unacceptable phase shifts.

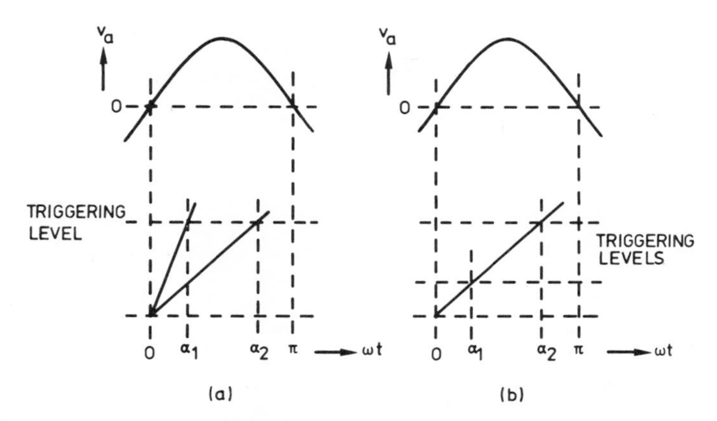

Figure 6.46 Firing-angle control by waveform intersection: (a) variable-slope ramp; (b) variable triggering level.

Digital systems

Like the waveform intersection method, digital systems normally depend on measuring off the required time interval from the supply-voltage zero, and considered simply as an alternative to the analogue approach to this particular requirement they do not appear to have any great advantage. They do, however, have considerable potential advantage as sections of wholly or mainly digital control systems, and lend themselves particularly well to microelectronic techniques.

Additional design features

Practical firing circuit design frequently has to take into account additional aspects beyond the basic function of firing-angle determination. These include:

(i) Firing-angle limitation: a positive means of limiting to a value that will not lead to loss of commutation is commonly required in systems involving fully controlled converters, and can be useful with a half-controlled bridge as an alternative to a free-wheel diode.

(ii) Pulse distribution or gating: depending on the design principles used, gating circuits may be needed to direct firing pulses to each thyristor in an appropriate region of the supply cycle and to inhibit them in regions where they might lead to a circuit malfunction or other undesirable effect.

(iii) Pulse duplication: three-phase circuits in which the main-circuit current paths include two thyristors in series, such as the fully controlled bridge converter, require that each thyristor should receive two firing pulses in each cycle, unless a single pulse is of sufficient duration to embrace both firing instants. This can be done by deriving the pulse for each thyristor from two appropriate phases.

APPENDIX 6(i)
BASIC HEAT-TRANSFER RELATIONSHIPS

Conduction

The rate of heat flow across an element of cross-sectional area A, length (thickness) d and thermal conductivity k, due to a temperature difference $\Delta\theta$ is

$$W_c = \frac{Ak\,\Delta\theta}{d}\ \mathrm{W}$$

and the thermal resistance is

$$R_{th} = \frac{\Delta\theta}{W_c} = \frac{d}{Ak}\ ^\circ\mathrm{C/W} \qquad (6.39)$$

Some representative thermal conductivities are

Copper:	385 W/m °C
Aluminium:	208 W/m °C
Mild steel:	45 W/m °C
Beryllia:	210 W/m °C
Boron nitride:	61 W/m °C
Alumina:	21 W/m °C
Mica:	0.5–0.8 W/m °C

Radiation

The nett rate of heat loss from a surface of area A and emissivity ε (maximum value 1, for a perfect 'black body') at an absolute temperature θ_{A1}, surrounded by black bodies at an absolute temperature θ_{A2}, is given by

$$W_r = 5.7\varepsilon A(\theta_{A1}^4 - \theta_{A2}^4) \times 10^{-8}\ \mathrm{W} \tag{6.40}$$

Some typical emissivities are:

Painted (not metallic) or matt-coated surfaces: 0.90–0.97
Oxidized aluminium: 0.11–0.19
Polished aluminium: 0.04.

Emissivity varies greatly according to material and surface condition and sometimes with temperature; the above figures are intended only as a rough illustration.

Natural convection

An empirically determined expression for the heat lost from a vertical plane surface of area A and height h in free air, at a uniform temperature $\Delta\theta$ above ambient temperature, is

$$W_n = 1.37\frac{A\,\Delta\theta^{1.25}}{h^{0.25}}\ \mathrm{W} \tag{6.41}$$

Forced convection

The rate of heat loss from a plane surface of area A and length d (in the direction of the air flow) in air moving with velocity V, at a uniform temperature $\Delta\theta$ above the air temperature, has been determined empirically as follows:

Laminar flow: $$W_f = 3.9A\,\Delta\theta\sqrt{\left(\frac{V}{d}\right)}\ \mathrm{W} \tag{6.42}$$

Turbulent flow: $$W_f = 6.0A\,\Delta\theta\frac{V^{0.8}}{d^{0.2}}\ \mathrm{W} \tag{6.43}$$

Thermal capacity

Thermal capacity is defined as the rate of change of thermal energy with temperature, and measured, for present purposes, in joules per °C. A body of mass M with specific heat c has a thermal capacity

$$C_{\text{th}} = cM \tag{6.44}$$

Some values of c are:

Water:	4200 J/kg.°C
Aluminium:	766 J/kg.°C
Copper:	396 J/kg.°C

APPENDIX 6(ii)
RESPONSE OF DAMPED *LCR* CIRCUITS

Voltage step input

The circuit representing the cases of interest here, namely dv/dt suppression and transient overvoltage suppression, is that of Figure 6.47. In Figure 6.47(a), a

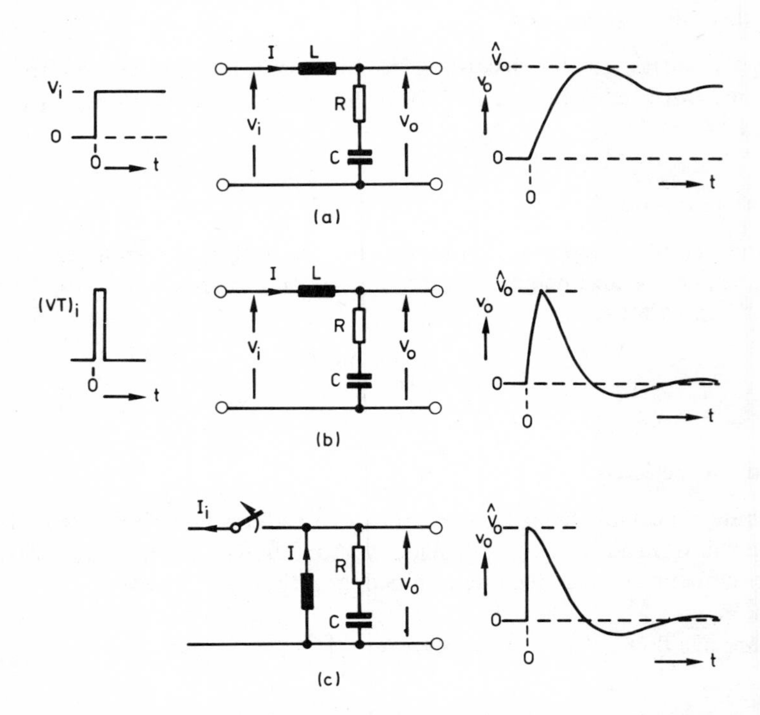

Figure 6.47 Response of damped resonant circuit ($\delta = 0.5$): (a) voltage step input; (b) voltage impulse input; (c) current step input.

voltage step V_i is applied to the input and the resulting output voltage is $v_o(t)$. The current that flows when V_i is applied when $t = 0$ is

$$i = \frac{V_i}{\omega_0 L} e^{-Rt/(2L)} \sin \omega_0 t \tag{6.45}$$

where ω_0 is the resonant frequency $\sqrt{\left(\frac{1}{LC} - \frac{R^2}{4L^2}\right)}$

From (6.45)

$$\frac{di}{dt} = \frac{V_i}{L} e^{-Rt/(2L)} \left(\cos \omega_0 t - \frac{R}{2\omega_0 L} \sin \omega_0 t\right) \tag{6.46}$$

Then

$$v_o = V_i - L\frac{di}{dt}$$

Putting

$$\frac{R}{2\omega_0 L} = d$$

$$\frac{v_o}{V_i} = 1 + e^{-d\omega_0 t} (d \sin \omega_0 t - \cos \omega_0 t) \tag{6.47}$$

and

$$\frac{dv_o/dt}{V_i} = \omega_0 e^{-d\omega_0 t} \{2d \cos \omega_0 t + (1 - d^2) \sin \omega_0 t\} \tag{6.48}$$

When v_o is at a maximum,

$$\frac{d}{dt}\left(\frac{v_o}{V_i}\right) = 0$$

and

$$\tan \omega_0 t = -\frac{2d}{1 - d^2}$$

From which

$$\cos \omega_0 t = -\frac{1 - d^2}{1 + d^2} \tag{6.49}$$

and

$$\sin \omega_0 t = \frac{2d}{1 + d^2}$$

Then, substituting from (6.49) in (6.47),

$$\frac{\hat{V}_o}{V_i} = 1 + e^{-d\omega_0 t}\left(\frac{2d^2}{1 + d^2} + \frac{1 - d^2}{1 + d^2}\right)$$

$$= 1 + e^{-d \tan^{-1}\{-2d/(1 - d^2)\}} \tag{6.50}$$

Differentiating (6.48), dv_o/dt is at a maximum when

$$(1 - 3d^2) \cos \omega_0 t - (3d - d^3) \sin \omega_0 t = 0 \tag{6.51}$$

Then

$$\left.\begin{aligned} \tan \omega_0 t &= \frac{1-3d^2}{3d-d^3} \\ \cos \omega_0 t &= \frac{3d-d^3}{(1+d^2)^{3/2}} \\ \sin \omega_0 t &= \frac{1-3d^2}{(1+d^2)^{3/2}} \end{aligned}\right\} \tag{6.52}$$

Substituting in (6.48) gives

$$\left(\frac{\mathrm{d}v_o/\mathrm{d}t}{V_i}\right)_{max} = \omega_0 \sqrt{(1+d^2)}\,\mathrm{e}^{-d\tan^{-1}\{(1-3d^2)/(3d-d^3)\}} \tag{6.53}$$

Equations (6.52) and (6.53) give a valid result so long as $(1-3d^2)$ is positive—i.e. if the value of $\omega_0 t$ for which $\mathrm{d}v_o/\mathrm{d}t$ is maximum is greater than zero. If d is greater than $1/\sqrt{3}$, the maximum value of $\mathrm{d}v_o/\mathrm{d}t$ occurs when $t = 0$, when

$$\frac{\mathrm{d}v_o/\mathrm{d}t}{V_i} = 2d\omega_0 = \frac{R}{L} \tag{6.54}$$

Voltage impulse input

The impulse response (Figure 6.47(b)) can be obtained simply from the previous result by treating the input voltage impulse $(VT)_i$ as a pulse of amplitude V_i and of a duration T which is short compared with π/ω_0. This pulse may in turn be regarded as a positive step of voltage V_i at $t = 0$ followed by a negative step $-V_i$ at $t = T$; then from (6.47)

$$\frac{v_o}{V_i} = \mathrm{e}^{-d\omega_0 t}(d \sin \omega_0 t - \cos \omega_0 t) - \mathrm{e}^{-d\omega_0 (t-T)}\{d \sin \omega_0 (t-T) - \cos \omega_0 (t-T)\} \tag{6.55}$$

Using the approximations

$$\cos \omega_0 T \approx 1, \qquad \sin \omega_0 T \approx \omega_0 T \qquad \text{and} \qquad \mathrm{e}^{-d\omega_0 T} \approx 1,$$

this reduces to

$$\frac{v_o}{V_i} \approx \omega_0 T \mathrm{e}^{-d\omega_0 t}\{(1-d^2)\sin \omega_0 t + 2d \cos \omega_0 t\}$$

or

$$\frac{v_o}{(VT)_i} \approx \omega_0 e^{-d\omega_0 t}\{(1-d^2)\sin \omega_0 t + 2d \cos \omega_0 t\} \tag{6.56}$$

This is the same expression as (6.48), and therefore yields the corresponding expressions for $\hat{V}_o$:

$$\left.\begin{aligned} d < \frac{1}{\sqrt{3}}: \frac{\hat{V}_o}{(VT)_i} &= \omega_0 \sqrt{(1+d^2)}\,\mathrm{e}^{-d\tan^{-1}\{(1-3d^2)/(3d-d^3)\}} \\ d > \frac{1}{\sqrt{3}}: \frac{\hat{V}_o}{(VT)_i} &= 2d\omega_0 = \frac{R}{L} \end{aligned}\right\} \tag{6.57}$$

Current step input

In Figure 6.47(c), the output voltage is produced by the interruption of a current I_i flowing in L. For convenience, use may be made of the previous results by regarding the current I_i as the result of a voltage impulse $(VT)_i = LI_i$. Then, from (6.57),

$$\left.\begin{aligned} d < \frac{1}{\sqrt{3}}: \frac{\hat{V}_o}{LI_i} &= \omega_0 \sqrt{(1+d^2)}\,\mathrm{e}^{-d\tan^{-1}\{(1-3d^2)/(3d-d^3)\}} \\ d > \frac{1}{\sqrt{3}}: \frac{\hat{V}_o}{LI_i} &= 2d\omega_0 \qquad \text{or} \qquad \hat{V}_o = I_i R \end{aligned}\right\} \qquad (6.58)$$

Damping factor

For greater convenience, a damped resonant circuit is usually characterized by a damping factor $\delta = R/2\sqrt{(L/C)}$ rather than the parameter d used above. The two values are related as follows

$$\left.\begin{aligned} \omega_0 &= \sqrt{\left(\frac{1}{LC} - \frac{R^2}{4L^2}\right)} = \frac{1}{\sqrt{(LC)}}\sqrt{(1-\delta^2)} \\ d &= \frac{R}{2\omega_0 L} = \frac{R\sqrt{(LC)}}{2L\sqrt{(1-\delta^2)}} = \frac{\delta}{\sqrt{(1-\delta^2)}} \\ \text{Conversely,} \qquad \delta &= \frac{d}{\sqrt{(1+d^2)}} \end{aligned}\right\} \qquad (6.59)$$

The break-point $d = 1/\sqrt{3}$ thus corresponds to $\delta = 0.5$.

Bibliography

1. Bedford, B. D. and Hoft, R. G. (1964). *Principles of Inverter Circuits*, Wiley, New York.
2. Blicher, A. (1976). *Thyristor Physics*, Springer, New York.
3. Davis, R. M. (1971). *Power Diode and Thyristor Circuits*, Peter Peregrinus, London.
4. Dewan, S. and Staughan, A. (1975). *Power Semiconductor Circuits*, Wiley, New York.
5. Finney, D. (1980). *The Power Thyristor and its Applications*, McGraw-Hill (U.K.).
6. Fishenden, M. and Saunders, O. A. (1982). *An Introduction to Heat Transfer*, Oxford University Press.
7. Grafham, D. R. and Golden, F. B. (Ed.) (1979). *SCR Manual*, General Electric Company, Auburn, New York.
8. Grove, A. S. (1967). *Physics and Technology of Semiconductor Devices*, Wiley, New York.
9. Gyugyi, L. and Pelly, B. R. (1976). *Static Power Frequency Changers*, Wiley, New York.
10. Hempel, H-P. (1980). *Power Semiconductor Handbook*, Semikron International, Nuremberg.
11. Hoffman, A. and Stocker, K. (Ed.) (1965). *Thyristor-Handbuch*, Siemens, Berlin.
12. Kloss, A. (1980). *Leistungselektronik ohne Ballast*,
13. Kusko, A. (1968). *Solid State D. C. Motor Drives*, The MIT Press.
14. Lander, C. W. (1981). *Power Electronics*, McGraw-Hill (U.K.).
15. Maggetto, G. (1971). *Le Thyristor; definitions- protections- commandes*, Presses Universitaires de Bruxelles.
16. McMurray, W. (1972). *The Theory and Design of Cycloconverters*, The MIT Press.
17. Möltgen, G. (1967). *Netzgeführte Stomrichter mit Thyristoren*, Siemens, Berlin.
18. Murphy, J. M. D. (1973). *Thyristor Control of A. C. Motors*, Pergamon Press, London.
19. Pelly, B. R. (1971). *Thyristor Phase-Controlled Converters and Cycloconverters*, Wiley, New York.
20. Rissik, H. (1941). *Mercury-Arc Current Converters*, Pitman, London.
21. Sanders, C. W. (1981). *Power Electronics*, McGraw-Hill,
22. Schaefer, J. (1965). *Rectifier Circuits*, Wiley, New York.
23. Smith, R. A. (1959). *Semiconductors*, Cambridge University Press.
24. Takuechi, T. J. (1968). *Theory of SCR Circuits and Application to Motor Control*, Tokyo Electrical Engineering College Press.
25. Wood, P. (1981). *Switching Power Converters*, Van Nostrand Reinhold, New York.

Index